D0553877

CONCEPTS IN BUILDING FIRESAFETY

CONCEPTS IN BUILDING FIRESAFETY

M. David Egan

College of Architecture
Clemson University

A Wiley-Interscience Publication
JOHN WILEY & SONS
New York Chichester Brisbane Toronto

Library of Congress Cataloging in Publication Data

Egan, M. David.
Concepts in building firesafety.

"A Wiley-Interscience publication."
Includes index.
1. Fire prevention. 2. Fire extinction.
3. Building, Fireproof. I. Title.

TH9145.E3 693.8 '2 77-12184
ISBN 0-471-02229-2

Printed in the United States of America

10 9 8 7 6 5 4 3 2 1

FOREWORD

The old slang expression "let's get our heads together" is just as apropos for architects, engineers, and fire protection practitioners as for any other group of professionals. Professor M. David Egan, looking forward to a closer relationship between those who design and construct buildings and those who enforce codes and suppress fires, has produced a book which I think has long been needed.

Fire departments and fire suppression personnel for many decades have been in the position of amassing information on what causes and contributes to fire spread in buildings. One of the biggest problems has been the lack of communication of this knowledge to the professionals who design and build fire susceptible structures. Much of that lack of communication has been the inability to get conceptual descriptions of the fire problem across disciplinary lines. In bridging this communications gap, Professor Egan has taken a great step forward by describing with clear illustrations the how and why of concepts fire suppression technicians have known for many years. I suspect that the design profession also has a similar store of construction information that they should communicate to fire fighters.

This book represents an important step across the line between architecture and fire protection. It is hoped that other steps will follow in both directions. Simply bemoaning the lack of communication, which often is just the inability to understand the different professional languages and esoteric acronyms, does not accomplish anything. Professor Egan has made a significant contribution by defining and illustrating firesafety standards and techniques which should be considered in any building plan.

JAMES H. SHERN

Fire Chief
Pasadena Fire Department
Pasadena, California
and
Vice President
International Association of Fire Chiefs

PREFACE

The goal of this book is to present in a graphical format the principles of design for building firesafety. As a consequence, the illustrations are not supplements to the text as is often the case with traditional presentations. Instead, the illustrations represent the core of the book's coverage of factors affecting fire ignition and spread in buildings, building site planning for fire suppression and occupant rescue operations, fire protection by building materials and constructions, fire detection and suppression systems for buildings, smoke and heat venting techniques, escape and refuge principles for occupant protection, and finally the special case of firesafety for tall buildings.

It is hoped that the illustrations will help students of building design and fire science to understand the theoretical bases of building codes and standards, and to develop the fundamental knowledge needed to achieve firesafety in the built environment. In addition, the illustrations should facilitate understanding of building firesafety concepts by architects, builders, interior designers, and urban planners who have not had exposure to building firesafety education or training programs and do not have time to digest lengthy written descriptions.

It should be noted, however, that the illustrations and the accompanying data tables have been included in the book to demonstrate principles, to describe laboratory fire test methods, and to show examples of accepted construction techniques. For specific design applications, the minimum requirements for firesafety must be obtained from the prevailing building codes, standards, and rulings of the authority having jurisdiction. Technical data such as flame spread ratings and fire-resistance ratings, along with the complete construction details, should only be taken from up-to-date test reports available from Underwriters Laboratories, U.S. National Bureau of Standards, and other nationally recognized testing laboratories. In addition, several example problems (e.g., sprinkler piping pressure drop analyses, visibility of exits in smoke-filled rooms) have been included to demonstrate principles to the reader. These presentations are not intended to replace the comprehensive information and computational procedures that can be obtained from the various references given throughout the text. For example, National Fire Protection Association and National Bureau of Standards publications represent excellent sources of materials specifically developed to help the design professional and the fire service safeguard the built environment from fire.

The book also contains selected listings of firesafety references and standards of particular interest to the building profession. An introduction to the concept of systems analysis applied to design for building firesafety is included and the appendix has an extensive metric system conversion table for common building firesafety units.

M. DAVID EGAN, P.E.
Anderson, South Carolina
November 1977

ACKNOWLEDGMENTS

Thanks are due to many people who helped me during the preparation of this book.

Gratitude is extended to Mr. Glen S. LeRoy, Architect who prepared all the illustrations and margin art. Glen LeRoy's sensitivity and dedication to effective graphic communication techniques is deeply appreciated.

Dean Harlan E. McClure, FAIA and the students at Clemson University College of Architecture made valuable suggestions and comments on the preliminary versions of this book used by me in teaching building firesafety courses to architecture and building construction students. Also, Professors Donald L. Collins and Charlie R. Mitchell offered many helpful suggestions in the areas of site planning and structures.

Buddy Moose and Ken Bolin, former Clemson University students, prepared drawings for the Clemson University versions.

Jeannie Egan carefully edited the manuscript and art text and typed the overall manuscript.

Special thanks are due to Dick Custer, Gus Degenkolb, Bud Nelson, Jim Shern, and Carrol Burtner for their in-depth reviews of the manuscript.

It is of course truly impossible to name all the people who influenced and contributed to the preparation of this book without overlooking someone. Nevertheless, recognition should be given to the following people for their valuable comments, criticisms, and suggestions:

DR. JOHN L. BRYAN
Chairman, Fire Protection Curriculum
University of Maryland

MR. CARROL E. BURTNER
Occupational Safety and Health Administration

MR. RICHARD L. P. CUSTER
Center for Fire Research
National Bureau of Standards

MR. JOHN G. DEGENKOLB
Fire Protection Engineer

MR. CLAUD FREDERICK
Steelcraft Manufacturing Company

PROFESSOR HARRY E. HICKEY
Fire Protection Curriculum
University of Maryland

DR. B. L. JOESTEN
Union Carbide Corporation

MR. NORMAN A. KOPLON, P.E.
City of Atlanta Building Department

MR. RONALD L. MACE, AIA
Barrier Free Environments

PROFESSOR MICHAEL J. MUNSON
School of Architecture and Urban Planning
Princeton University

Mr. Harold E. Nelson
Center for Fire Research
National Bureau of Standards

Mr. J. L. Pauls
Division of Building Research
National Research Council of Canada

Dr. Anne W. Phillips
Executive Director, National Smoke, Fire and Burn Institute

Chief James H. Shern
Pasadena Fire Department

Dr. Edwin E. Smith
Chemical Research
Ohio State University

Dr. John A. Templer
Director, Pedestrian Research Laboratory
Georgia Institute of Technology

Mr. Dennis A. Yosick
Architectural Construction Manager
United States Gypsum Company

Mr. Calvin H. Yuill
Consultant

M.D.E.

CONTENTS

CONCEPTS IN BUILDING FIRESAFETY

BASIC THEORY

BASIC THEORY—BASIC FACTORS OF FIRE

Fire is the rapid oxidation of combustible materials and gases, producing heat and light. If a combustible material (called "fuel"), oxygen, or heat can be removed or prevented from coming together, there will be no fire. Listed below are the basic factors required for fire along with the general methods of extinguishing fire.

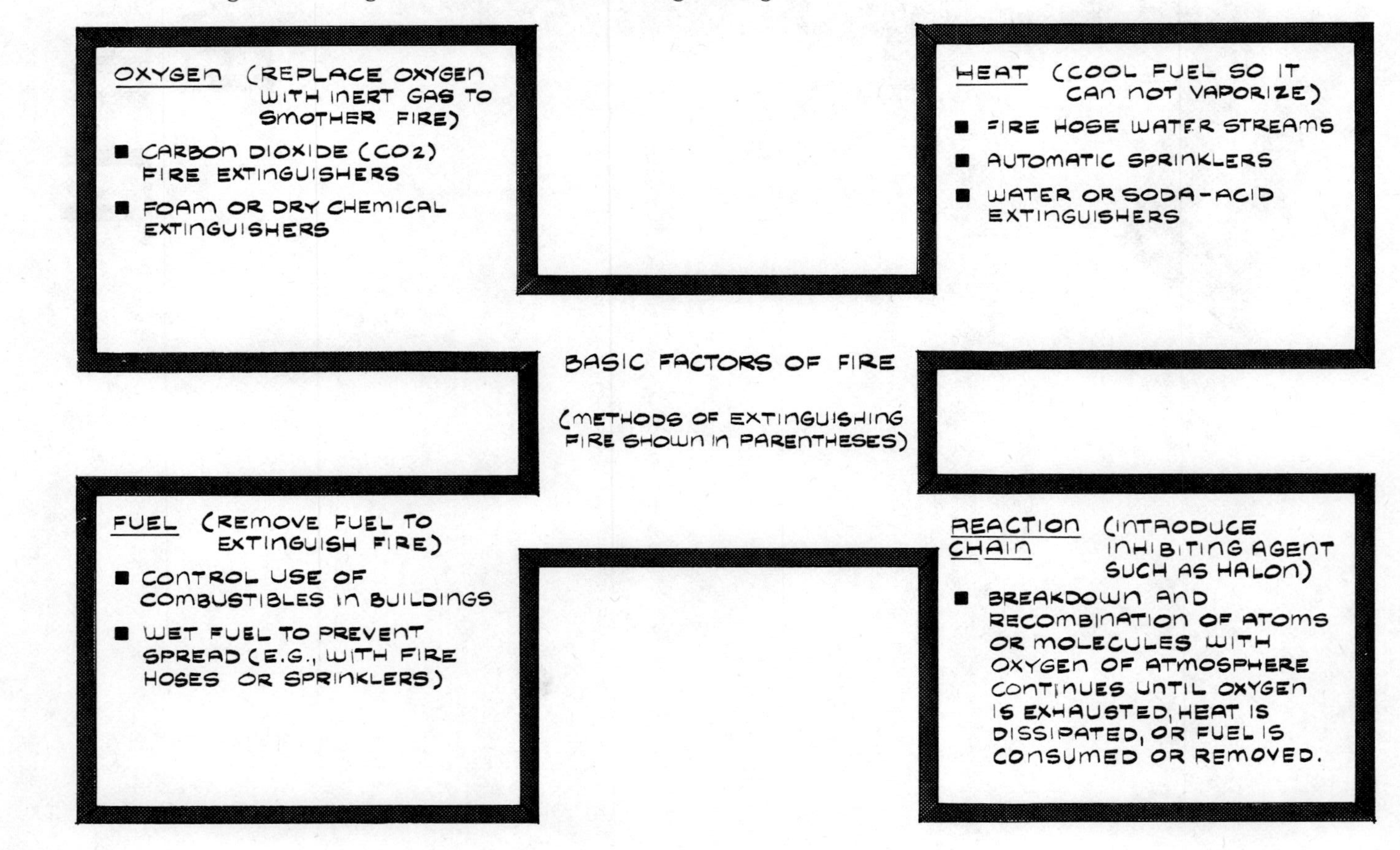

BASIC THEORY—FIRE BEHAVIOR IN CONFINED SPACES

Fire moves upward rapidly by convection and can spread laterally along ceilings. Combustion generates gases as well as heat and smoke. Hot gases, if confined, will tend to rise (or "mushroom") filling the entire room. In the example wastebasket corner fire shown below, the temperatures in the plume can be as high as 1200 to 1800°F. Although temperatures in the rest of the room may only increase slightly, rescue of occupants may be impossible unless fire gases are removed quickly. The majority of fire deaths in buildings are caused by smoke and toxic gases. The mushrooming hot gases and heated ceiling and upper wall surfaces radiate energy down to the unignited materials below in the room. The closer the ceiling is to the burning materials, the more heat it will radiate. Ceiling and wall finish materials that are combustible will have a significant influence on the fire growth and spread.

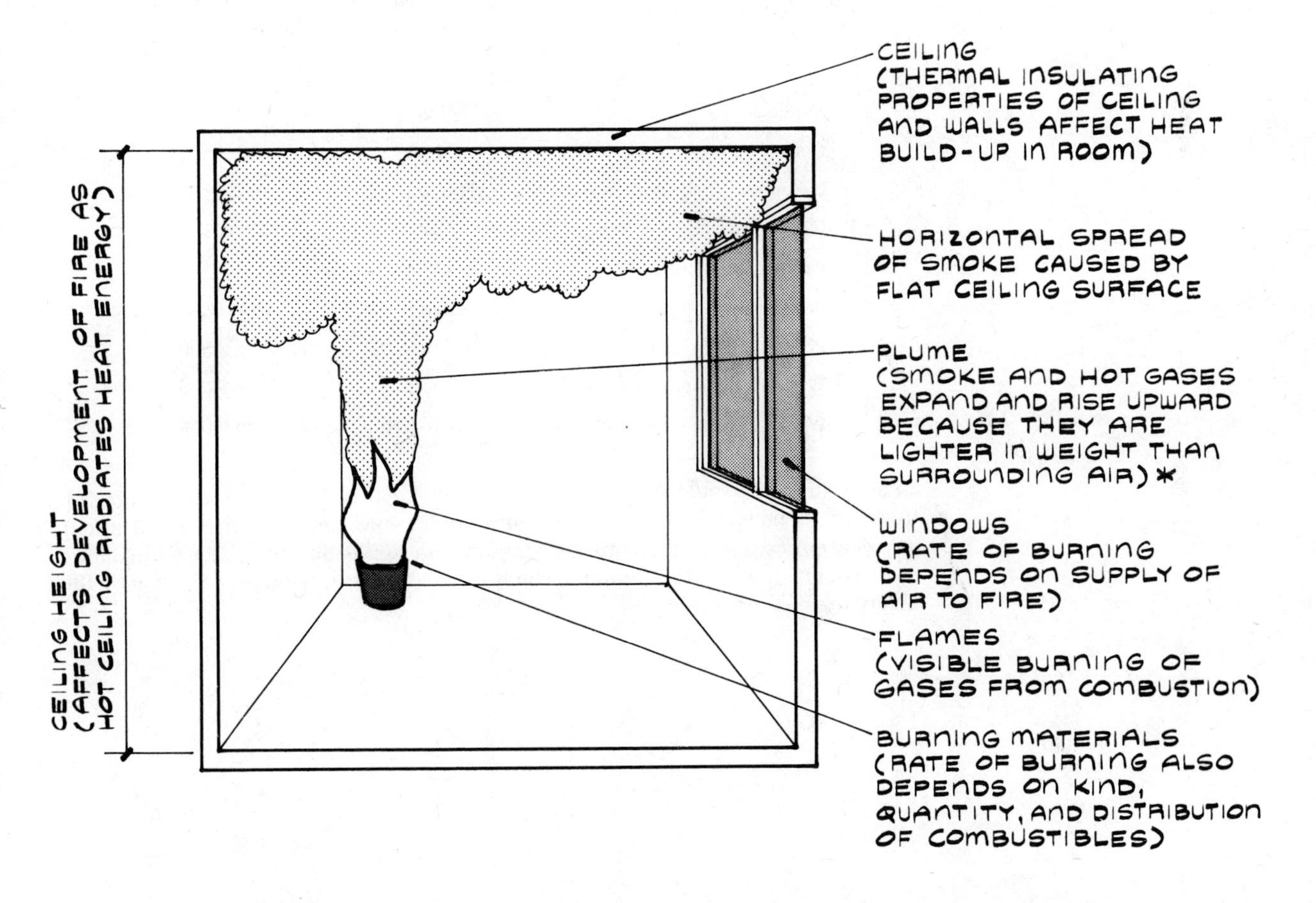

* THE BUOYANT FORCE IN THE IMMEDIATE VICINITY OF FIRE IS DUE TO A 3 TO 1 (OR MORE) THERMAL EXPANSION OF AIR VOLUME.

Note: Corner fire tests have been developed by Underwriters Laboratories, Factory Mutual System, and other laboratories to evaluate the flame spread characteristics of various finish materials. In a corner fire, wall surfaces are subjected to intense heat from radiation and confined hot gases.

IGNITION TEMPERATURE AND FUEL CONTRIBUTION

Ignition temperature is the minimum temperature to which a material must be heated in air to initiate self-sustained burning (or combustion).* Fuel contribution is a measure of the released heat from burning materials. Ignition temperature in degrees Fahrenheit (°F) and fuel contribution in British thermal units per pound (Btu/lb) of material are listed below for various materials. (A Btu is the amount of heat required to raise 1 lb of water 1°F.) Ignition temperatures in the table are approximate as ignition varies with moisture content (energy is required to evaporate moisture), shape, duration of exposure to heat, and other factors. However, they clearly indicate that building materials can be easily ignited in fully developed fires where temperatures well above 1000°F occur.

Material	*Ignition Temperature (°F)*	*Fuel Contribution (Btu/lb)*
Asphalt	905	17,150
Cotton batting	450	7,100
Gasoline	500 to 850	19,250
Oil, cottonseed	650	17,100
Paper, newsprint (cuts)	450	7,900
Polystyrene	900 to 950	18,000
Polyvinyl chloride (rigid)	800 to 900	7,500 to 9,500
Wood (sawdust and shavings)	400 to 500	7,500 to 9,050
Wood (fir, oak, pine, etc.)	450 to 500	7,500 to 9,050

*The term combustible refers to a material or structure that can burn. Be careful, however, as materials that may not burn under one condition may burn under another. For example, structural steel is noncombustible, but steel in the form of fine steel wool with a film of oil on it is combustible (due to increased surface area and ability to absorb heat).

BASIC THEORY—HEAT FROM FIRE

Building occupants cannot remain in superheated air for more than a few minutes. Shown below are examples of human response to elevated air temperatures from fire. For example, air temperatures 10 ft ahead of the flames can reach 300°F which, as indicated by the shaded portion of the thermometer, is tolerable for less than 5 minutes.

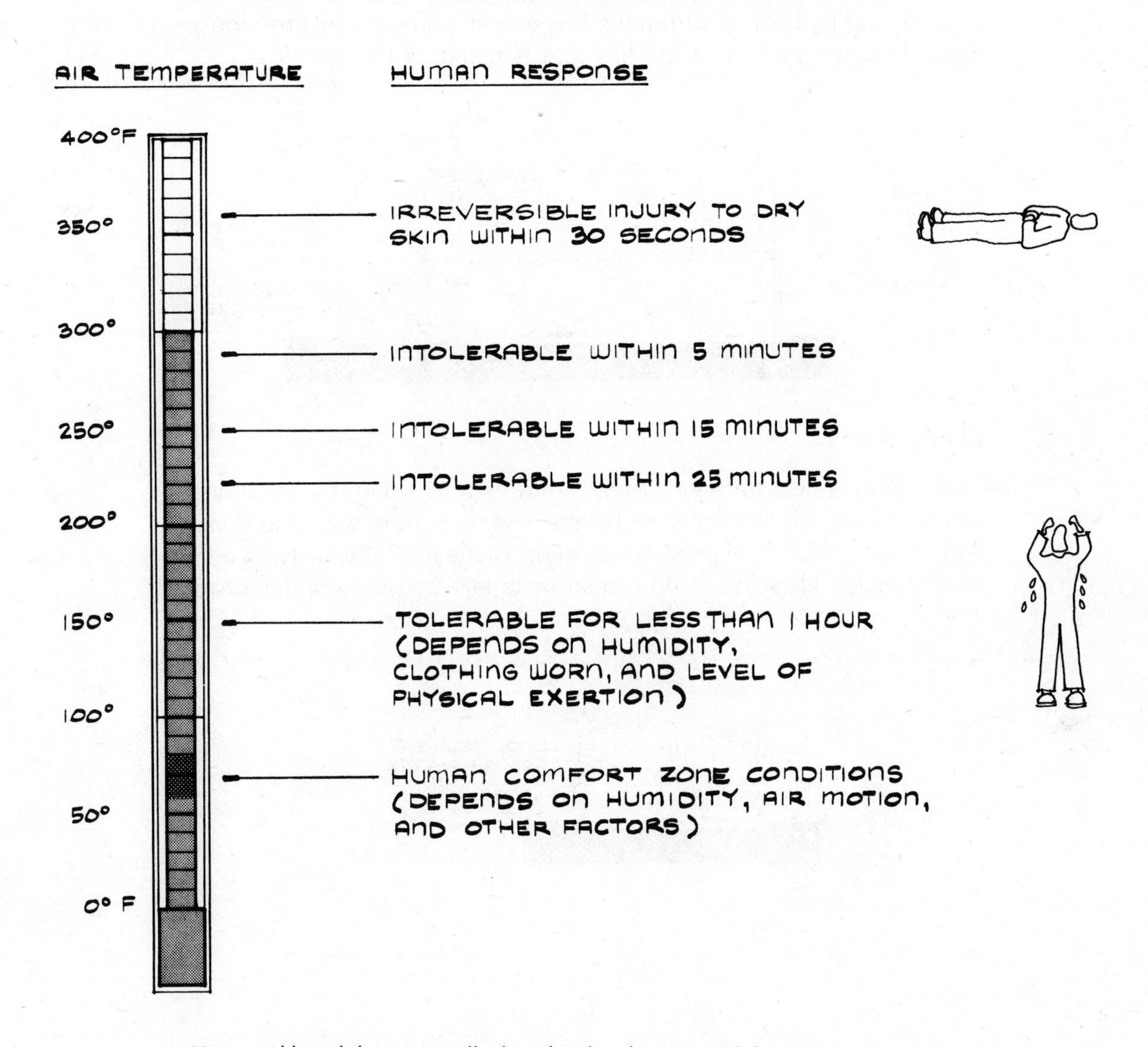

Note: Although heat is usually thought of as the principal danger from fire, it often is only the final blow to an already dying victim!

BASIC THEORY—HEAT TRANSFER

Shown below are examples of heat transfer within and between buildings by convection, conduction, and radiation.

Convection

Convection is heat transfer by air motion. During a fire heated air expands and moves away from the fire, exerting pressure against doors, dampers in air ducts, and the like and penetrating the openings.

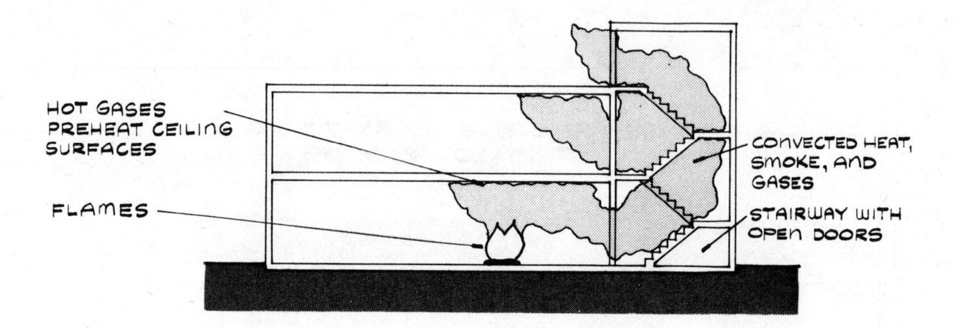

Conduction

Conduction is heat transfer through solid materials. During a fire heat can be transferred through steel beams, metal conduit, wire and ducts, and so on which are good conductors of heat. Conversely, wood, mineral wool, glass-fiber, and similar materials are poor conductors of heat.

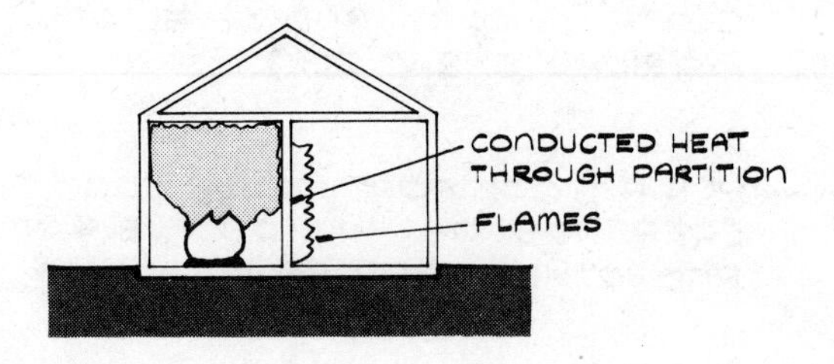

BASIC THEORY—HEAT TRANSFER (Continued)

Radiation

Radiation is heat transfer by electromagnetic waves. During a fire hot surfaces can radiate heat, igniting combustibles considerable distances away. For example, fire can leapfrog across wide malls in shopping centers by radiation of heat energy as shown by the sketch below. Radiant heat energy increases rapidly as the source becomes hotter. The radiated energy is proportional to the fourth power of the absolute temperature (e.g., °F + 460) of the source.

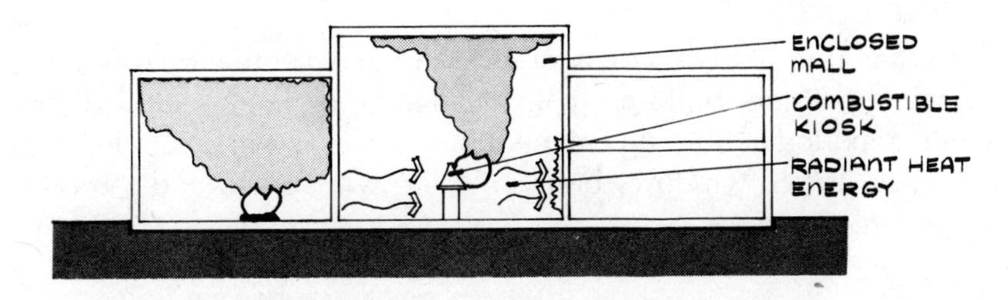

Note: For comprehensive tables of heat flow properties of building materials and example problem solutions, see M. D. Egan, *Concepts in Thermal Comfort*, Englewood Cliffs, N. J.: Prentice-Hall (1975), pp. 47–55. In rooms with low ceilings that are insulated by materials having low thermal conductivity (k) values, dangerous fire conditions can be reached in only a few minutes after ignition.

BASIC THEORY—VENTILATION AND FUEL CONTROLLED FIRES

The major factors that determine the rate of burning in a building compartment or confined area are the fuel load and the ventilation. The fuel load from building furnishings can vary from room to room and change with time. It is, therefore, difficult to predict if a fire will be ventilation or fuel controlled. Consequently, fire-resistance requirements for buildings should be determined by assuming that the more destructive ventilation controlled fire will occur.

Ventilation Controlled

Where the fuel load is considerable and the ventilation poor (e.g., basements, theaters, buildings with small areas of fixed, sealed glazing), the rate or period of burning can be prolonged as it will be controlled by ventilation. When windows break, the fire will spread and the rate of burning will increase.

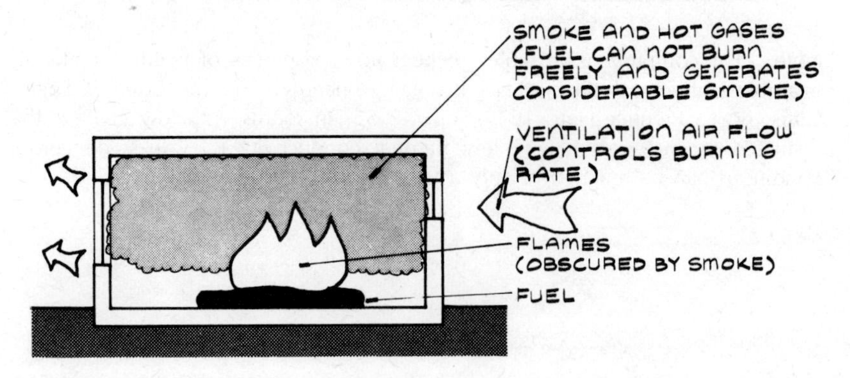

Fuel Controlled

Where the fuel load is small and the ventilation sufficient (e.g., buildings with very large window openings), the fire will be controlled by the surface area of the fuel. Fuel controlled fires are of short duration and the room temperature is not excessively high due to infiltration of cooler outdoor air. In multistory buildings with low ceilings, however, flames can spread from floor to floor through exterior openings. With high ceilings, flames would be confined to the room.

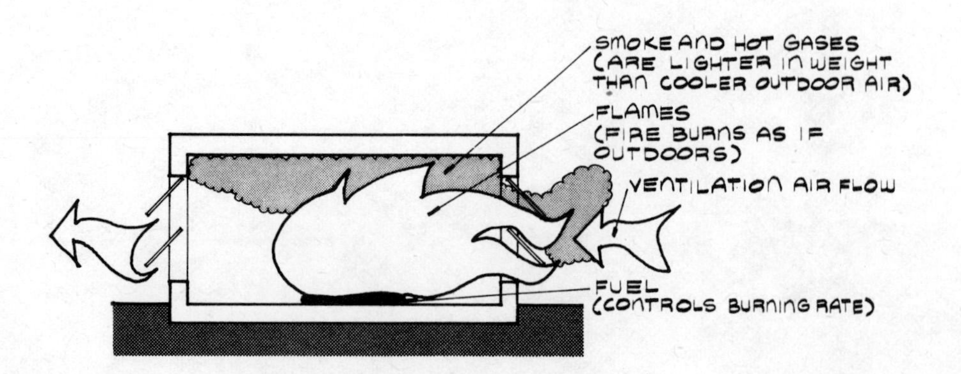

BASIC THEORY—TIME-TEMPERATURE CURVE FOR FIRES

An example time-temperature curve for an actual building compartment fire is shown below. During the "fire growth" time period, escape from buildings is usually possible if occupants are familiar with the egress routes and the exits and access to exits are protected (see Chapter 6). The time of burning preceding fully developed fire conditions is also critical for effective fire suppression operations. It is essential, therefore, that fires be detected early so that time will be available for escape by occupants and response by fire fighters (see Chapter 4).

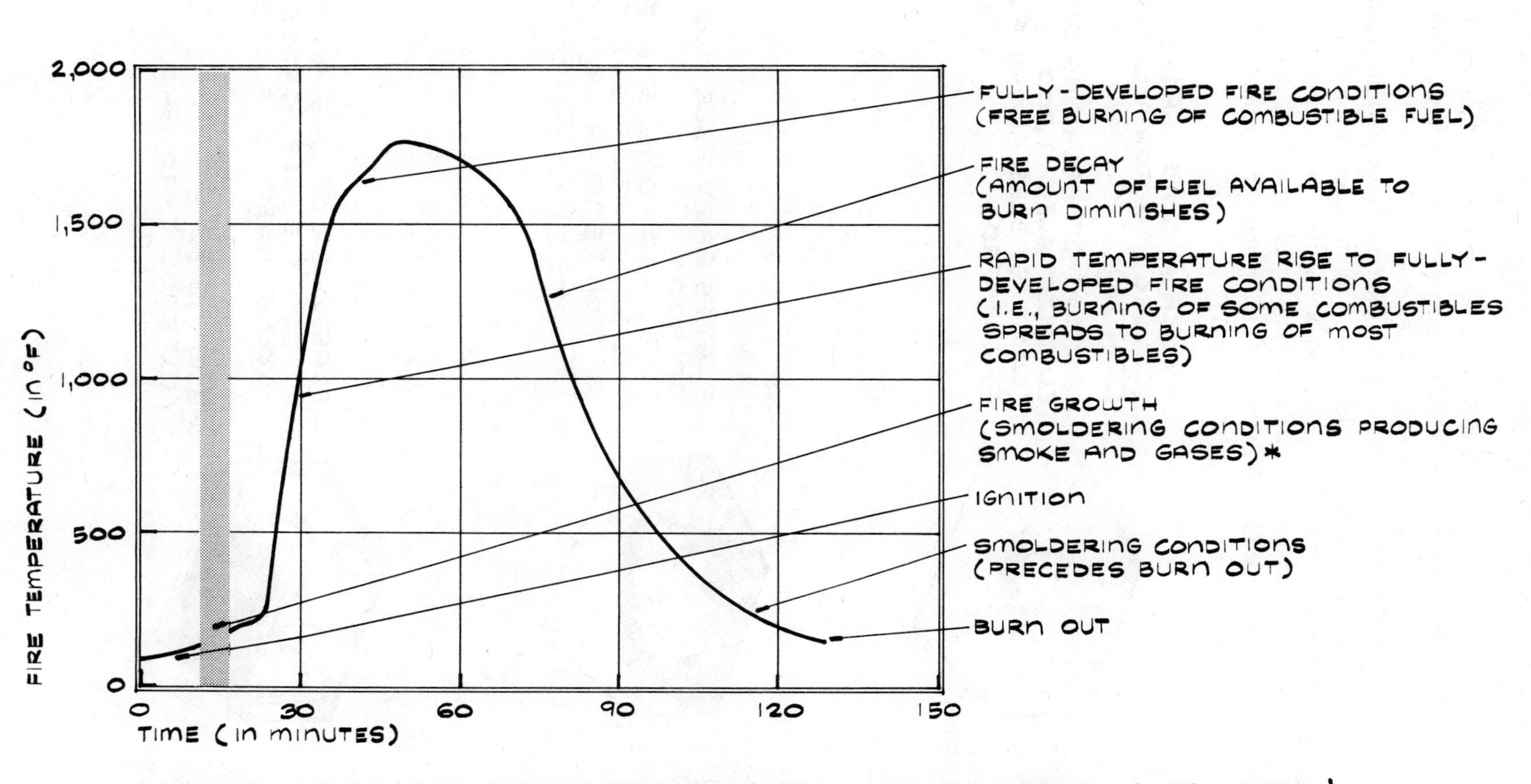

BASIC THEORY—FLASHOVER

Flashover is usually defined as the sudden and dramatic simultaneous ignition of most combustible materials and gases in a room or area. It occurs when room temperatures near the ceiling rapidly rise to 800 to 1200°F. The time between the ignition of fire in a room and flashover is critical to the safe evacuation of room occupants and to effective fire fighter rescue and suppression operations. The sketches below show the development of a typical fire in a small room. Note that well-insulated walls retard heat flow from the room and hence would reduce the time between ignition and flashover.

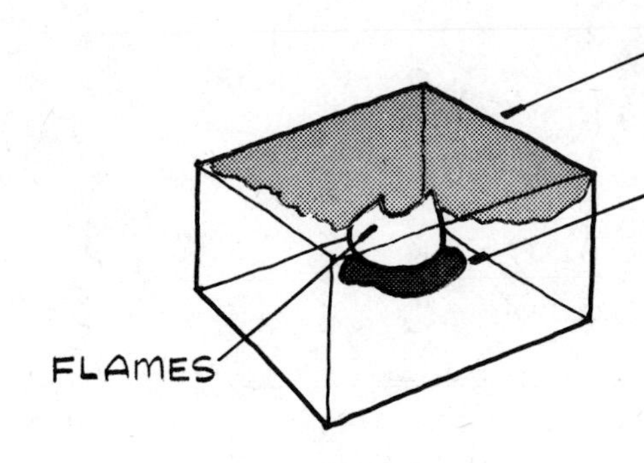

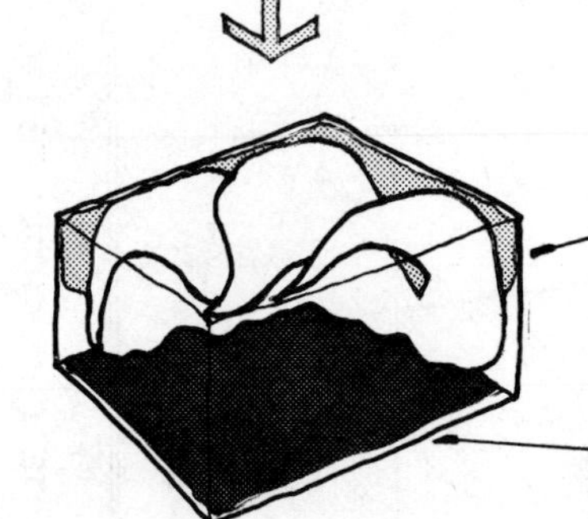

BASIC THEORY—RELEASE RATE TEST APPARATUS

The release rate test apparatus (developed at Ohio State University) shown below consists of a small test chamber containing a radiant panel and a pilot flame burner to ignite the test sample. Test samples can be tested in vertical and horizontal orientations. Air flows through the chamber at a controlled rate and thermocouples measure the rise in air temperature in the chamber. A photocell, mounted at the air outlet, measures the optical density of smoke. The apparatus is calibrated to measure rate of heat release, rate of smoke release, and concentration of toxic gases. Rate of release data (e.g., time to reach hazardous conditions) are extremely vital information as escape to safety depends on how much time is available before escape is blocked by heat, smoke, or toxic gases.

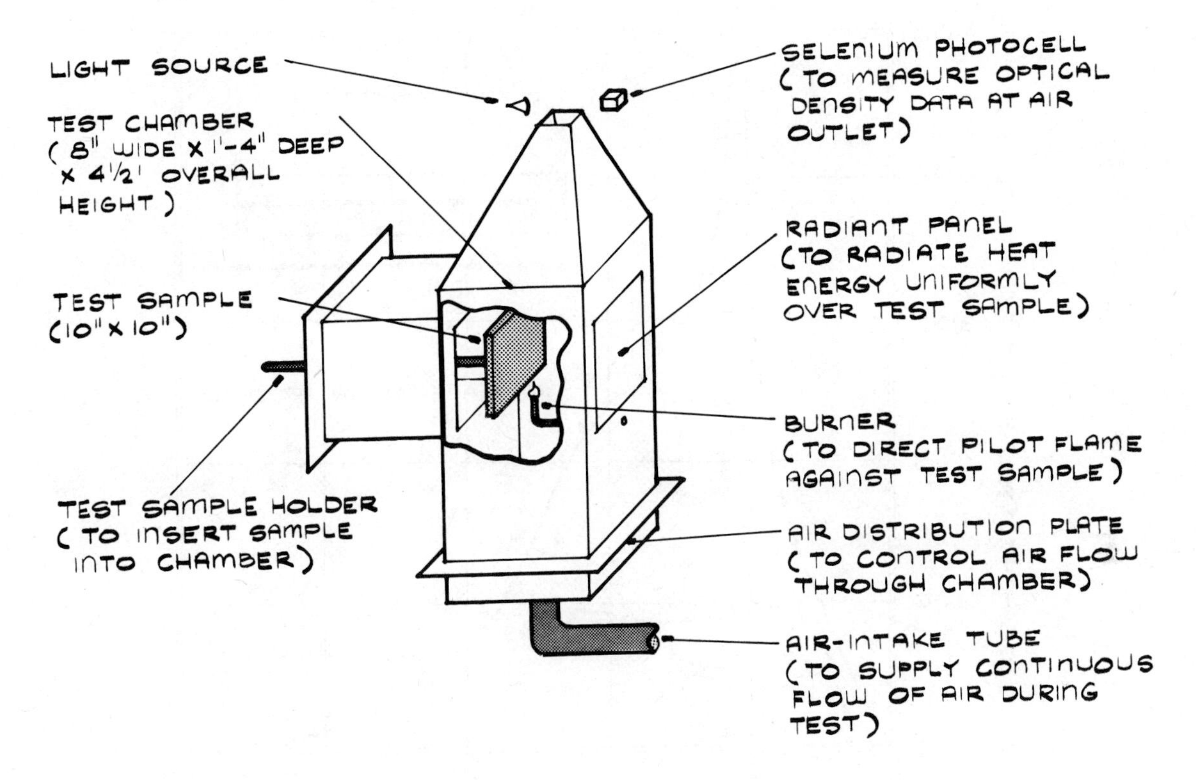

Note: Rate of heat release also can be measured by a combustion chamber developed by the National Bureau of Standards (cf., W. J. Parker, and M. E. Long, "Development of a Heat Release Rate Calorimeter at NBS," U. S. National Bureau of Standards, NBS Report 10462, March 1972).

REFERENCE

Smith, E. E., "Fire Hazard Characteristics of Duct Materials," *ASHRAE Journal*, July 1972.

BASIC THEORY—HEAT RELEASE RATE

The weight of building materials and furnishings determines the quantity of heat released during a fire. The higher the heat release rate of building materials, the more rapid the fire growth. Surface area, texture, orientation, and fire exposure conditions will determine the heat release rate for furnishings with identical chemical composition. The curves below show rate of heat release in Btu per minute at room temperature (to represent the ignition and smoldering stages of fire) of identical weight samples of nylon upholstery materials padded with urethane foam and cotton batting. Note that there will be considerable variations in rate of heat release among various urethane foams due to physical property and chemical composition differences.

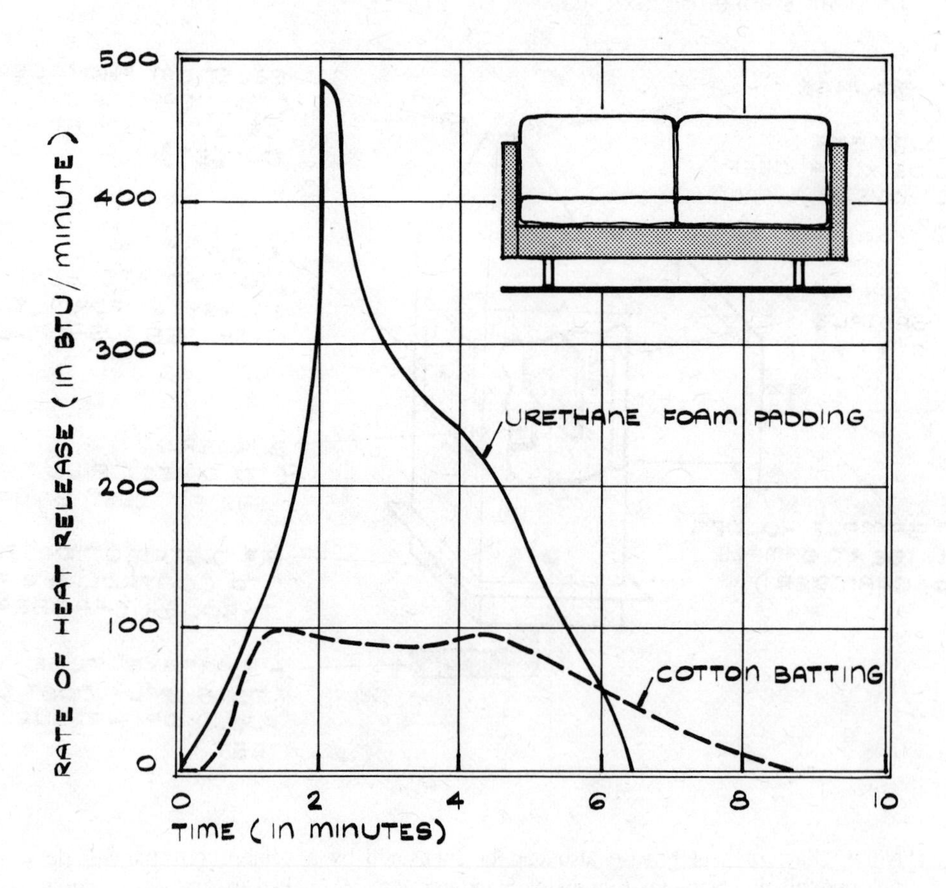

Note: At exposure temperatures of 970°F and above, there would be little difference in heat release for the urethane and cotton paddings. Consequently, a material's performance should be based on test conditions that represent anticipated fire conditions. For example, acoustical ceiling materials should have high test temperature exposure to represent fully developed fire conditions.

REFERENCE

Smith, E. E., "Measuring Rate of Heat, Smoke, and Toxic Gas Release," *Fire Technology*, Vol. 8, No. 3, August 1972.

BASIC THEORY—SMOKE RELEASE FROM BURNING MATERIALS

The curves below show the effects on visibility of burning wood and vinyl covered urethane padding. At 6 minutes after ignition, smoke from the vinyl-urethane reduces visibility of a standard exit sign to less than 1 ft. Smoke from the wood, however, only reduces visibility to about 14 ft (see dashed lines on graph). Levels of smoke produced by burning materials are vital considerations for firesafety design. Tests for smoke release must represent anticipated fire conditions as identical burning materials can produce different smoke (i.e., chemical composition and quantity) under different fire exposures.

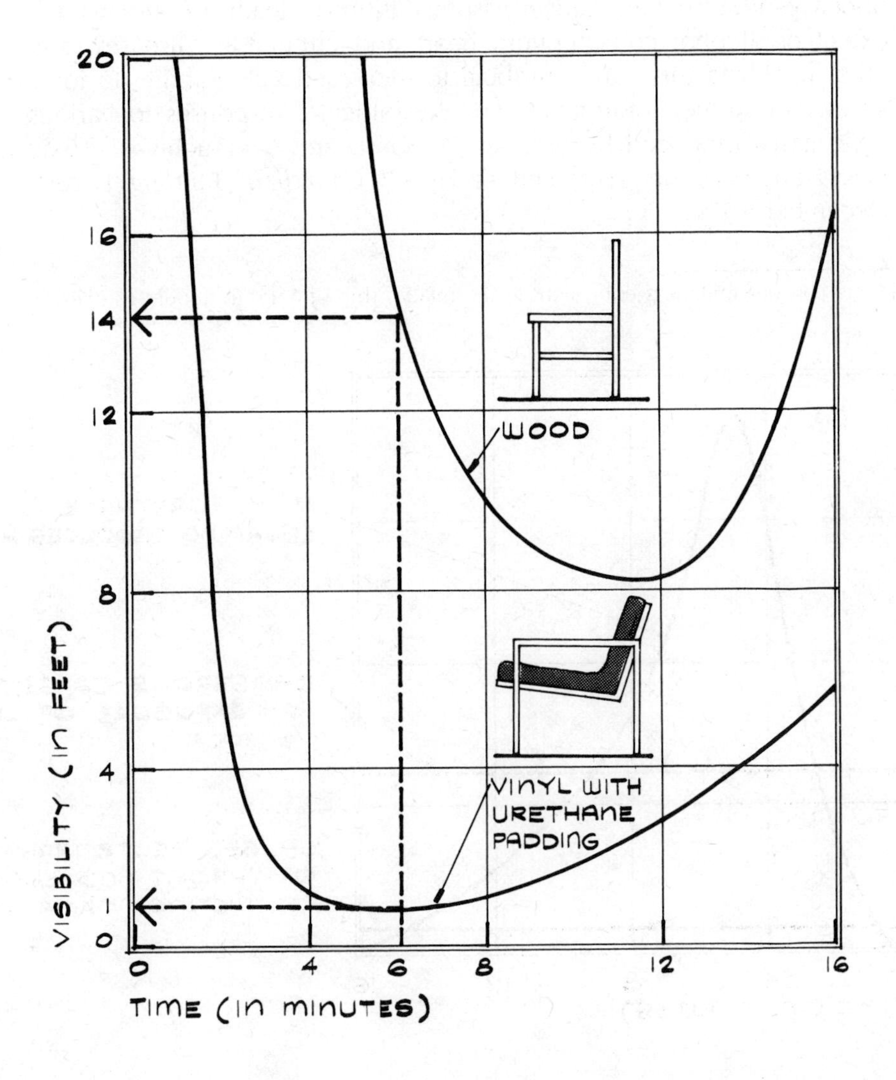

REFERENCE

"Tests Show Combustion Products of Some Synthetics to be Lethal," *News in Engineering*, Ohio State University, May 1973.

BASIC THEORY—TOXIC GASES FROM BURNING MATERIALS

Toxic gases from burning materials can cause fatalities if they are present in sufficient concentrations or as dangerous compounds with other gases. The graphs below show concentrations, in parts per million (ppm)* of air, for hydrogen chloride (HCl) produced by burning test samples of vinyl-urethane and odorless carbon monoxide (CO) produced by burning test samples of wood. Dangerous threshold concentration levels are also indicated on the graphs. Within only 3 minutes after ignition, the vinyl-urethane sample produced a dangerous HCl concentration. At low concentrations, HCl causes coughing, choking, and irritation of the eyes and at high concentrations can damage the upper respiratory tract, causing asphyxiation or death. In humans, the extent of alcohol consumption, heart and circulatory diseases, and other health factors can contribute to increased vulnerability to toxic gases. For further information on physiological responses to various toxic gases produced by fire, see K. Sumi, and Y. Tsuchiya, "Toxic Gases and Vapours Produced at Fires," *Canadian Building Digest*, December 1971.

*One part per million is equivalent to the relationship of 1 in. to about 16 miles.

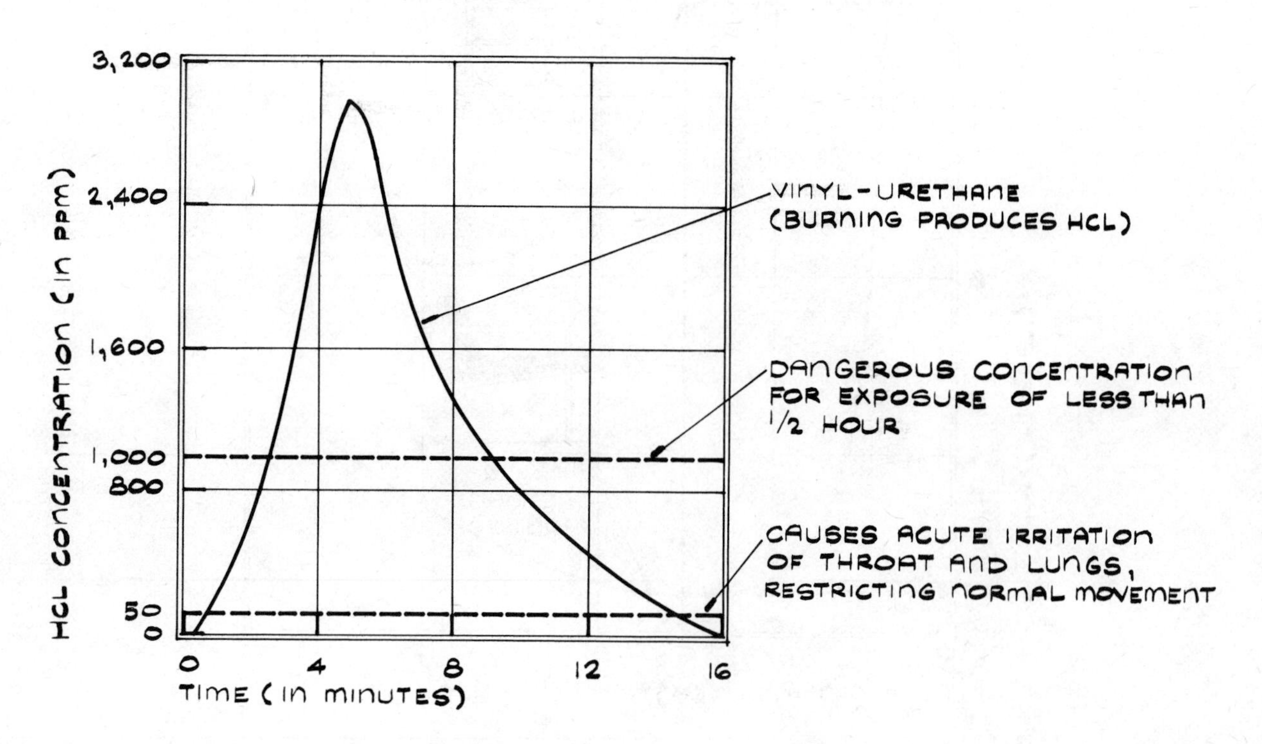

BASIC THEORY—TOXIC GASES FROM BURNING MATERIALS (Continued)

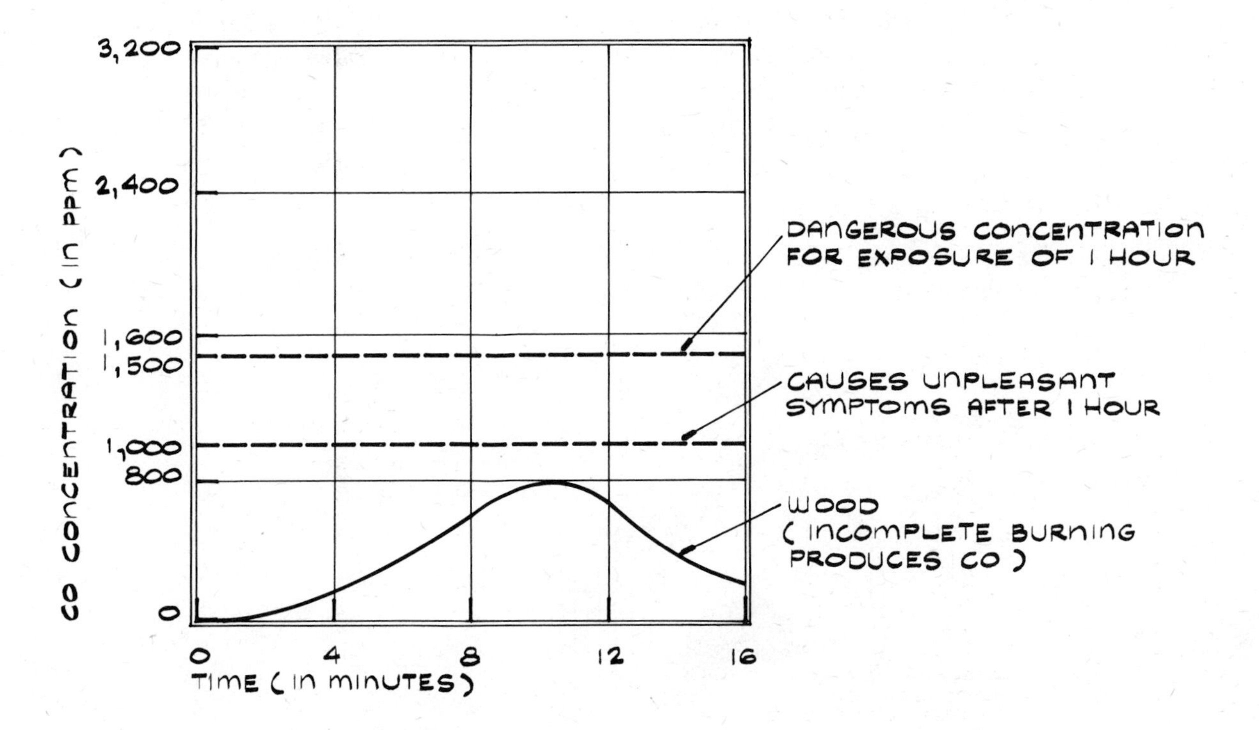

Note: HCl also is produced by the burning of many common plastics (e.g., polyvinyl chloride). Burning materials can produce many other toxic gases in dangerous concentrations: for example, hydrogen sulfide (H_2S) from burning leather and wool, sulfur dioxide (SO_2) from burning rubber and wood, and hydrogen cyanide (HCN) from burning polyamides and modacrylics.

REFERENCE

"Tests Show Combustion Products of Some Synthetics to be Lethal," *News in Engineering*, Ohio State University, May 1973.

BASIC THEORY—FLAME SPREAD

Often the ability to escape a building fire depends on the rate of flame spread along combustible surfaces. Shown below are examples of flame spread along thin, combustible textile curtains and plywood wall finish. Note that fabrics can be chemically treated to prevent (or retard) ignition.

Vertical Spread

Flame spread in an upward direction can be rapid as convected heat accelerates burning by preheating fuel above the flames.

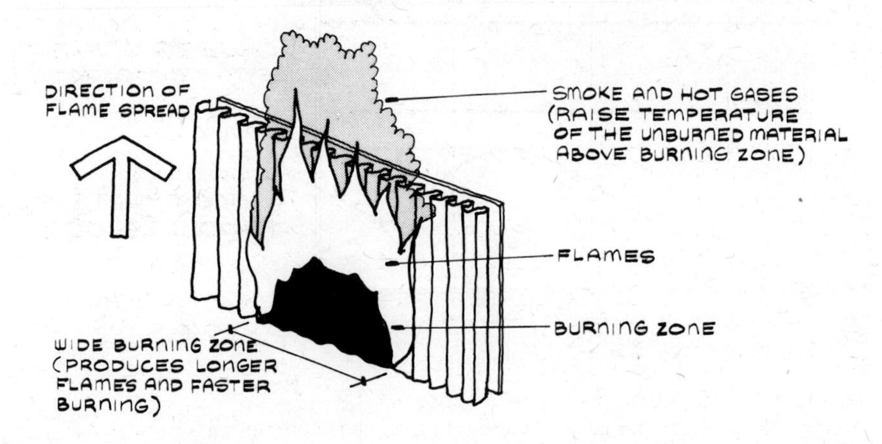

Horizontal Spread

Flame spread in the horizontal direction is less rapid due to heat being convected away from unburned material. As a consequence, flame spread along ceilings generally is more rapid than along floors.

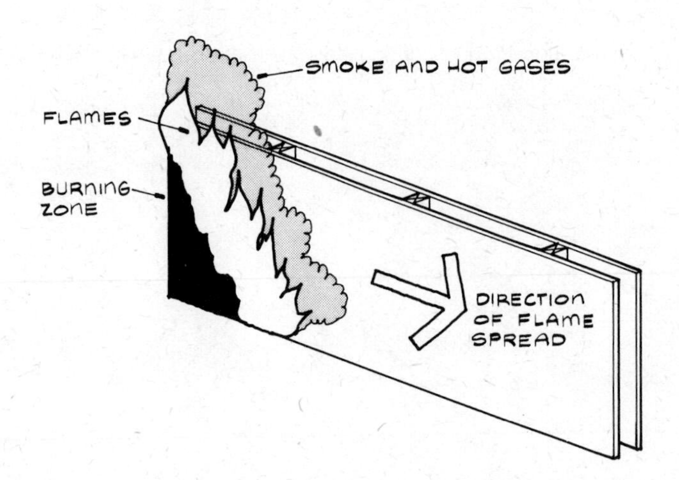

BASIC THEORY—FLAME SPREAD TUNNEL APPARATUS

In the flame spread tunnel test (called the Steiner tunnel), a 20 in. wide by 25 ft long test sample is placed on the underside of a removable top panel as shown by the sketch. One end of the sample is ignited by gas burners and the surface burning characteristics are observed through side window openings. The test is designed to simulate a fire exposure of about 1400°F in the area of the flames. Flame spread test data (i.e., the time for flame to travel down the length of the tunnel or until it ceases or recedes) is compared both to asbestos cement board which is rated at 0 and to select grade red oak rated at 100 for the same test conditions. Asbestos cement board and red oak also are used as reference points for "fuel contributed" and "smoke developed" ratings determined by the tunnel test.

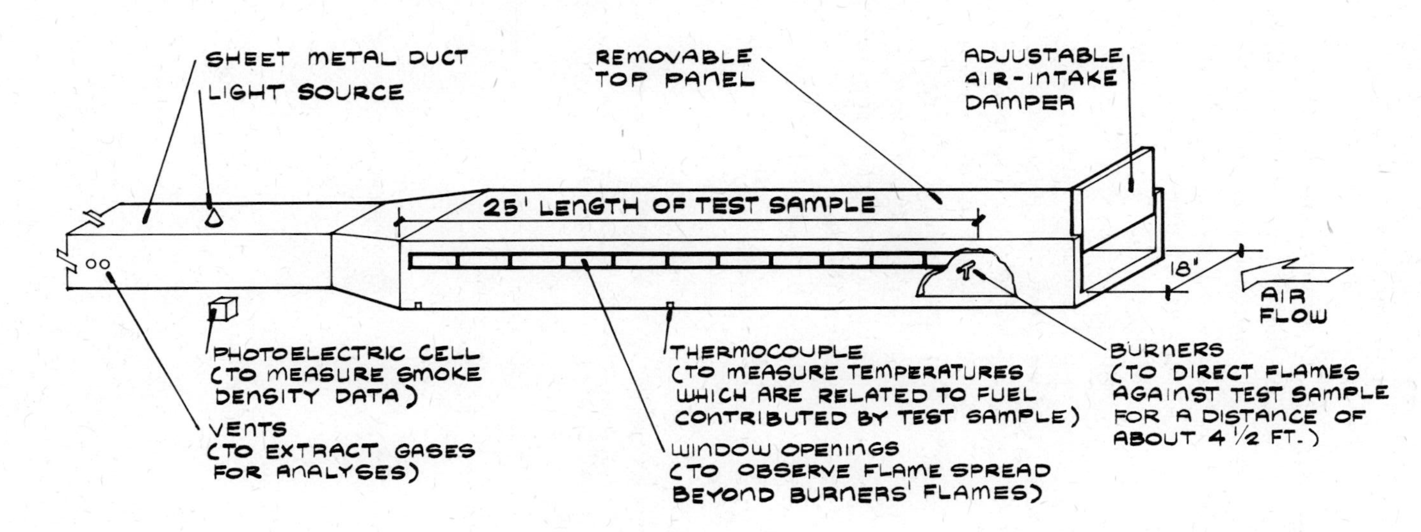

TEST REFERENCE

"Test for Surface Burning Characteristics of Building Materials," ASTM Method E 84.

FLAME SPREAD TUNNEL TEST DATA

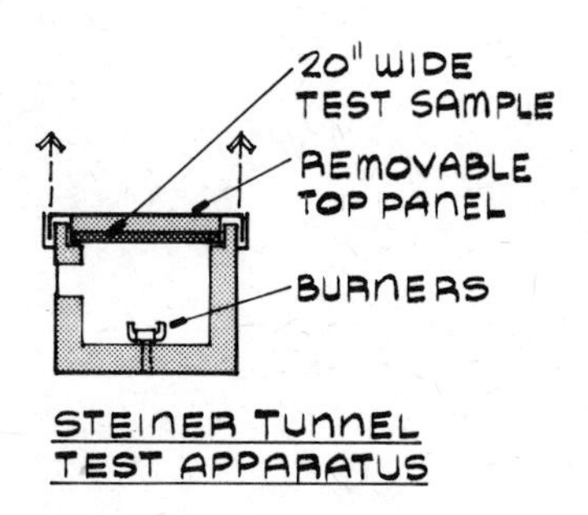

STEINER TUNNEL TEST APPARATUS

Flame spread ratings are listed opposite for various interior finish materials. Although the values given are typical, flame spread ratings for design purposes should be based on data from up-to-date tests by recognized fire testing laboratories (e.g., Underwriters Laboratories, National Bureau of Standards, Southwest Research Institute). The flame spread rating is determined in the tunnel test by observing surface burning along a horizontal test sample exposed to flames from underneath. For comparison purposes, a scale from 0 (asbestos cement board) to 100 (red oak) is used. For red oak, flame will travel 19½ ft in 5½ minutes. Consequently, a test sample that has a flame travel of 19½ ft in 4 minutes will have a flame spread rating of (5.5/4) × 100 = 138. Tests should rate materials installed in the same manner as they will be used in buildings. The back-up surface (called "substrate") to which finish materials are applied can have a significant effect on flame spread. For example, materials with low flame spread ratings when attached to asbestos cement board will have higher ratings when attached to painted plywood (see D. Waksman, and J. B. Ferguson, "Fire Tests of Building Interior Covering Systems," *Fire Technology,* Vol. 10, No. 3, August 1974). Synthetic materials (e.g., polyurethane, polystyrene) supported on the underside of the tunnel's top panel by metal rods (or bars, wire mesh) soften, melt, and drip when tested. As a consequence, the actual fire hazard of synthetic materials may not be measured by the tunnel test.

FLAME SPREAD TUNNEL TEST DATA (Continued)

Material	*Flame Spread Rating*
Ceilings	
Glass-fiber sound-absorbing blankets	15 to 30
Gypsum board (with paper surface on both sides)	10 to 25
Mineral-fiber sound-absorbing panels	10 to 25
Shredded wood fiberboard (chemically treated)	20 to 25
Sprayed cellulose fibers (chemically treated)	20
Walls	
Aluminum (with baked enamel finish on one side)	5 to 10
Asbestos cement board	0
Brick or concrete block	0
Cork	175
Gypsum board	10 to 25
Pine, southern*	130 to 190
Plywood paneling*	70 to 275
Red oak	100
Floors	
Carpeting**	10 to 600
Concrete or terrazzo	0
Linoleum	190 to 300
Vinyl asbestos tile	10 to 50

*Plywood and treatable species of wood (e.g., southern pine, white fir, redwood) can be impregnated with salt solutions by a pressure process to achieve low flame spread ratings. Fire-retardant mastic paints also can be used to reduce flame spread ratings. However, the fuel contribution will not be reduced by these processes.

**Carpets installed on underlayments with good thermal insulation properties generally will have higher flame spread ratings due to the slower dissipation of heat.

Note: For a comprehensive listing of flame spread ratings for various materials, see current edition of the Underwriters Laboratories publication, "Building Materials Directory."

FLAME SPREAD RATINGS FOR INTERIOR FINISHES

Flame spread requirements for interior finish materials should be based on the occupancy and the location within a given occupancy. For example, a finish with a high flame spread rating generally would be more hazardous in an exit stairway or corridor (which can be a path of fire spread through buildings) than in an isolated room. Typical maximum flame spread rating restrictions are given in the table. For an example rating proposal, see "Fire Protection Through Modern Building Codes," American Iron and Steel Institute, 1971, p. 278. Variations in fire severity and distribution of finish materials often can obscure the distinction between flame spread ratings of 26 and 75 or of 76 and 200. For example, refer to "Full-Scale Fires Used to Test Interior Finishes," *Architectural Record,* November 1970.

MAXIMUM FLAME SPREAD RATING

Occupancy	*Enclosed Stairways*	*Corridors*	*Rooms (< 1500 sq ft)*	*Rooms (> 1500 sq ft)*
Residential	0 to 25	26 to 200	76 to 300	26 to 75
Educational	0 to 25	26 to 75	26 to 75	26 to 75
Institutional	0 to 25	0 to 25	0 to 75	0 to 25
Assembly	0 to 25	0 to 25	26 to 75	0 to 25
Office	0 to 25	26 to 75	76 to 200	26 to 75
Mercantile	0 to 25	26 to 75	76 to 200	26 to 75
Industrial	0 to 25	26 to 75	76 to 200	26 to 75
Storage	0 to 25	26 to 75	76 to 200	26 to 75
Hazardous	0 to 25	26 to 75	26 to 75	26 to 75

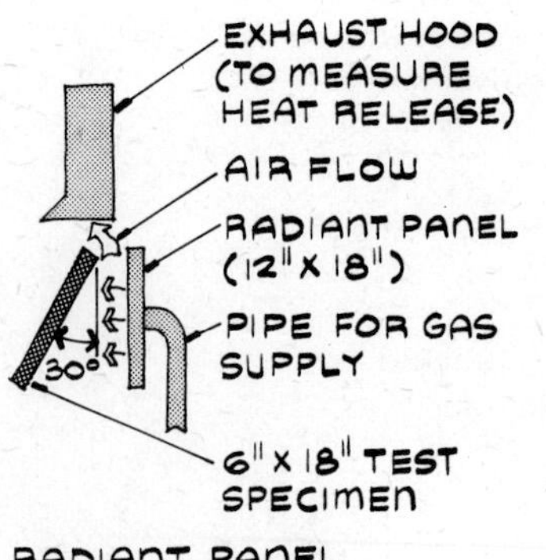

RADIANT PANEL TEST APPARATUS

Note: The ASTM E 162 radiant panel test method was developed for research and quality control purposes. The test, similar to the flooring radiant panel test described on page 23, subjects a 6 in. x 18 in. test specimen mounted on a 30 degree angle to a vertical radiant panel heat source. Maximum heat release and downward flame propagation rate from ignition at upper edge of test specimen are combined to give a flame spread index for the tested material. The index results are presented on a numerical scale that is similar to the results from the tunnel test.

BASIC THEORY—EXAMPLE SURFACE TEMPERATURES DURING ROOM FIRES

The graph below depicts typical ceiling and floor temperatures in simulated fire situations. As indicated by the large temperature differential during the 10 minute time period shown, floor covering materials are potentially less hazardous than comparable ceiling (or wall) surface materials. Nevertheless, carpeted floors can act as a path for fire spread in buildings (refer to J. G. Quintiere, and C. Huggett, "An Evaluation of Flame Spread Test Methods for Floor Covering Materials," in M. J. Butler, and J. A. Slater (eds.), "Fire Safety Research," U.S. National Bureau of Standards, NBS Special Publication 411, November 1974).

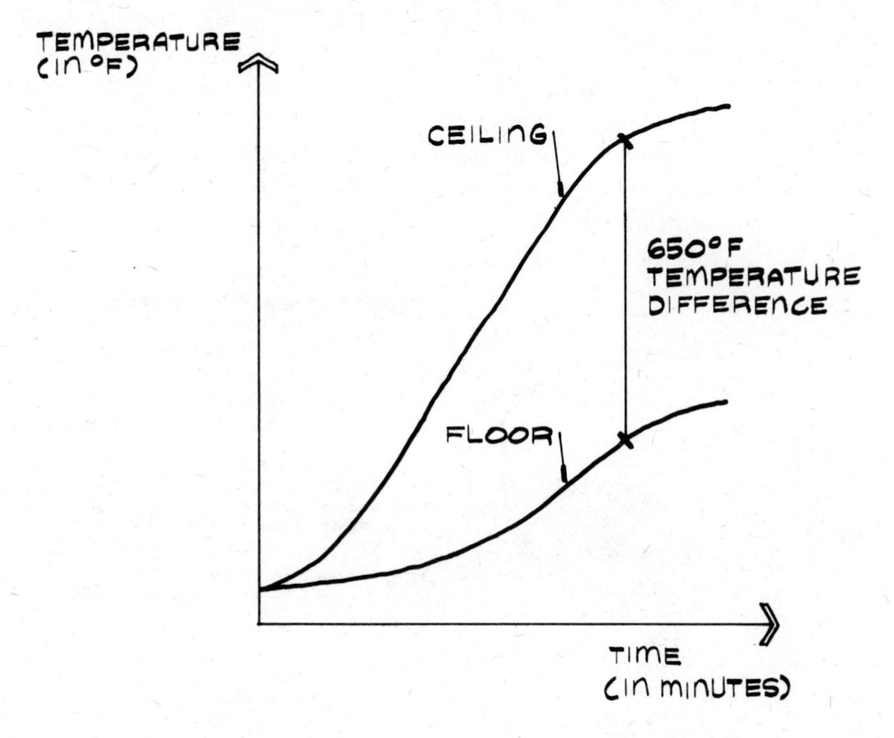

REFERENCE

Yuill, C. H., "The Flammability of Floor Coverings," *Journal of Fire & Flammability,* Vol. 1, January 1970, p. 66.

BASIC THEORY—CARPETS AND RUGS

Actual carpet flooring applications in buildings cannot be simulated in the tunnel test where carpeting is mounted in an abnormal upside down position (bonded to asbestos board or held in place by a wire mesh). Shown below are tests used to determine the surface flammability and fire propagation potential of carpets and rugs.

Pill Test

The ease of ignition and surface flammability of carpets can be evaluated by a methenamine pill to represent a small ignition source such as a slow burning cigarette, glowing ember from a fireplace, and so on. A 9 in. x 9 in. carpet test specimen, held down by a steel flattening frame with an 8 in. circular cutout, is placed in a 12 in. x 12 in. x 12 in. open top test chamber as depicted below. The pill is ignited and the test continues until the specimen burns out (passes test) or the charred portion reaches to within 1 in. of the circular cutout edge of the flattening frame (fails test).

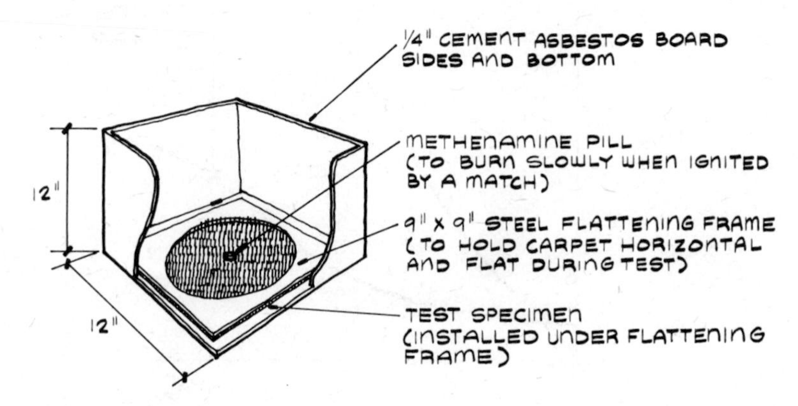

TEST REFERENCE

"Standard for the Surface Flammability of Small Carpets and Rugs (Pill Test)," DOC-FF-2-70, *Federal Register, Vol. 35, No. 251, December 29, 1970.*

Flooring Radiant Panel Test

The fire propagation potential of carpets in corridors and exitways can be evaluated by the flooring radiant panel test. An 8 in. x 39 in. carpet test specimen, installed in its normal orientation, is exposed to a source of varying radiant heat energy. The specimen is ignited by a small flame at the high energy end. The distance to where the specimen extinguishes itself determines its rating. The higher the radiant energy needed to propagate the fire, the more resistant the carpet will be to the spread of fire.

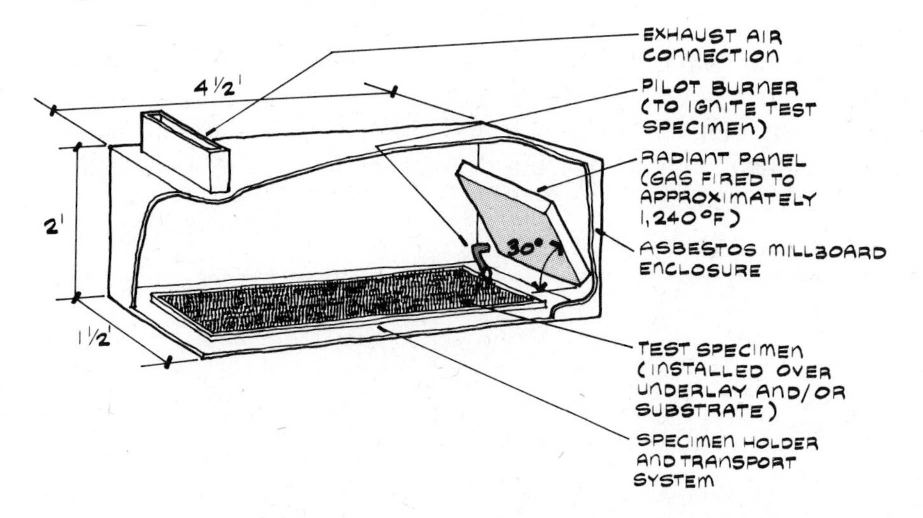

Note: A chamber furnace (similar to the tunnel apparatus, but with the test specimen mounted along the bottom of the chamber) has been developed by Underwriters Laboratories to evaluate surface flame propagation characteristics of floor coverings. In the test, an 8 ft long by 2 ft wide test specimen is exposed for a given time period to flames from overhead gas jets at one end of the chamber. The extent of surface burning is observed through side vision ports and a flame propagation index is calculated in inches/minute.

REFERENCE

Benjamin, I. A. and C. H. Adams, "Proposed Criteria for Use of the Critical Radiant Flux Test Method," U.S. National Bureau of Standards, NBSIR 75-950, December 1975.

BASIC THEORY—FIRE ENDURANCE

In the 1920s, S. H. Ingberg proposed that fire endurance requirements for office buildings and similar occupancies be prescribed on the basis of fire load (i.e., weight of combustible materials per square foot of floor area). The weight of fire load is determined by converting the fuel contribution from the various combustibles in a room to a weight of wood that would contribute an equivalent amount of heat energy. The relationship between fire load and fire endurance, shown by the graph below, was determined by a series of burn-out tests in existing buildings. For a fire load of 20 psf (pounds per square foot), a fire endurance of 2 hours would be anticipated. (See dashed lines on graph.) Fire endurance requirements, determined by fire load, assume that all fires burn approximately at the same rate and that heat is absorbed by the surrounding building surfaces at the same rate. However, in modern buildings the fire endurance also depends on the kind of combustibles and on the ventilation.

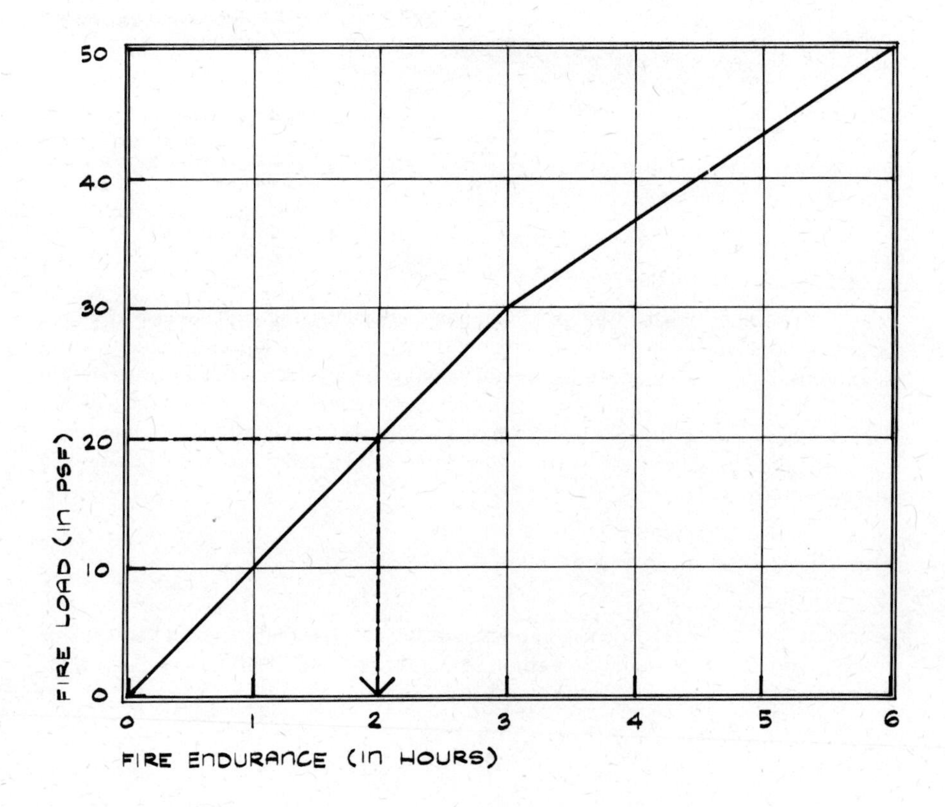

REFERENCE

Ingberg, S. H., "Tests of the Severity of Building Fires," *Quarterly of the NFPA*, Vol. 22, 1928.

FIRE LOADS

The table below gives typical "fire load" data in pounds per square foot (psf) of floor area for various occupancies. Fire loads are likely to increase with time and can vary considerably within buildings of the same type occupancy as indicated by the table. Additionally, changes in building occupancy and in the distribution of combustibles within buildings can significantly alter the fire load.

Type of Space	*Fire Load (psf)*
Apartments*	8 to 10
Classrooms, typical	7
File rooms**	4 to 86
Library stacks, typical	36
Offices**	2 to 45
Reception areas	3 to 9
Toilet rooms	2 to 10

*Average for whole dwelling unit is given in table. Closets can exceed 50 psf!

**Fire loads will be reduced considerably when combustibles are stored in steel filing cabinets, safes, vaults, or steel desks. For example, where one-half to three-fourths of ordinary combustibles are stored in steel filing cabinets and desks, the fire load can be reduced by 80%, and where greater than three-fourths, by 90%.

SOURCES

Culver, C. G., "Survey Results for Fire Loads and Live Loads in Office Buildings," U.S. National Bureau of Standards, Building Science Series 85, May 1976.

Ingberg, S. H. et al., "Combustible Contents in Buildings," U.S. National Bureau of Standards, BMS Report 149, July 1957.

BASIC THEORY—FIRE-RESISTANCE TESTS

In a fire-resistance test, building constructions are exposed to heat from a test furnace. The standard fire exposure is defined by the time-temperature curve shown below. The curve represents the heat that would be produced by a fire load of about 10 psf/hour of test. Specific conditions of acceptance vary according to the type of construction element being tested and its intended use. In general, however, if the test specimen withstands the simulated fire exposure for a certain time period supporting the applied load (for loadbearing constructions), does not allow flames or hot gases to ignite the cotton waste on the unexposed side, and, in some tests, withstands a fire hose stream of cold water, then it is fire-resistance rated. In addition, the test specimen may be required to prevent the average surface temperature, measured on the laboratory side, from exceeding 250°F above its initial temperature (or 325°F rise at any point). The test results are expressed in terms of time of fire exposure during which the above endpoint criteria are satisfied (e.g., $\frac{1}{2}$ hour, 1 hour, 2 hour, etc.).

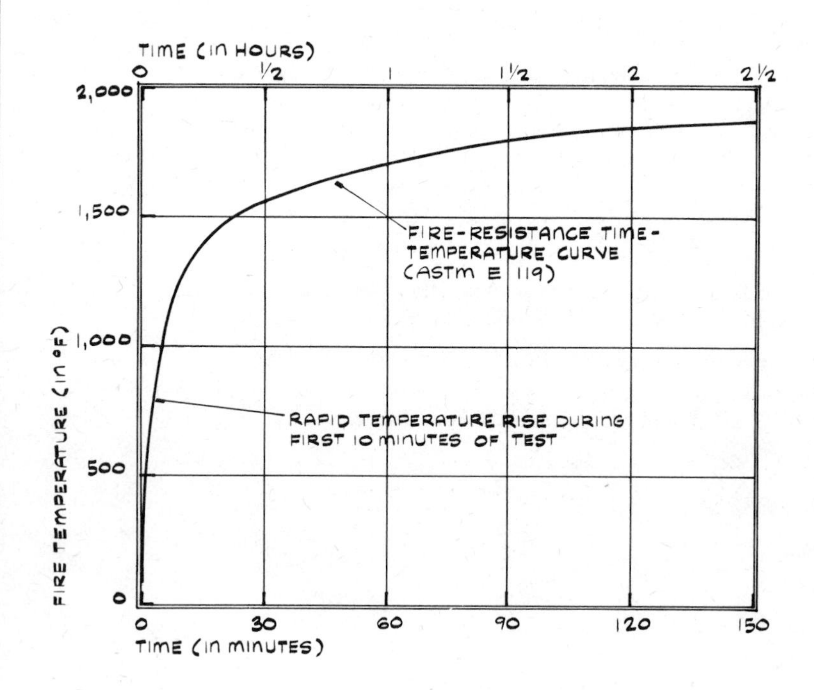

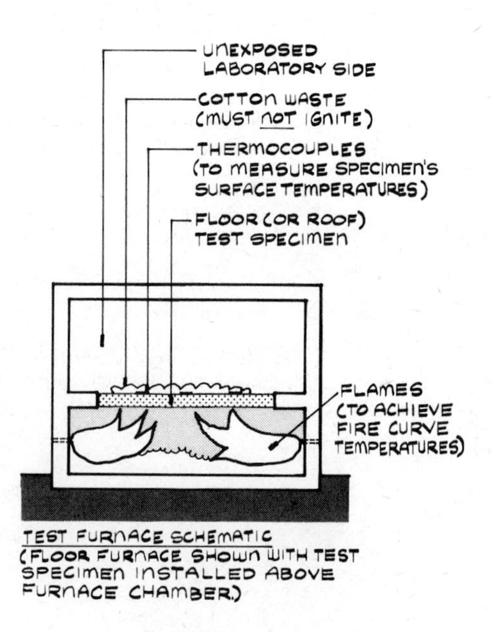

TEST REFERENCE

"Fire Tests of Building Construction and Materials," ASTM Method E 119.

BASIC THEORY—FIRE SEVERITY

The area under the time-temperature curve for a fire can be used to represent its destructive potential, called "fire severity." The severity of actual fires can be related to the standard ASTM time-temperature curve by determining the area under the standard curve which is equal to the area under the actual fire curve. For example, the actual fire shown below at the left has a burning time of t_1, whereas the equivalent severity under the standard ASTM curve occurs after a longer burning time of t_2. As a consequence, the fire-resistance time reported from the standard laboratory test should be considered a fictitious time as failure could occur in buildings well before the time reported (i.e., closer to t_1 than t_2 in the example curves below).

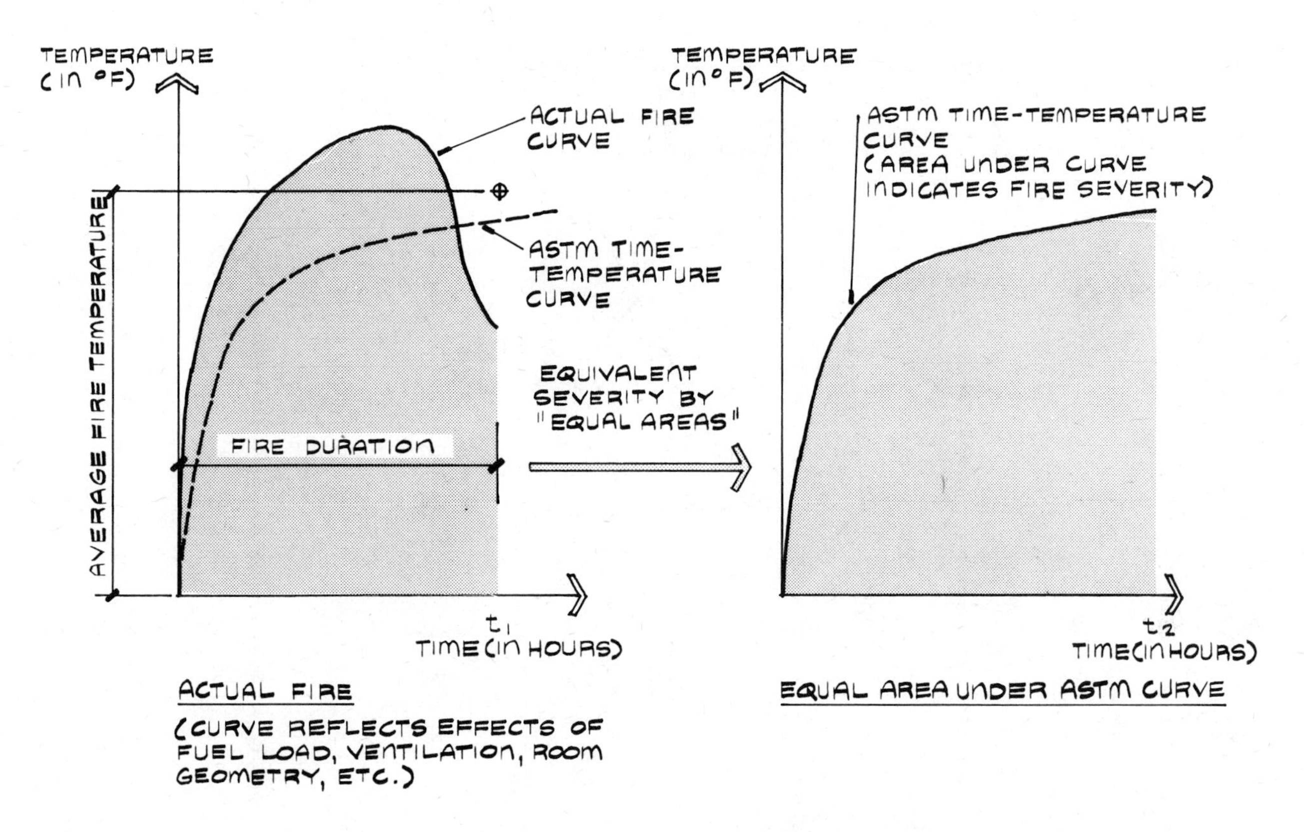

Note: In addition to fire duration and fire temperature of fully developed fires, fire severity depends on the thermal insulation properties of the surrounding building surfaces. Where rooms are well insulated, heat will be contained in the room increasing the fire intensity and growth.

BASIC THEORY—EXAMPLE FIRE GROWTH RATES

Shown below are time-temperature curves for two fires. Five minutes after ignition, the stairway fire temperature is 500°F. (See dashed lines on graph.) In just under 2 minutes, however, the storage room fire is 1000°F. Fire load, surface area, and surface texture of the burning materials determine the rate of heat release. For example, a block of wood will eventually release the same heat as an equal weight of wood shavings. The greater fire hazard would be from the shavings as the rate of heat release would be greater. Although the standard time-temperature curve does not duplicate the growth pattern of every building fire, it is a valuable method for establishing a common basis of comparison for fire-resistance tests.

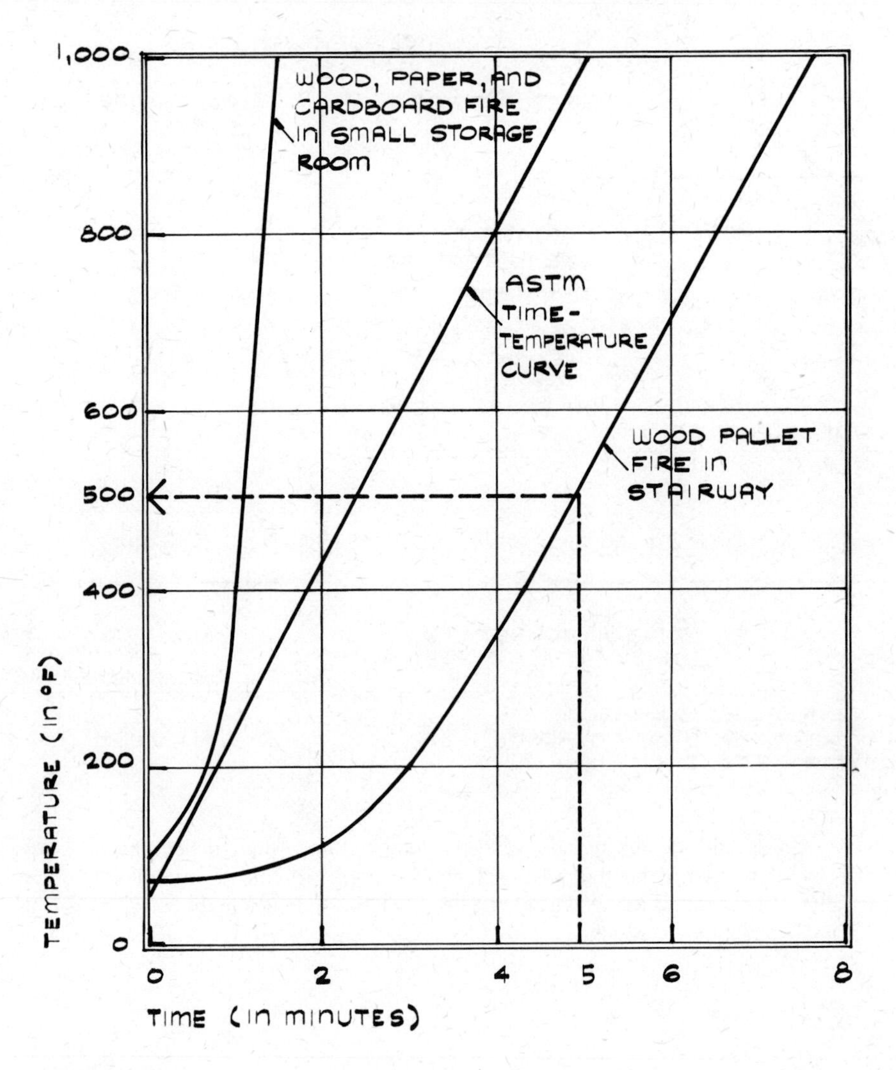

REFERENCE

Wilson, R., "Time! The Yardstick of Fire Control," *NFPA Firemen*, September 1962.

CLASSIFICATION OF BUILDING CONSTRUCTIONS

Common classifications for building constructions are given in the table below in order of decreasing fire resistance. For detailed descriptions of the various building constructions classifications, refer to building codes and fire insurance organizations' publications.

Classification	*Description*
Fire resistive	Loadbearing walls and principal structural members have a fire-resistance rating of 4 or 3 hours (depends on "fire resistive" category).
	Nonloadbearing partitions and secondary structural members enclosing stairways and other vertical openings have a fire-resistance rating of 3 or 2 hours.
	Other nonloadbearing partitions must be constructed of noncombustible or limited-combustible materials.
Noncombustible	Constructed of materials which do not contribute to fire.
	Protected by fire-resistance rated bearing walls, floor, and roof constructions.
Heavy timber	Detailed requirements are given for size of timber members and their connections (generally larger than structural design considerations alone would dictate).
	Loadbearing walls have a fire-resistance rating of 2 hours.
Protected combustible (or "ordinary")	Structural members are partly or entirely combustible.
	Exterior walls are noncombustible.
Unprotected combustible (or "wood frame")	Similar to "protected combustible," but exterior walls can be combustible.

Note: "Unprotected" noncombustible is similar to "protected" noncombustible, but has lower degree of fire resistance.

EXAMPLE FIRE-RESISTANCE RATING REQUIREMENTS

Fire ratings for construction elements are based on the occupancy, building height, and other considerations. Example building code fire-resistance rating requirements in hours for apartment construction elements used in buildings of various construction types are given in the table below.

FIRE-RESISTANCE RATINGS FOR VARIOUS CONSTRUCTION TYPES (IN HOURS)

Construction Element	*Fire Resistive*	*Noncombustible*	*Protected Combustible*	*Wood Frame*
Fire-resistive walls	4	4	4	4
Exterior loadbearing walls	4	3	3	3
Exterior walls (nonloadbearing)	2	2	2	2
Interior loadbearing walls	4	2	1	1
Stairs, elevator shafts, pipe and duct shafts, and corridors	2	2	2	1

Note: Where mixed occupancies (i.e., two or more occupancies so intermingled that separate firesafety measures cannot be used) occur, the most restrictive code provisions must be applied.

BUILDING HEIGHTS

Building heights (in stories or feet) are restricted by codes to assure that fire hose streams will be able to reach upper story fires from the outside and to help assure safe egress by occupants. In addition, fires in multistory buildings may involve more than one floor making fire-fighting operations more difficult. Nevertheless, greater allowable heights are permitted where occupancies use automatic fire extinguishing systems (e.g., sprinklers). Allowable heights in stories for buildings of various construction types are given in the table. The data represent the average allowable story height from the four model codes in the United States. Most codes classify buildings having similar functions into occupancy groups such as "residential" for hotels, motels, apartments; "assembly" for auditoriums, churches; and so on.

ALLOWABLE HEIGHTS FOR VARIOUS CONSTRUCTION TYPES
(IN STORIES)

Occupancy	*Fire Resistive*	*Noncombustible*	*Protected Combustible*	*Heavy Timber*
Residential	No limit	3	4	3
Educational	No limit	2	2	2
Assembly	No limit	2	2	2
Mercantile	No limit	2	4	3
Industrial	No limit	2	3	3

BASIC THEORY—BUILDING HEIGHT COMPUTATIONS

To find the story height of a building, measure vertically from the first story at the lowest grade (or ground level) elevation to the roof. A story is the portion of a building between the upper surface of a floor and the upper surface of the floor (or roof) above. Example buildings are shown below.

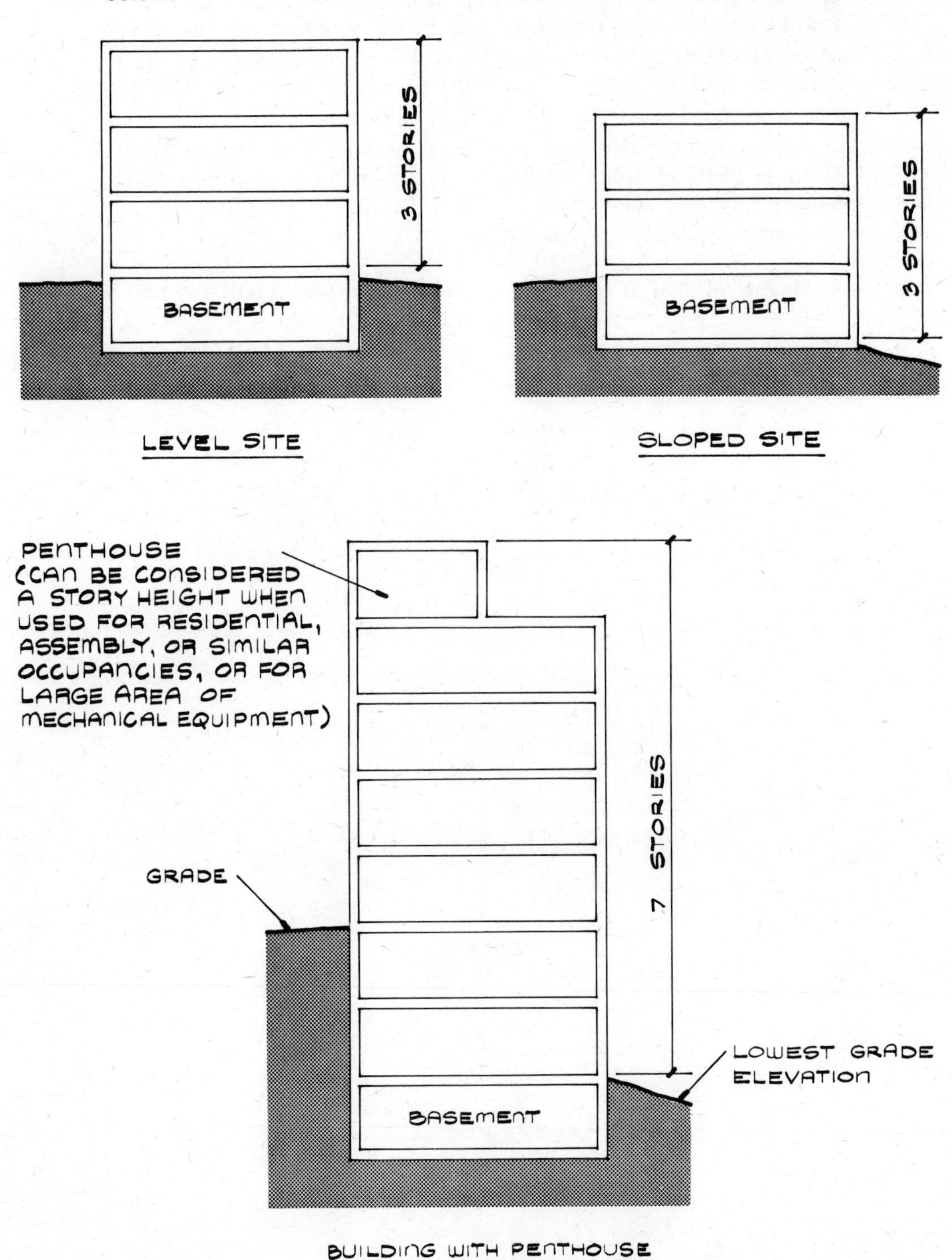

Note: If the upper floor surface of a story is less than 6 ft above grade, it usually can be considered a basement.

BUILDING AREAS

Building areas (in square feet per floor) are limited by codes since large areas generally contribute to larger fires, cannot be effectively penetrated by fire hose streams from the outside, and are more difficult to evacuate. Nevertheless, larger areas are often permitted where occupancies involve low fire hazards, complete automatic sprinkler protection, relatively narrow shapes with access from the long sides, or where buildings front on more than one street providing better access for fighting fires. Example allowable areas in square feet for single story buildings of various construction types are given in the table (cf., "Fire Protection Through Modern Building Codes," American Iron and Steel Institute, 1971). For buildings where heights are limited by codes, allowable individual floor areas will be less than those for single story buildings.

ALLOWABLE AREAS FOR SINGLE STORY BUILDINGS OF VARIOUS CONSTRUCTION TYPES (IN SQ FT)

Occupancy	*Fire Resistive*	*Noncombustible*	*Protected Combustible*	*Heavy Timber*
Residential	No limit	13,200	11,000	8,800
Educational	No limit	13,200	11,000	8,800
Assembly	No limit	13,200	11,000	8,800
Mercantile	No limit	8,800	7,800	6,600

Note: The limits in the table do not necessarily restrict the overall size of a building as fire-rated walls can be used to divide the building into areas not exceeding the limits.

BASIC THEORY—BUILDING AREA COMPUTATIONS

To find the total building area, sum the area of each floor, basement, "penthouse," finished attic with sufficient, useable headroom, and so on. The total area of all floors must not exceed code limits. Shown below, on a sketch with exaggerated enclosure thicknesses, is an example residential occupancy.

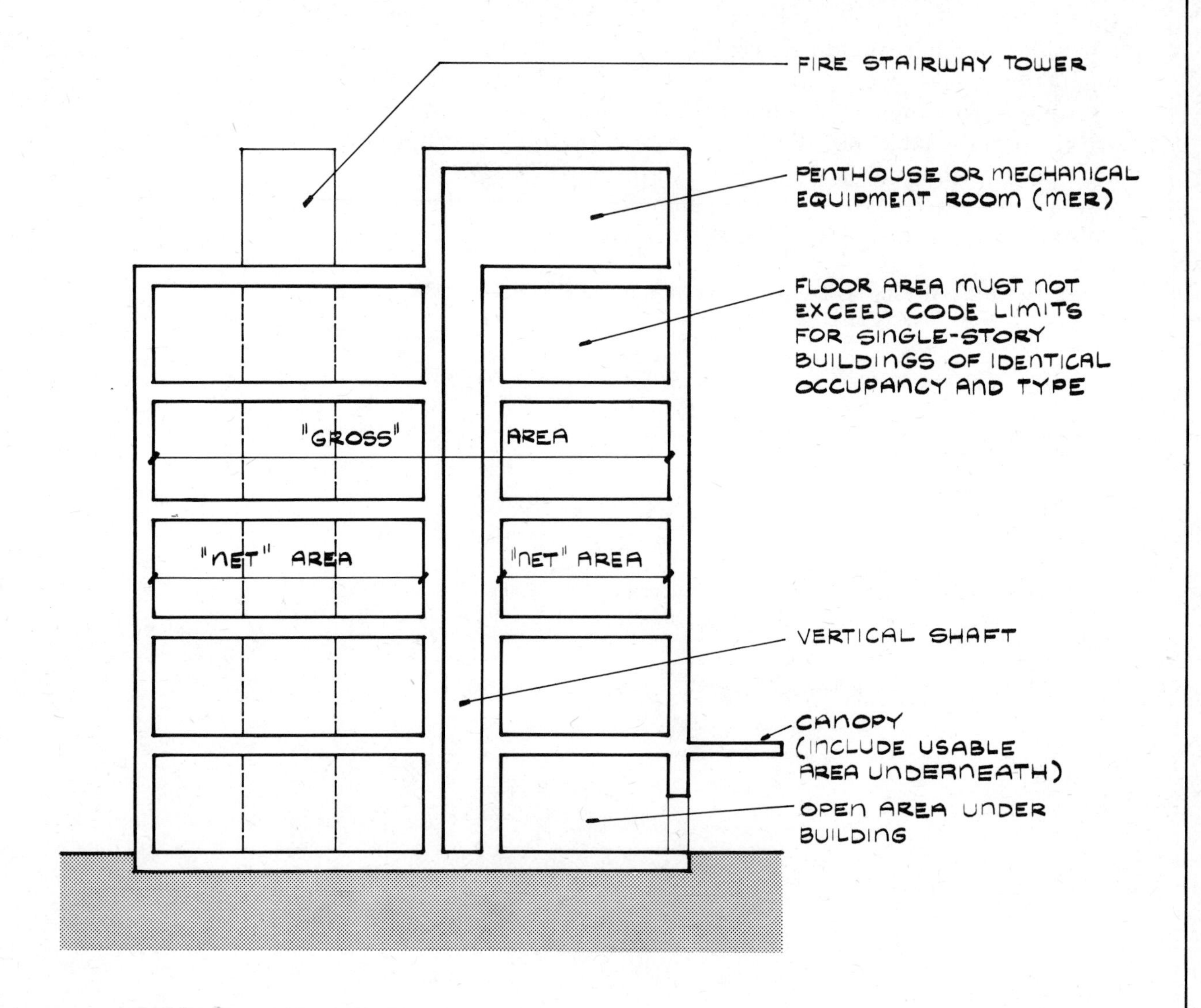

BASIC THEORY—DECISION TREES

A "decision tree" is a network diagram consisting of two basic parts: decisions and results or events. The tree (possible events branch out like the limbs of a tree) shows the potential consequences of a series of decisions. It can be applied to building firesafety situations by identifying fire factors and emphasizing their interrelationships. For example, the diagram below is concerned with deciding what are the alternate methods for controlling a building fire (i.e., control combustion process, suppress fire, or control by constructions).

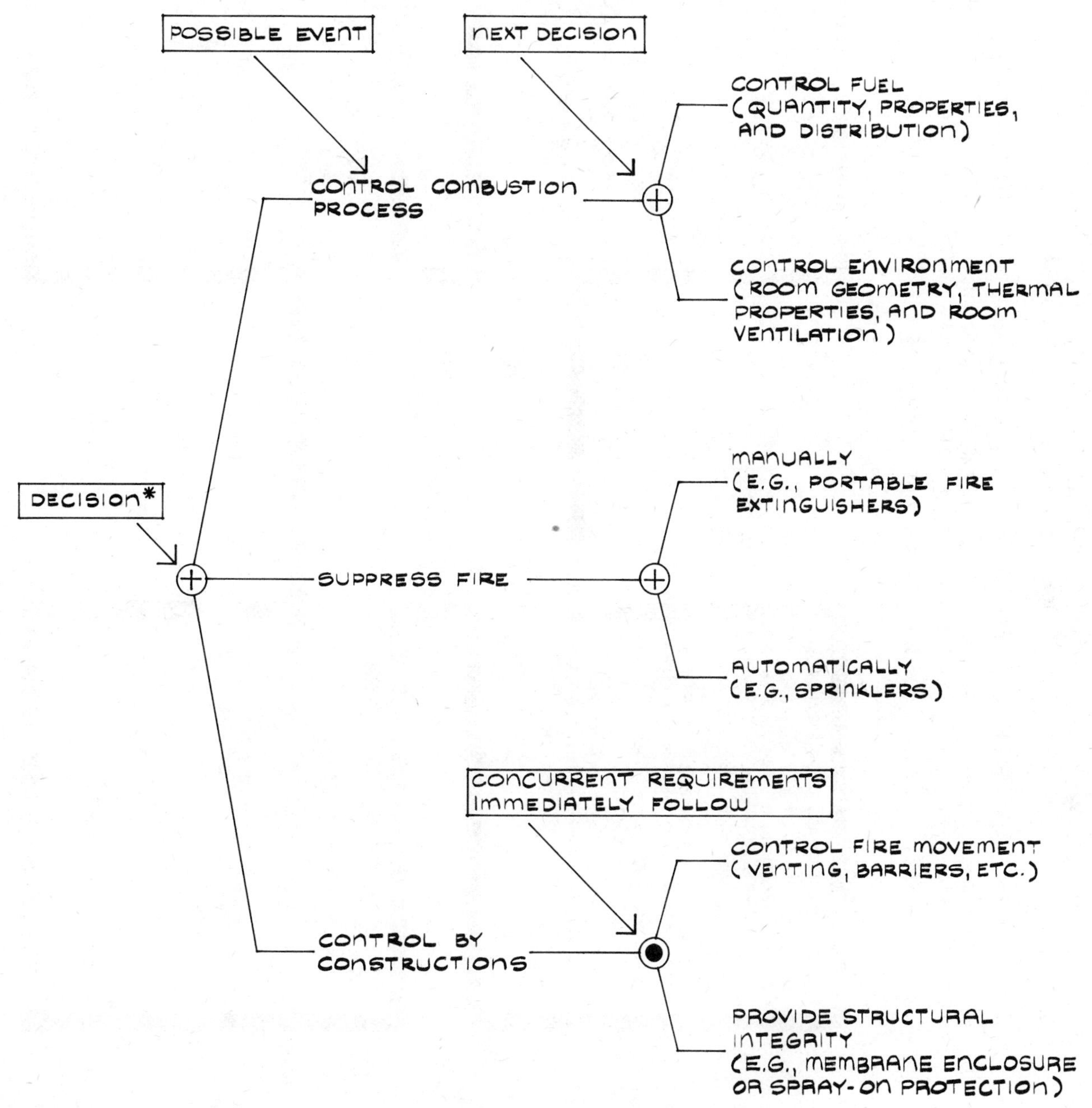

Note: A detailed building firesafety decision tree is included in the appendix.

BASIC THEORY—METHODS OF FIRE CONTROL

Alternate approaches to controlling fires in buildings are shown below. Basically, the alternate methods are to control the combustion process (how fire burns), to control the building constructions, and to suppress the fire (automatically or manually). Effective fire control can utilize one or parts of the four approaches listed below.

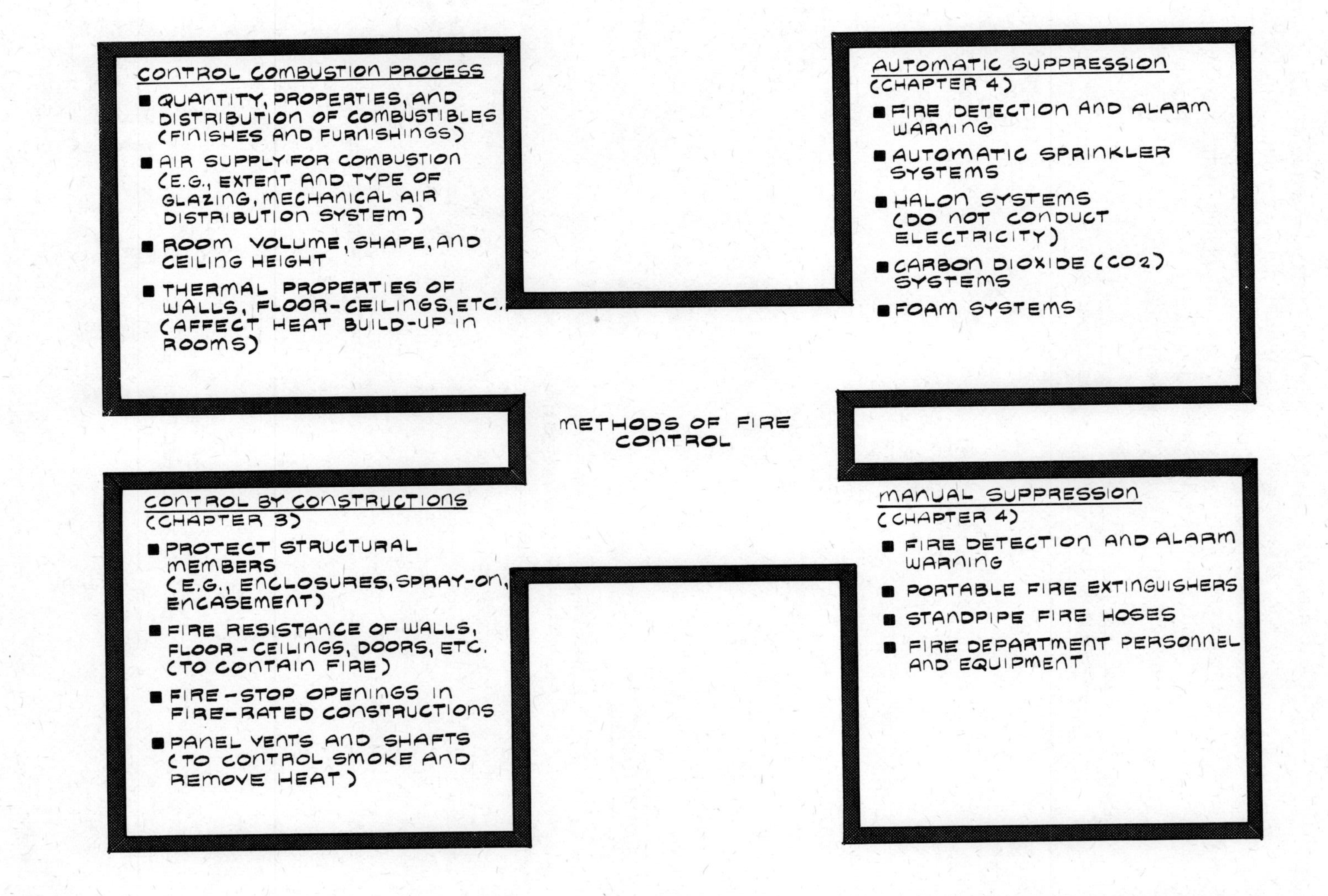

SITE PLANNING

SITE PLANNING—SITE ACCESS FOR FIRE APPARATUS

The building site plan should provide adequate driveway widths, turning radii, and parking space on firm, level surfaces for fire apparatus. Avoid man-made and natural barriers that could interfere with movement of fire vehicles. Fire apparatus turning radii (*R*) typically vary from 28 to 40 ft and vehicle length (*L*) from 40 to 65 ft for ladder trucks and from 20 to 40 ft for pumpers.

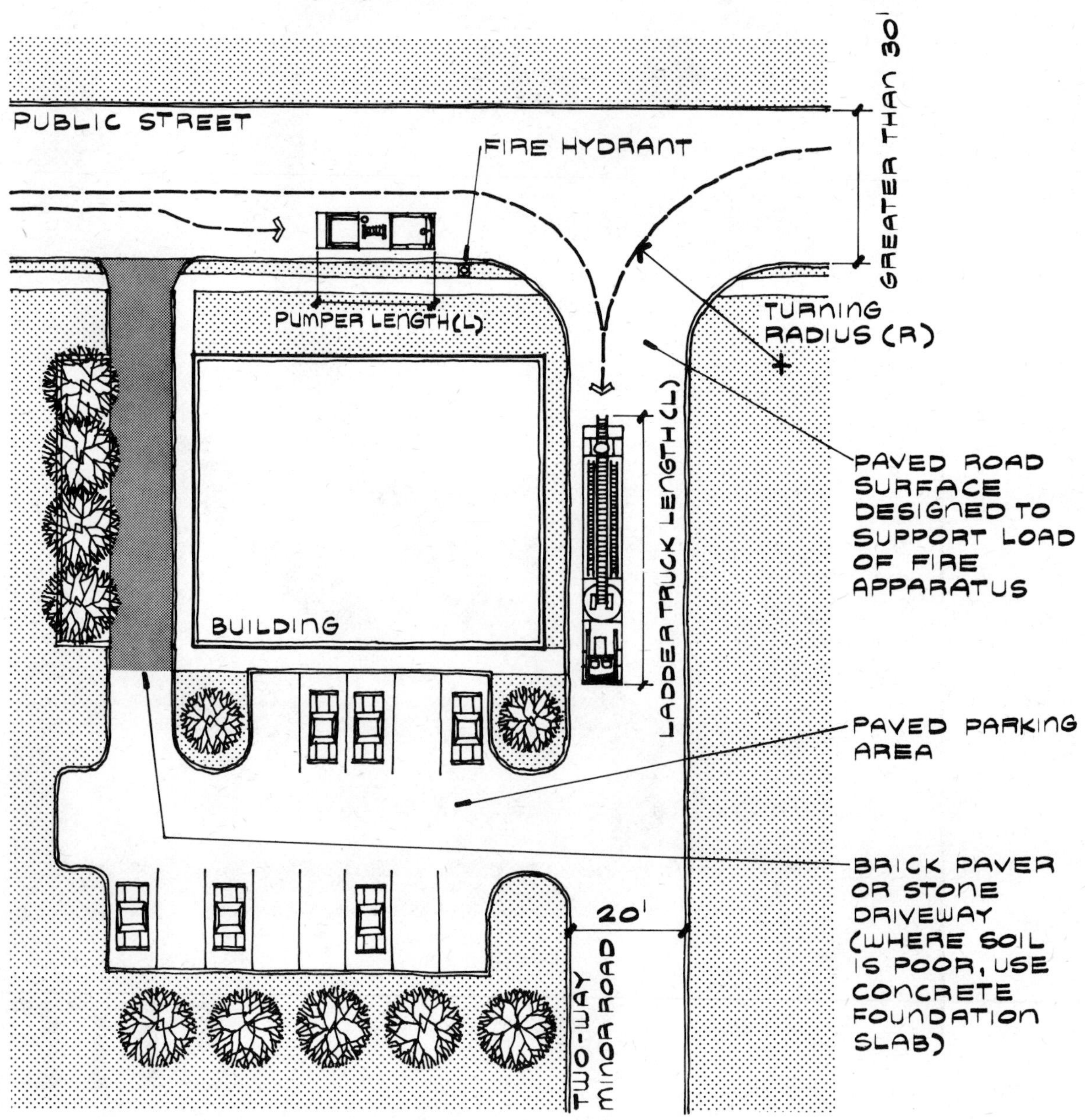

Note: Vertical clearance into below-grade parking areas, under pedestrian bridges, and the like should be at least 9 to 12 ft to provide access for fire apparatus.

SITE PLANNING—BARRIERS TO SITE ACCESS

Site layouts should not restrict fire apparatus access to buildings. Example potential barriers are shown below.

Fence

Fences for security should have gates of sufficient width (*W*) when open to allow fire apparatus access to the protected buildings.

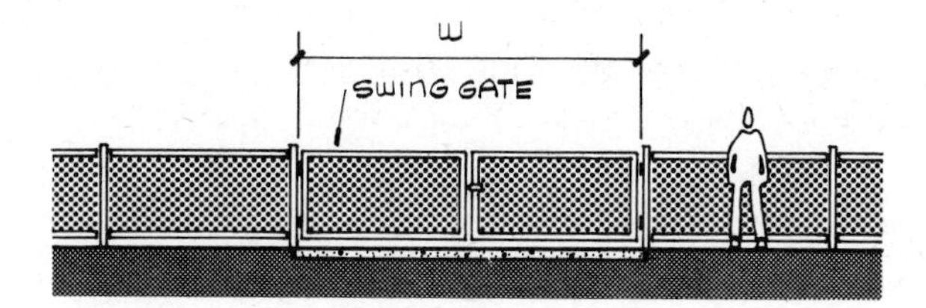

Bollard

Obstructions to prevent motor vehicles from entering an area (e.g., pedestrian street shopping malls, private drives) should be removable for fire apparatus access. Bollards, as shown below, can be secured by locks that can be opened by fire department keys.

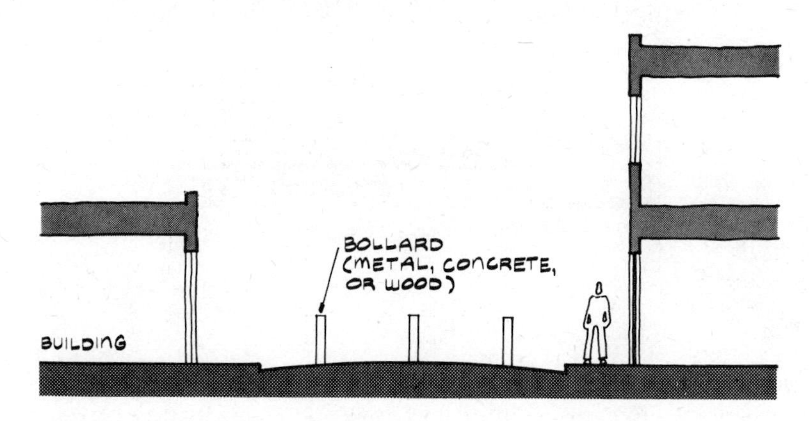

SITE PLANNING—FIRE STATIONS AND FIRE APPARATUS

Fire stations should be located near major roads, close to the center of the area being served, and set well back from the curb. Stations should be distributed in a community so they will be able to provide effective fire apparatus response times to existing areas requiring protection and be able to serve future patterns of growth. Avoid locations that can easily become jammed with vehicular traffic (e.g., near large parking garages, major intersections). Traffic light control from the station is often desirable. Consider the effects of barriers (e.g., railroad tracks, draw bridges, highways without sufficient cross streets) which can seriously reduce response times. Example fire apparatus sizes are shown below. Actual dimensions vary according to the manufacturer and to the needs and specifications of the local fire department.

Ladder Truck

This type of apparatus carries a hydraulically operated aerial ladder for rescue and fire suppression operations. It also carries an assortment of ground ladders, tools, and rescue devices. Aerial ladder apparatus are either two- or three-axle tractor drawn as shown below. Tractor drawn apparatus have greater maneuverability on narrow streets or in heavy traffic, but require an additional driver to steer the trailer's rear wheels.

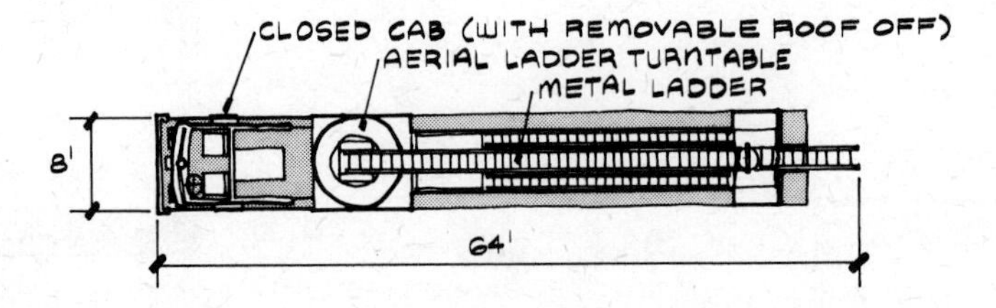

Pumper

A pumper provides adequate water pressure from fire hydrants and other water sources to control and suppress fires. It also carries hose, ladders, forcible entry tools, self-contained breathing units, portable fire extinguishers, and other equipment. Pump sizes typically vary from 500 to 1500 gpm at 150 psi. Combination pumpers have a pump, hose compartment, and a water tank on one chassis.

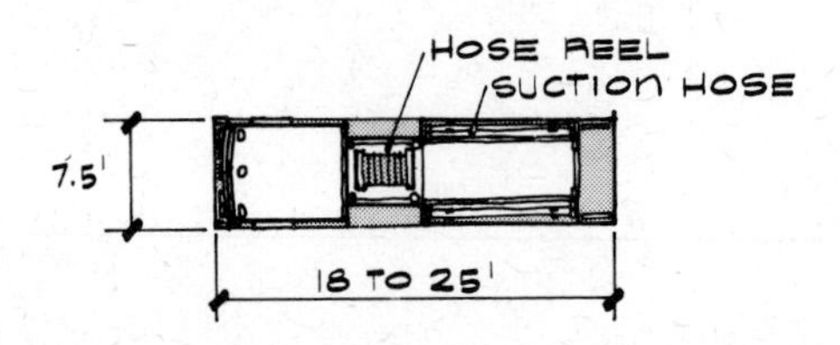

SITE PLANNING—FIRE STATIONS AND FIRE APPARATUS (Continued)

Tanker

Tankers transport water to areas without hydrant or other protection for fire suppression purposes. Tank sizes typically vary from 500 to 5000 gal.

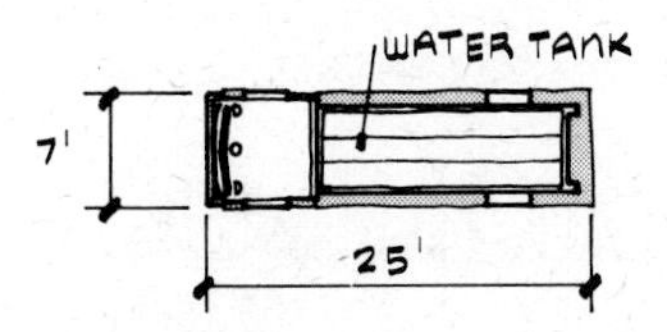

Note: Other fire apparatus include floodlight truck, salvage truck ("salvage" means to cover or remove furnishings or goods which could be damaged by fire or water), emergency rescue truck, aerial platform (i.e., cherry picker), and squad truck (with special tools and provisions for carrying several fire fighters).

SITE PLANNING—TURNING CLEARANCES FOR NARROW DRIVEWAYS

The table below gives the preferred minimum distance (d) in feet for equal legs of the right triangle-shaped open area for "turning clearance." Table values are presented for private driveway width (W) and vehicle length (L).

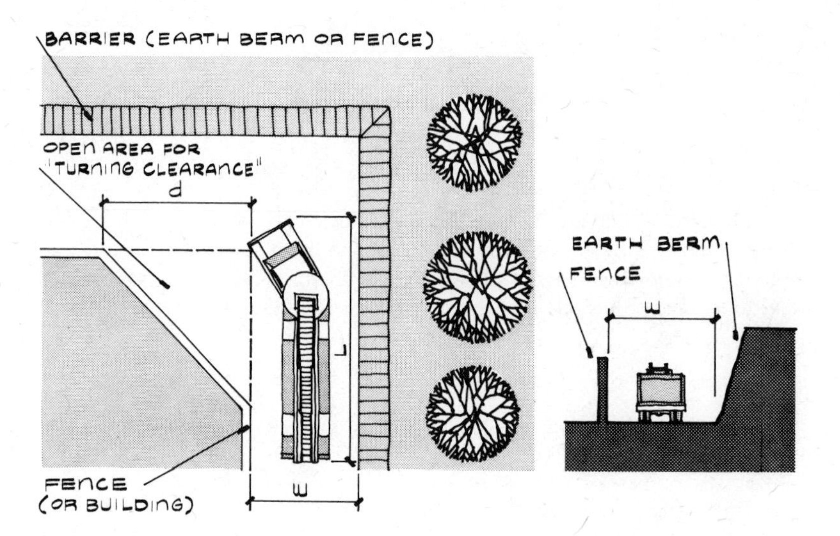

d Value Table

L \ W	10	12	14	16
35	36	31	25	19
40	38	32	28	25
45	47	37	34	30

L AND W ARE IN FEET

Note: The table values may be used for preliminary clearance estimates; however, consult with local fire department for specific operating requirements of their fire apparatus.

REFERENCE

Callender, J. H. (ed.), *Time-Saver Standards*, New York: McGraw-Hill (1966), p. 1204.

SITE PLANNING—DRIVEWAY LAYOUTS

To provide for ease in maneuvering fire apparatus, avoid long (greater than 150 ft), narrow driveways that do not have adequate turnarounds. Dead ends should be avoided. Shown below are example driveway layouts presented in order of increased provisions for ease of fire apparatus maneuvers.

Dead End

Can cause time consuming, difficult back-up maneuvers.

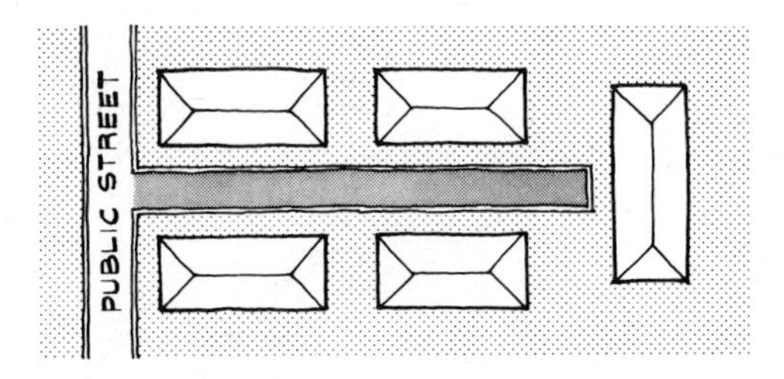

T-Turn or Hammer Head

Provides means to maneuver and change direction without lengthy back-up maneuvers.

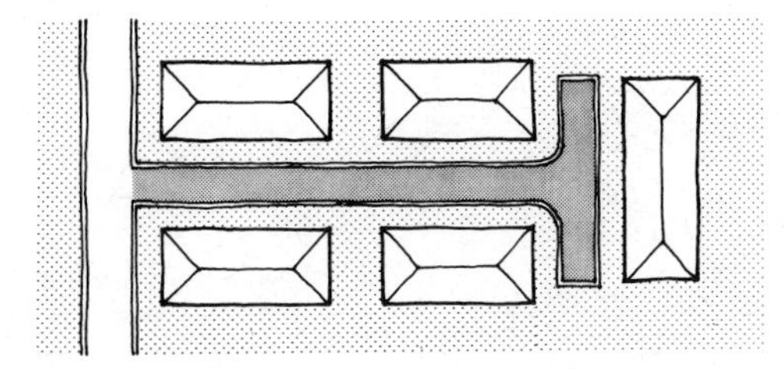

Cul-de-Sac

Clear turning radius should be at least 40 ft.

Curved Driveway

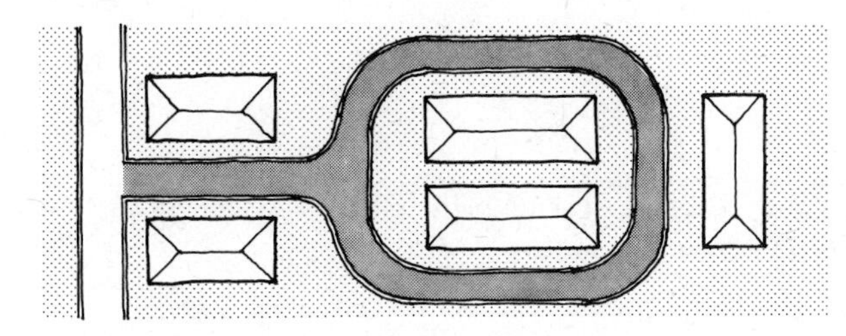

POOR

BETTER

Note: Consult local authority having fire jurisdiction to determine the layout best suited to local conditions.

SITE PLANNING—TURNAROUNDS

Turning space for fire apparatus should be provided at dead ends to avoid dangerous back-up maneuvers. Shown below are example turnarounds that allow expeditious changes of direction. Parking should not be permitted in T-turns or shunts. The examples are given for a road width of 20 ft, the generally accepted minimum width for a two-way minor road without parking.

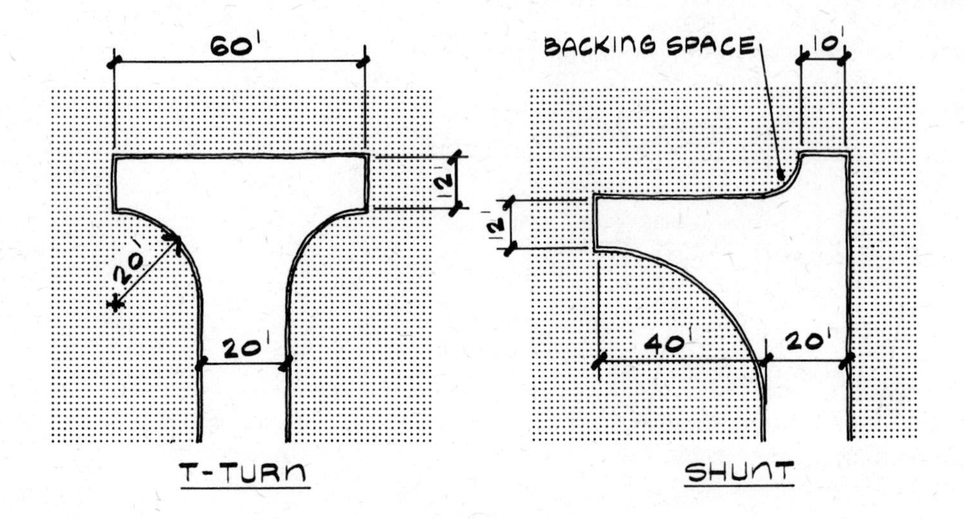

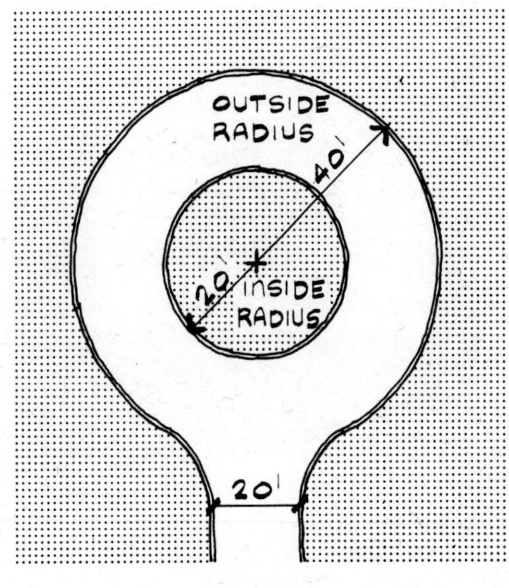

CIRCLE
(WITHOUT PARKING)

NOTE: IN GENERAL, PROVIDE GREATER OUTSIDE RADIUS FOR COMMERCIAL AND INDUSTRIAL AREAS.

REFERENCE

"Geometric Design Guide for Local Roads and Streets," American Association of State Highway Officials, Washington, D.C., 1971.

SITE PLANNING—DRIVEWAY WIDTHS FOR AERIAL APPARATUS

To allow full extension of aerial ladders at a safe climbing or elevation angle (Θ) of 60 to 80 degrees, sufficient space is needed to position (or "spot") the apparatus. The graph below shows typical width (*W*) in feet to allow operational space in driveways to reach buildings of various height (*H*) in feet. For example, to reach the roof of a building 60 ft high, a width of at least 24 ft is needed for a preferred elevation angle of 70 degrees (see dashed lines on graph).

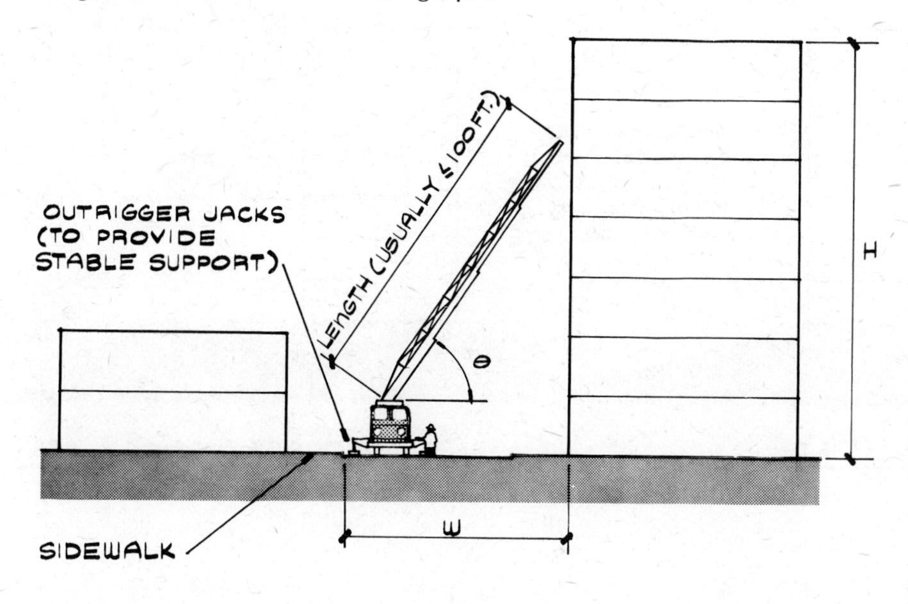

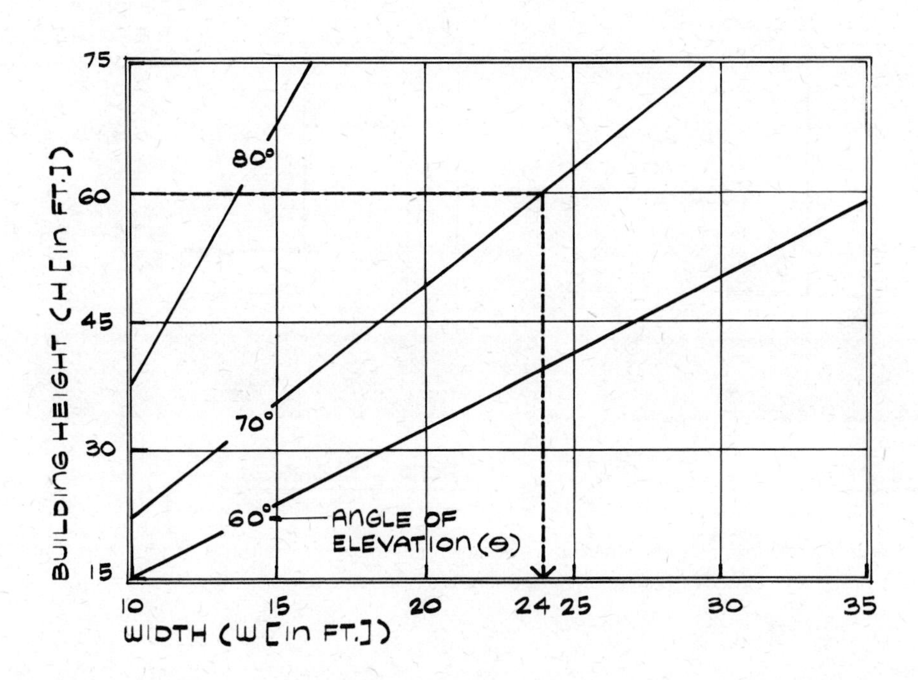

SITE PLANNING—OBSTRUCTIONS TO AERIAL APPARATUS OPERATIONS

Site characteristics can impede effective fire-fighting and rescue operations. Shown below are example site and building features that can restrict the placement of aerial and other fire apparatus.

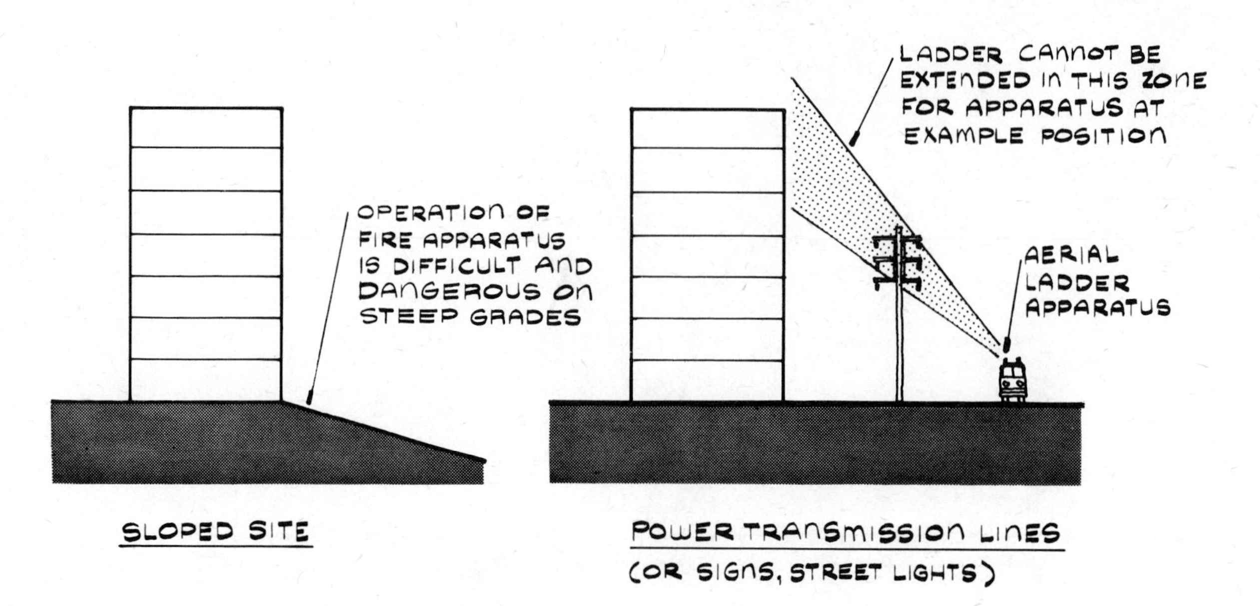

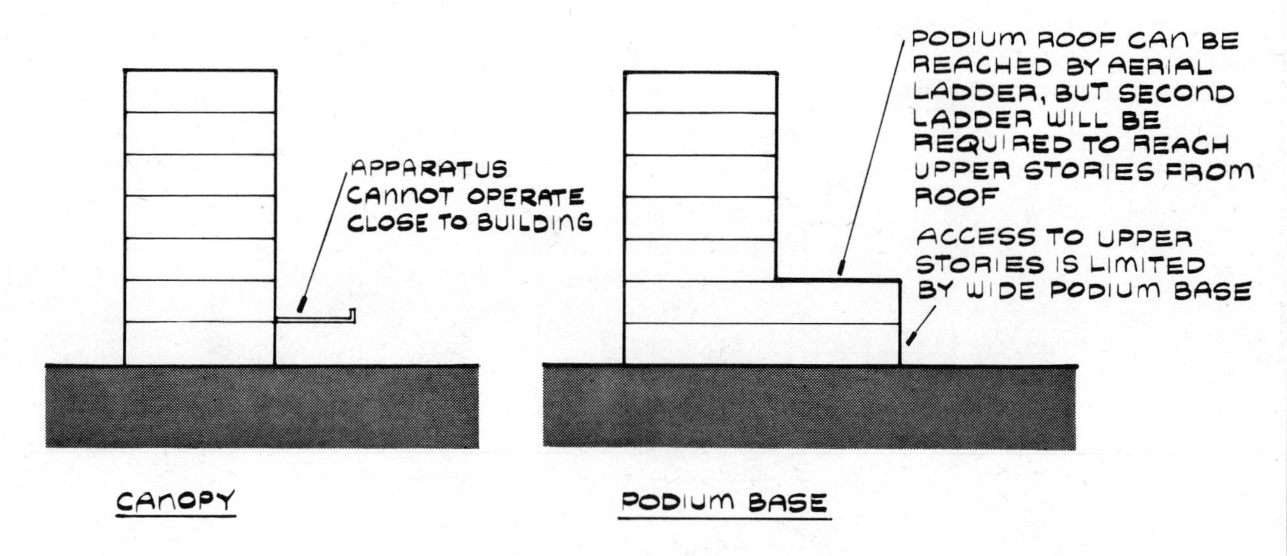

Note: Site designs must discourage vehicle parking near buildings at locations that will prevent effective placement of apparatus.

SITE PLANNING—FIRE HYDRANT LAYOUTS

Shown below is an example layout for fire hydrants serving a row of dwelling units. Locate hydrants at street intersections and at midpoints along streets where the distance between intersections exceeds 400 ft. The need for long fire hose lines will cause delay and will require high water pressures from fire apparatus pumps. Do not place hydrants within 50 ft of buildings, unless the construction is fire resistive or the facing wall is masonry without openings. For remote site locations, be sure that hydrants are not further than 300 ft away from the buildings to be protected.

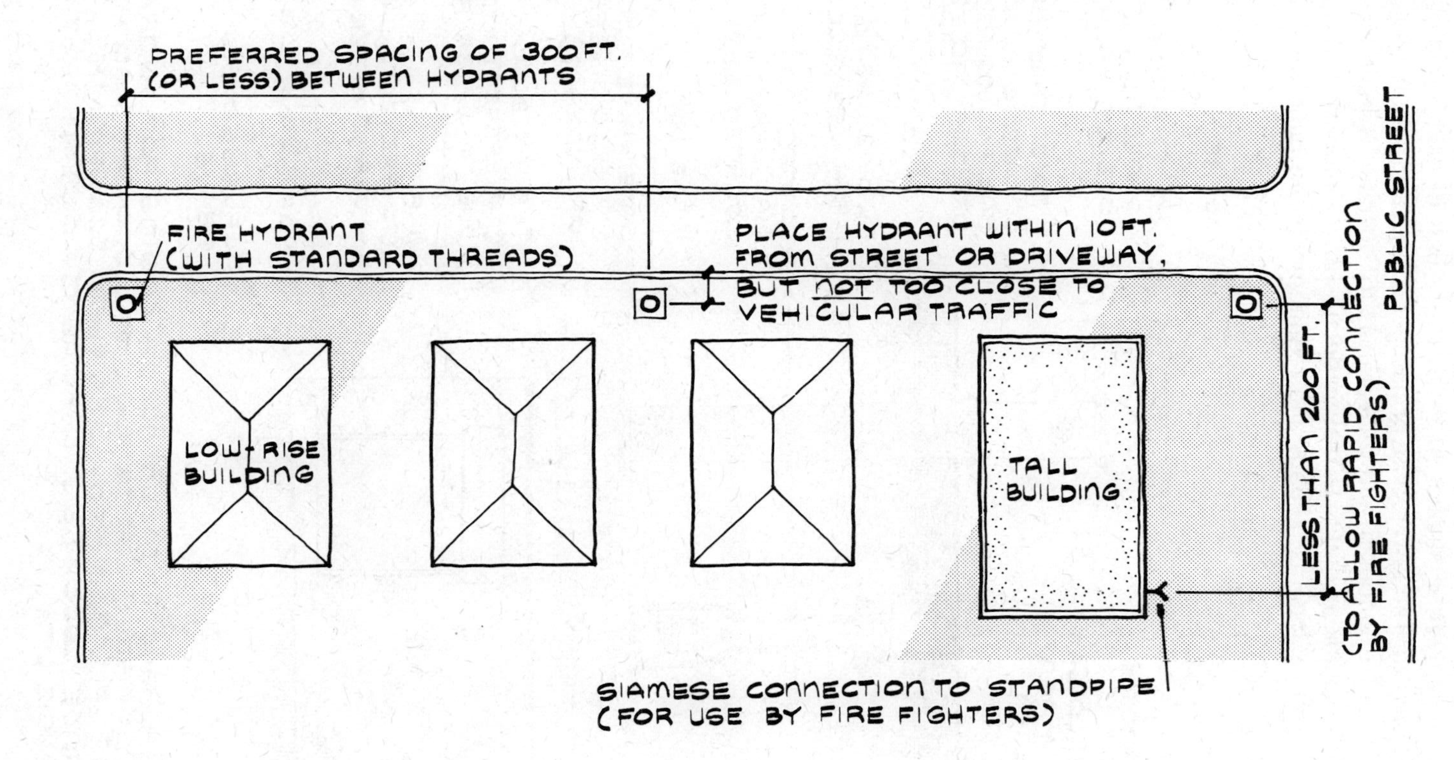

REFERENCES

"Fire Safety in Housing," U.S. Department of Housing and Urban Development, Washington, D.C., July 1975.

Municipal Fire Administration, Washington, D.C.: International City Managers' Association (1967).

SITE PLANNING—FIRE HYDRANTS

Adequate water flow in gallons per minute (gpm) to fire hydrants is required to provide sufficient water to suppress fires. In addition, adequate water pressure in pounds per square inch (psi) is needed to enable hose streams to reach every part of each building to be protected. The lowest hose or pumper connection should be at least 15 in. above grade to allow clearance to operate hydrant wrenches. The example hydrant installation detail shown below should be adapted to satisfy local conditions and requirements of municipal governments.

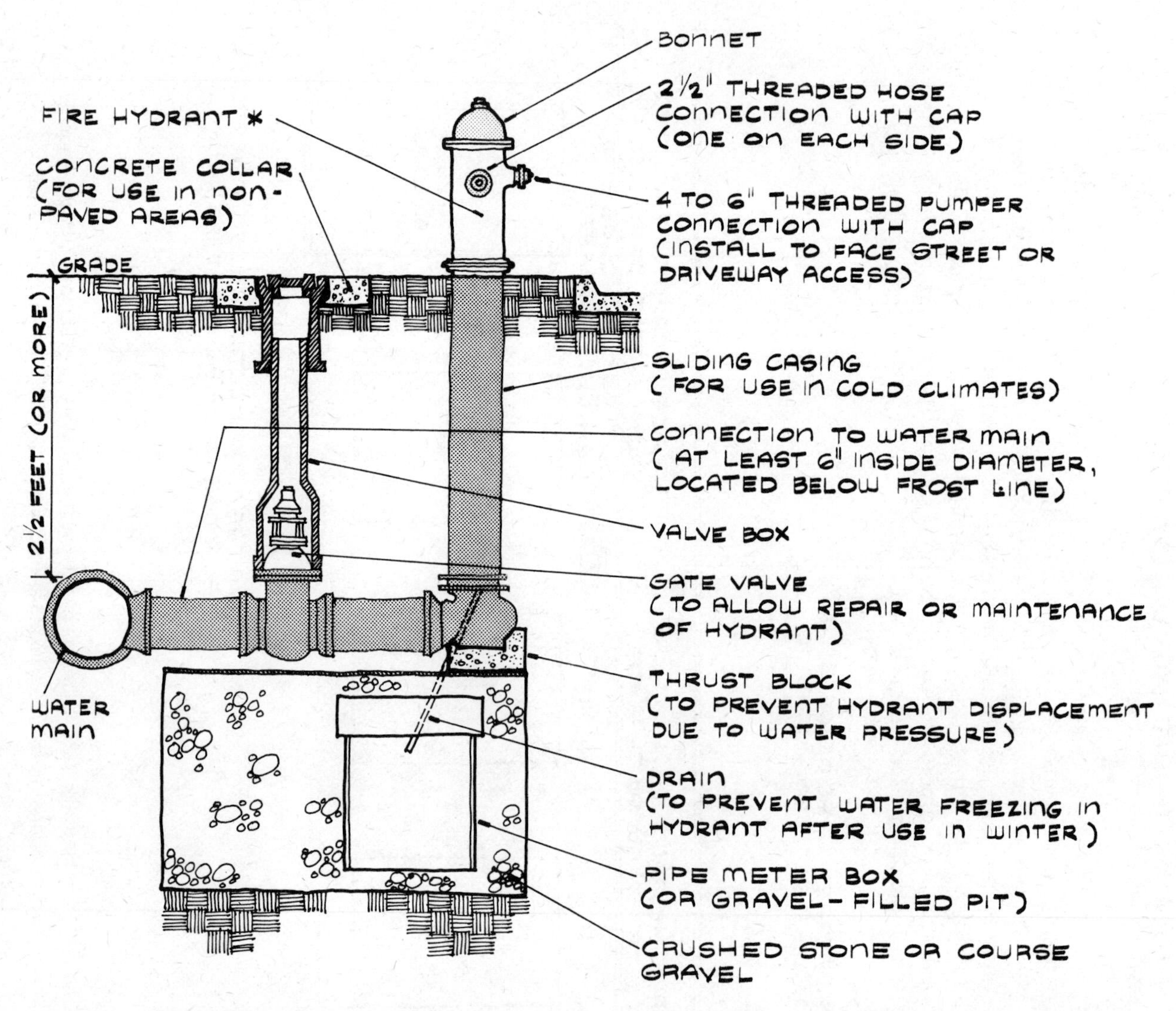

* THE NFPA RECOMMENDED COLOR CODING FOR HYDRANT CAPS AND BONNETS IS: GREEN-WATER FLOW OF 1,000 GPM AND ABOVE, ORANGE-500 TO 999 GPM, AND RED-LESS THAN 500 GPM. USE A HIGH LIGHT-REFLECTANCE COLOR (E.G., CHROME YELLOW) ON REST OF HYDRANT FOR MAXIMUM VISIBILITY.

REFERENCE

Brzezinski, F. T., "Utility Design" in *Handbook of Landscape Architectural Construction,* American Society of Landscape Architects Foundation, McLean, Va., 1975.

SITE PLANNING—FIRE HYDRANT PLACEMENT

Fire hydrants should be located within 10 ft from street or driveway access to allow rapid connection of fire apparatus suction hose lines. In addition, to avoid damage from vehicular traffic, hydrants should not be located too close to the street or driveway.

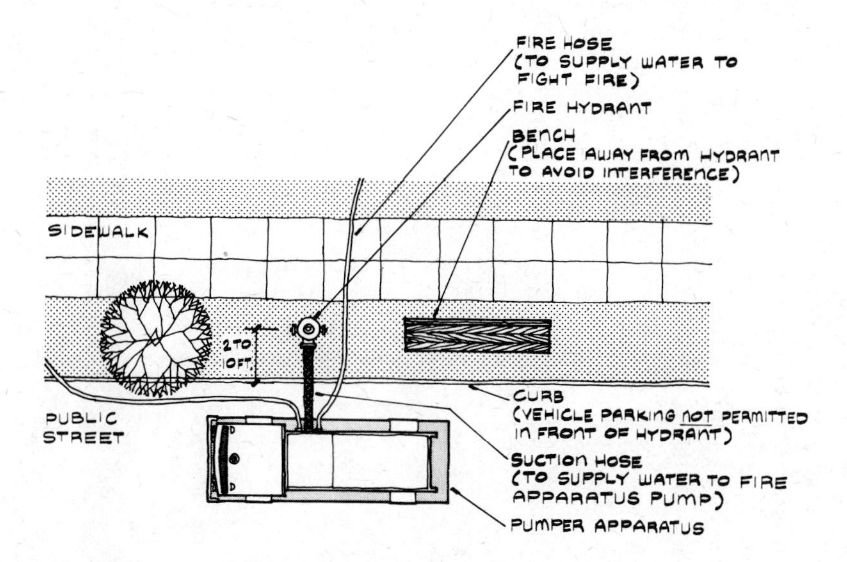

Shown below are examples of site barriers to the effective use of fire hydrants. Hydrants that are behind shrubs, or buried in the ground as shown, will be difficult to locate, particularly at night.

SITE PLANNING—EXTERIOR WALL OPENINGS

Heat radiated through exterior wall openings can ignite adjacent combustible materials and nearby buildings. Shown below are factors (e.g., separation distance, wall openings, nearby combustible materials) that influence fire spread between buildings. Operable windows are assumed to be open as this may be the situation during a fire. Allowable exterior wall openings (expressed as a percentage of exterior wall) can be restricted by building codes to reduce the possibility of fire spread from one building to another. For example, the "National Building Code" of the American Insurance Association restricts allowable percentage of windows in exterior bearing-wall areas for fire-resistive construction to 0% for 0 to 3 ft horizontal separation distance; 20% for over 3 to 20 ft; 30% for over 20 to 30 ft; and 40% for over 30 ft.

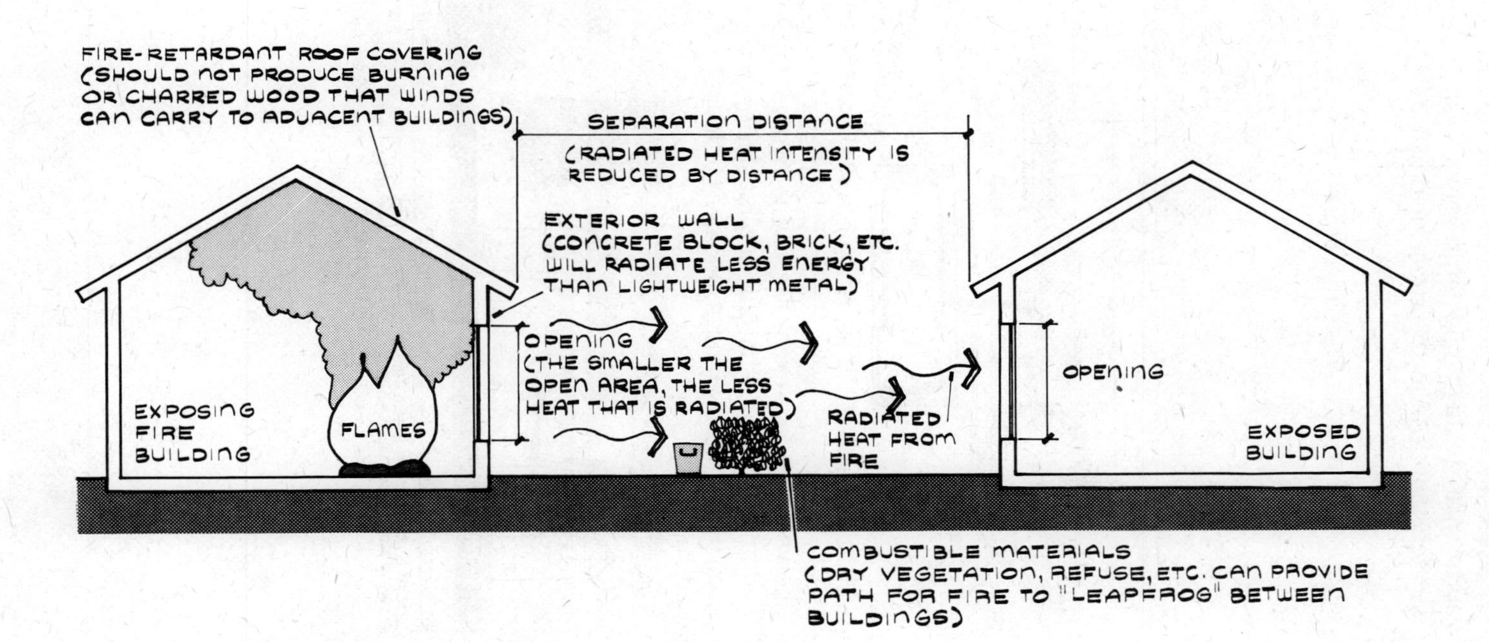

Note: Fire-retardant roof coverings should not spread fire, produce burning brands, or slip from position. They are classified by Underwriters Laboratories as class A (effective against severe fire exposure), class B (moderate exposure), and class C (light exposure). Steep pitched roofs are more susceptible to ignition and sustained burning than are flat or low pitched roofs.

SITE PLANNING—ROOF COVERINGS

The fire test apparatus used to determine a classification for roof coverings is shown below. Class A roof coverings are considered effective against severe fire exposure, class B against moderate exposure, and class C against light exposure.

Intermittent Flame Exposure Test

For classes A and B, the test specimen is exposed to a gas flame at a temperature of 1350 to 1450°F for a cycle of 2 minutes on and 2 minutes off. Class A coverings must withstand 15 cycles and class B, 8 cycles. During and after the test, the specimen must not (1) produce flaming or glowing brands that leave the test deck, (2) exhibit sustained flaming on the underside of the deck, or (3) expose the deck by displacement or falling away. Specimens generally are tested at a slope (or pitch) of 5 in. per horizontal foot or at the maximum slope recommended by the roof covering manufacturer.

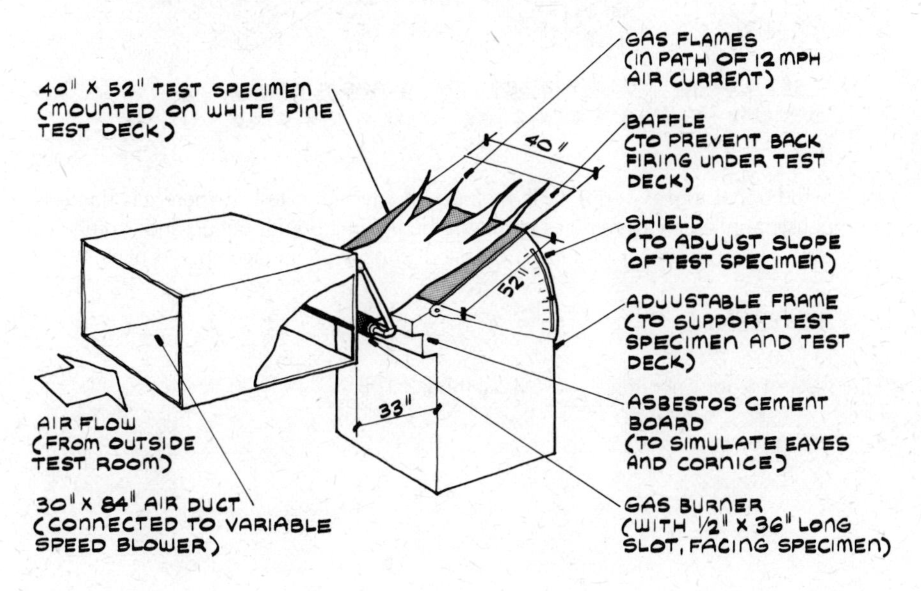

Burning Brand Test

In the burning brand test, the specimen is placed 60 in. away from the air duct. All sides of the brands are exposed so they will be burning freely when placed on the roof covering specimen. For class A coverings, 1 class A brand is used; for class B, 2 class B brands; and for class C, 25 class C brands. Failure criteria are similar to the intermittent flame exposure test.

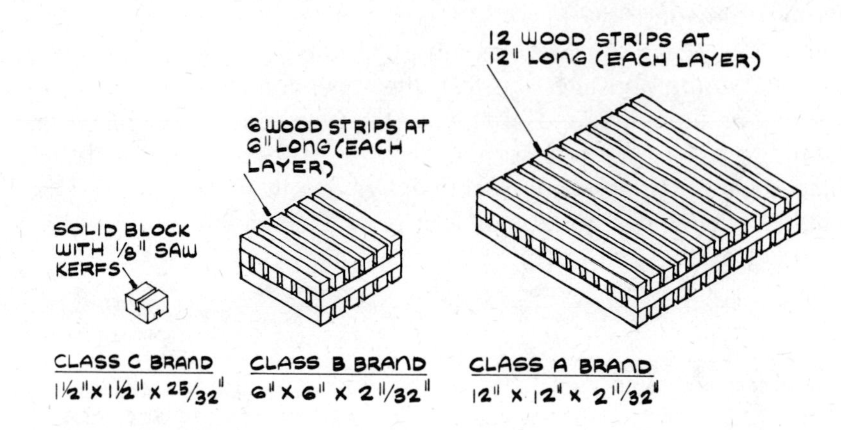

Note: Additional roof covering tests include "flying brand test" (where gas flame is continuously applied for 14 minutes to determine if flying brands occur) and "spread of flame test" (where a longer specimen is tested to observe surface spread of flame).

TEST REFERENCE

"Fire Tests of Roof Coverings," ASTM Method E 108.

SITE PLANNING—METHODS OF EXPOSURE PROTECTION

Buildings located near a burning building are exposed to radiant and convected heat energy. Radiated heat energy can be reduced by (1) increasing the separation distance, (2) using outside sprinklers, (3) self-supporting barrier walls, and (4) decreasing (or eliminating) the area of the wall openings. Example methods of exposure protection are shown below.

Separation Distance

Depends on fire hazard, site characteristics, and the like.

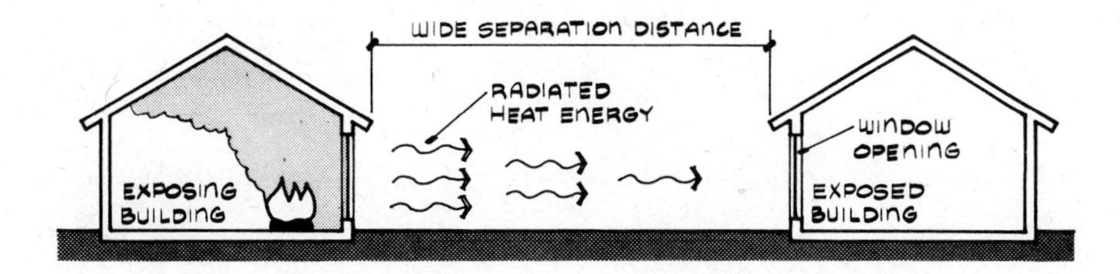

Outside Sprinklers

Prevent ignition of a building's exterior surfaces, interior finishes, and contents.

Barriers

Concrete, brick, or block walls shield adjacent buildings.

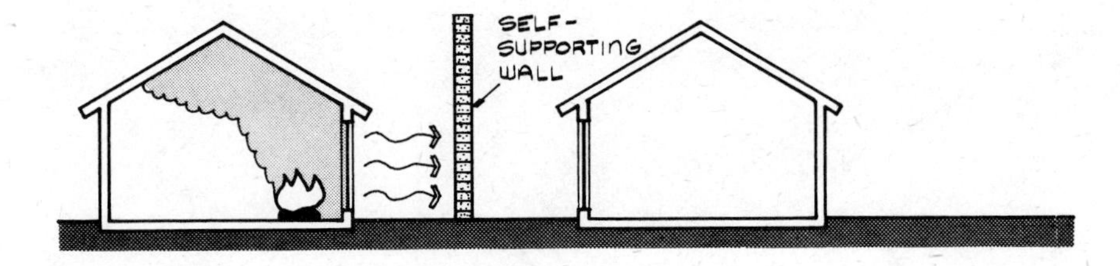

No Openings

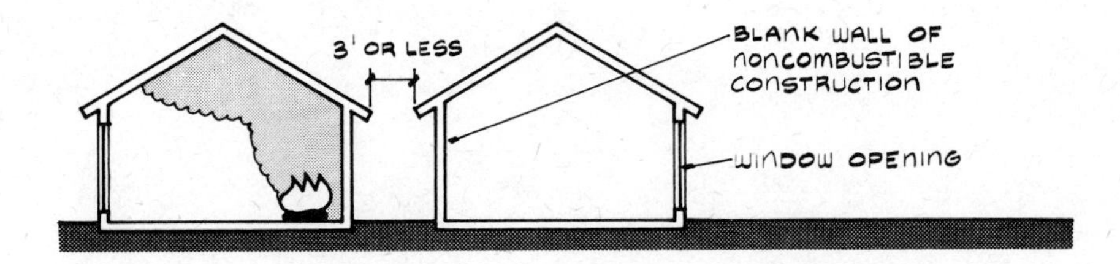

FIRE EXPOSURE SEVERITY

The probable exposure severity between two buildings of equal height can be estimated by the table below. Refer to "Protection of Buildings from Exterior Fire Exposures," NFPA No. 80A, 1975 for a discussion of radiation hazard theory and a comprehensive list of related publications.

Exposure Severity	*Construction*	*Interior Finish*
Severe	Wood frame	Wood paneling
Moderate	Wood frame	Gypsum board (or plaster)
	Noncombustible	Wood paneling
	Fire resistive	Wood paneling
Light	Noncombustible	Gypsum board (or plaster)
	Fire resistive	Gypsum board (or plaster)

SITE PLANNING—SEPARATION DISTANCES

The table below lists guide numbers (N) for determining separation distances to protect exposed buildings from fire spread through equally distributed windows. Separation distance can be found by the formula:

$$d = FN + 5$$

where d = distance between buildings in ft
F = width (W) or height (H) in ft
N = guide number from table (no units)

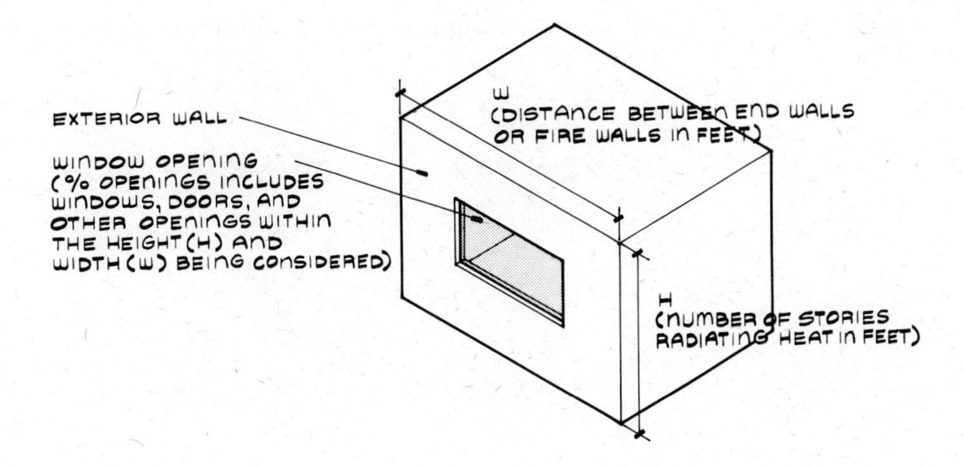

Openings (%) for Exposure Severity of:			*Guide Number (N) for Shape Ratio of W/H or H/W*									
Light	*Moderate*	*Severe*	*1.0*	*1.3*	*1.6*	*2.0*	*3.2*	*5.0*	*8.0*	*13.0*	*20.0*	*32.0*
20	10	5	0.36	0.40	0.44	0.46	0.49	0.51	0.51	0.51	0.51	0.51
30	15	7.5	0.60	0.66	0.73	0.79	0.88	0.92	0.94	0.95	0.95	0.95
40	20	10	0.76	0.85	0.94	1.02	1.17	1.27	1.32	1.33	1.34	1.34
50	25	12.5	0.90	1.00	1.11	1.22	1.42	1.58	1.66	1.70	1.71	1.71
60	30	15	1.02	1.14	1.26	1.39	1.64	1.85	1.99	2.05	2.08	2.08
80	40	20	1.22	1.37	1.52	1.68	2.02	2.34	2.59	2.73	2.79	2.81
100	50	25	1.39	1.56	1.74	1.93	2.34	2.76	3.12	3.36	3.48	3.52
	60	30	1.55	1.73	1.94	2.15	2.63	3.13	3.60	3.95	4.15	4.22
	80	40	1.82	2.04	2.28	2.54	3.12	3.77	4.43	5.01	5.41	5.60
	100	50	2.05	2.30	2.57	2.87	3.55	4.33	5.16	5.95	6.56	6.92
		60	2.26	2.54	2.84	3.17	3.93	4.82	5.80	6.78	7.63	8.18
		80	2.63	2.95	3.31	3.70	4.61	5.68	6.91	8.24	9.51	10.50
		100	2.96	3.32	3.72	4.16	5.19	6.43	7.88	9.50	11.15	12.59

Note: To find guide numbers for shape ratios not given in the above shortened table, see "Protection of Buildings from Exterior Fire Exposures," NFPA No. 80A, 1975.

EXAMPLE PROBLEM—SEPARATION DISTANCE

Given: The facing walls for adjacent buildings are 60 ft wide by 30 ft high. The maximum percentage of equally spaced openings is 60%. Find the minimum separation distance between the two buildings of fire-resistive construction with gypsum board interior finish.

Solution Process: Find shape ratio (W/H): $W/H = 60/30 = 2.0$ and guide number (N) of 1.39 for light exposure severity. (See tables on preceding pages for probable severity and guide number.)

Evaluate distance requirements using the smaller wall dimension of 30 ft.

$d = FN + 5 = (30)1.39 + 5 = \boxed{47 \text{ ft}}$

SITE PLANNING—EXTERIOR SHADING DEVICES

The building examples shown below have one or more sides covered by exterior, solar-shading devices. These exterior barriers can prevent rescue of building occupants through openings located behind them. Additionally, they can block the effective application of fire hose streams directed into the building during fire-fighting operations.

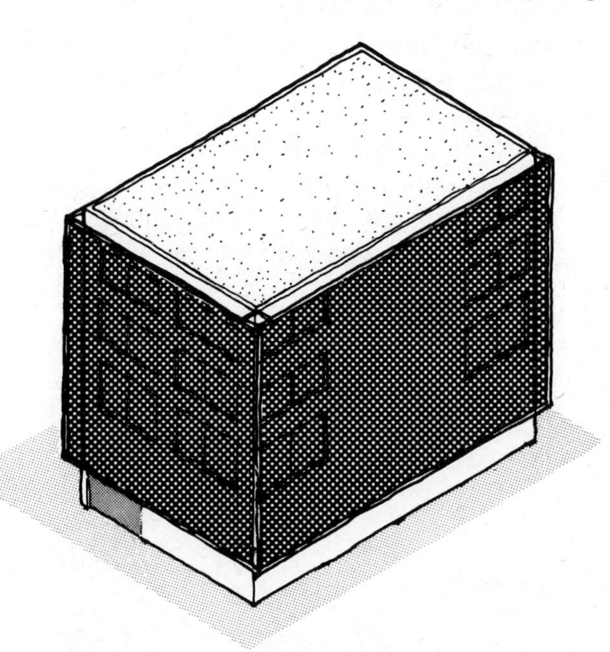

EXPANDED (OR PERFORATED) METAL DECORATIVE SCREEN

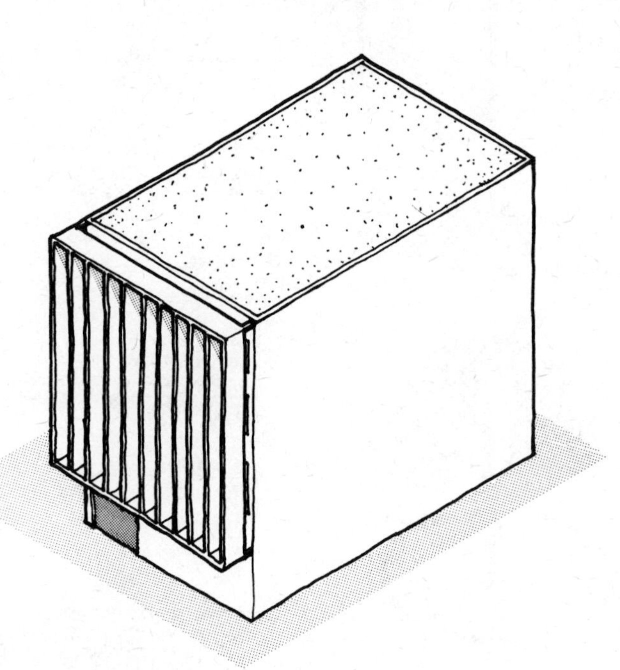

SUN SCREEN

Note: Screens should have areas that are removable by quick-release devices to permit normal fire-fighting operations and window access for rescue operations.

SITE PLANNING—LIGHTNING PROTECTION

Lightning is a frequent cause of fire. During thunderstorms the electron build-up in clouds can rapidly discharge as lightning to a lower voltage such as trees, buildings, ground or water, causing damage and fire. The need for building lightning protection should be determined by careful evaluation of the frequency and severity of thunderstorms at the site, the exposure of the building (e.g., slope of the site, location of adjacent tall buildings, etc.), and the probable cost of building repairs or replacement. Farm buildings, isolated buildings, and tall structures such as church steeples and chimneys often require lightning protection. The down conductor cable (i.e., low-resistance electrical path to the ground) in the examples below is shown exaggerated in size for clarity of presentation. Down conductors can be concealed under roofing, behind column facings, or within wall cavity spaces.

Isolated Building

Alternate ground techniques (plate, rod, water pipe) for a large, isolated building are shown.

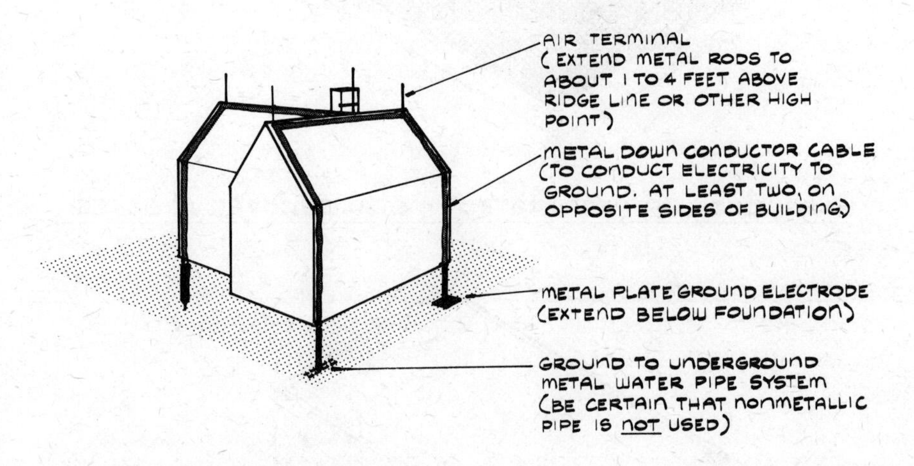

Tree of Historical Value

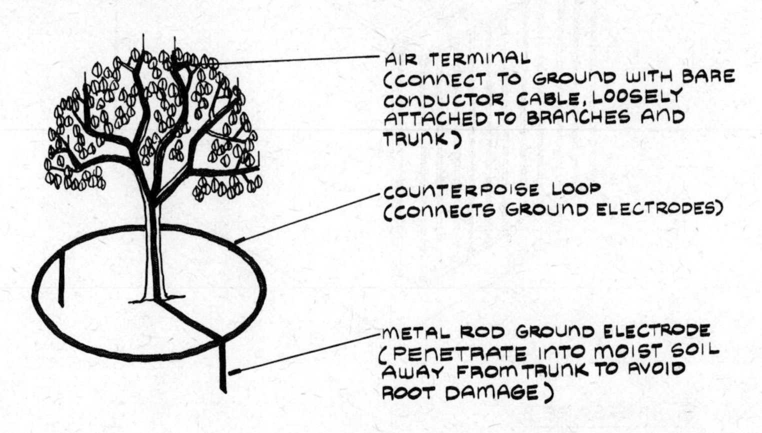

Note: For detailed information on lightning protection theory and applications, see R. H. Golde, *Lightning Protection,* London, England: Edward Arnold Ltd. (1973) or J. L. Marshall, *Lightning Protection,* New York: John Wiley & Sons (1973).

SITE PLANNING—FREQUENCY OF THUNDERSTORMS

The contours on the map of the United States indicate the frequency of thunderstorm occurrence in average number of days per year. Thunder must occur during a day for it to be recorded by National Weather Service stations. A day is recorded regardless of the number of occurrences during the day and regardless of the severity of the lightning storm. Thunderstorm data for various locations throughout the United States are available from publications of the National Oceanic and Atmospheric Administration, U.S. Department of Commerce, Rockville, Md. 20852.

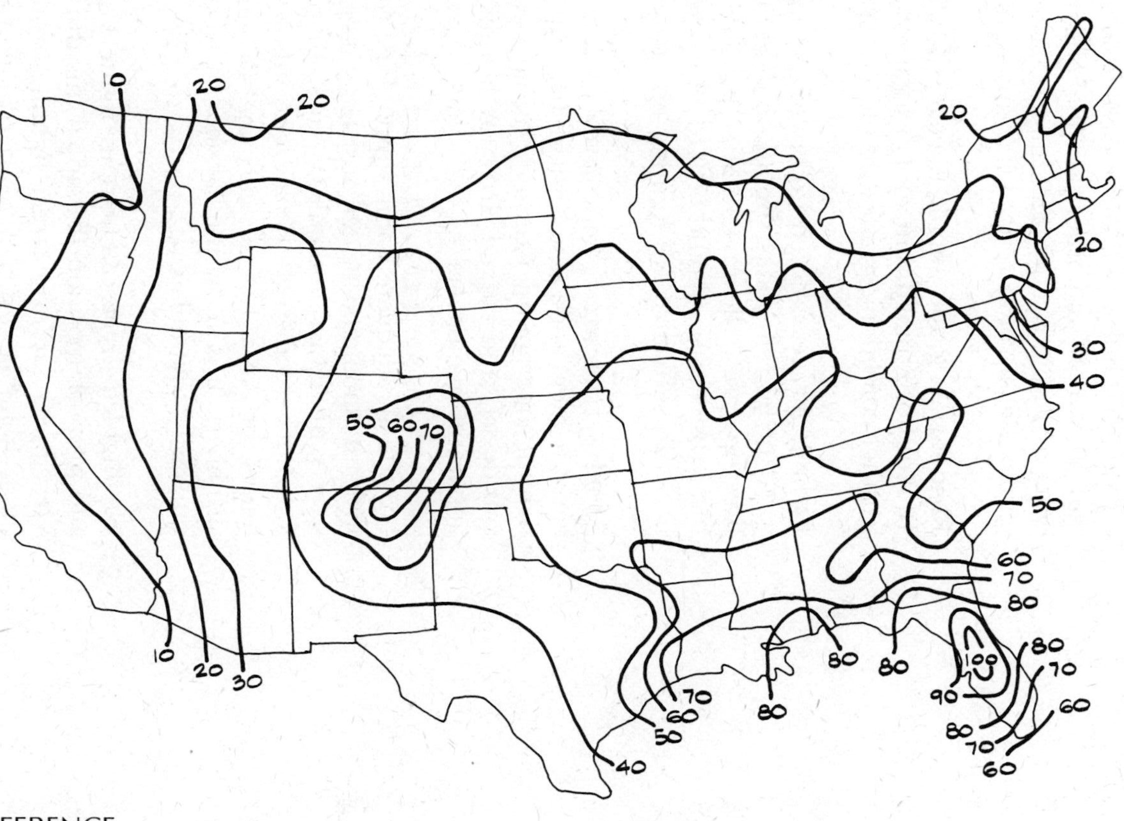

REFERENCE

Handbook of Industrial Loss Prevention, New York: McGraw-Hill (1967), p. 29–1.

SITE PLANNING CHECKLIST

Fire apparatus should have unobstructed street or driveway access to buildings, with adequate clear road widths and turning radii. For example, unobstructed access should be to within 75 ft of low-rise, multifamily housing and to within 50 ft of high-rise buildings.

Site access to buildings should be coordinated with the placement of lighting and utility poles, kiosks, outdoor sculpture, fountains, trees and vegetation, and other potential obstructions to fire-fighting operations. Avoid locating buildings at the edge of cliffs or other steep grades which could restrict fire apparatus access to only one side.

Road surface grades should not exceed 10% (i.e., 1 ft of vertical rise to 10 ft of horizontal distance). For locations with frequently wet or icy weather conditions, grades should not exceed 8%.

Parking garages adjacent to buildings, access ramps, and wide elevated pedestrian walks should be designed to support the load of fire apparatus. Clearance under pedestrian walks and other overhead obstructions should be 9 to 12 ft, depending on the apparatus needs for effective fire-fighting operations.

Avoid narrow streets and driveways that prevent full extension of aerial ladders and cause difficult back-up maneuvering during fire-fighting operations. Consider using T-turns (or hammer heads), culs-de-sac, and curved driveway layouts to allow expeditious fire apparatus maneuvers.

Provide an adequate water supply for fire-fighting operations. Fire hydrants should be located at street intersections and at points along streets, driveways, and culs-de-sac so that spacing between hydrants does not exceed 300 ft. Locate hydrants away from trees, posts, signs, fences, or any other obstructions. Do not locate hydrants behind shrubbery or other visual barriers, or where runoff water and snow can easily accumulate.

Well-lighted streets can help fire fighters quickly locate fire hydrants and properly position fire apparatus during nighttime operations. To achieve uniform street lighting from pole-mounted fixtures, there should be sufficient overlap of light from the adjacent fixtures. In general, the greatest horizontal ground dimension that can be covered by a single lighting pole is about two times the fixture mounting height above grade. In some situations, lighting fixtures required for street lighting can be integrated into the exterior of buildings or other structures.

Where vacant, useable land is adjacent to a proposed building project, site planning should consider the potential fire hazard that could occur if the land is used.

SITE PLANNING CHECKLIST (Continued)

Provide adequate separation distance between buildings to prevent the spread of fire. Fire spread between adjacent buildings also depends on the severity of the fire (fuel load), exterior wall constructions, extent of openings (doors and windows) in facing walls, combustibles between buildings, and so on.

The receiving, storage, and trash removal of combustible materials can become a source of fires. Effective site layouts should encourage the use of protected locations. For trash removal, steel containers with steel covers and pick-up locations remote from buildings are preferred.

Standard symbols used on a building's exterior to indicate the fire hazard of its contents can be integrated with the overall graphics design of the building. The fire hazard to fire fighters at chemical processing plants, warehouses, laboratories, and so on can be identified by the NFPA diamond-shaped symbol which visually presents coded information on health, flammability, and self-reactivity hazards as well as special hazard information (e.g., reaction with water, radioactive materials).

BUILDING MATERIALS AND CONSTRUCTIONS

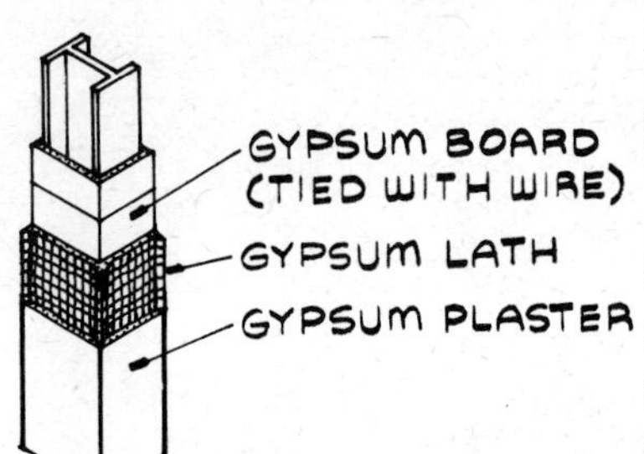

PROTECTED STEEL COLUMN

BUILDING MATERIALS AND CONSTRUCTIONS

For effective fire protection, building constructions must be able to support their structural loads during a fire, contain the spread of smoke and fire gases, and prevent excessive heat flow for a reasonable time period.

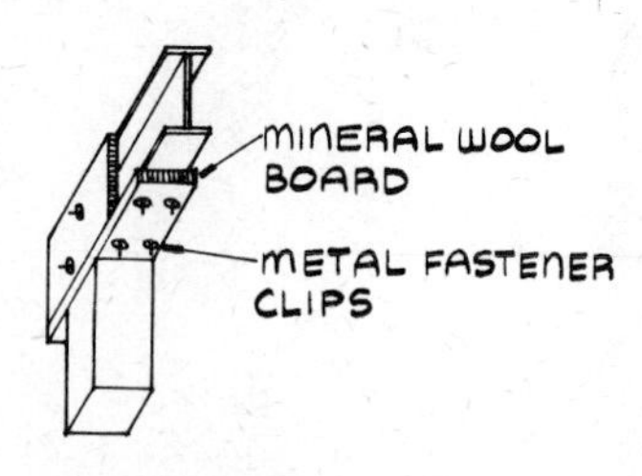

PROTECTED STEEL BEAM

An example protected steel column, enclosed in gypsum plaster on gypsum lath, is shown in the margin. When exposed to fire, gypsum releases water of crystallization which retards heat flow to the steel. Several other materials are available that can be used to protect steel columns and beams from fire temperatures. Examples of materials with high thermal insulation properties are vermiculite (i.e., weathered mica) and perlite (i.e., volcanic rock) used in plaster and in concrete; mineral wool; and concrete with lightweight aggregate (when exposed to fire, water is released which retards heat flow). An example protected steel beam, enclosed in mineral wool, is also shown in the margin.

LIQUID-FILLED TUBULAR COLUMN

Additional techniques for structural steel protection are intumescent mastic coatings (when exposed to fire, they expand to form a thick, lightweight heat barrier) and liquid-filled members (when exposed to fire, the liquid removes heat by convection).

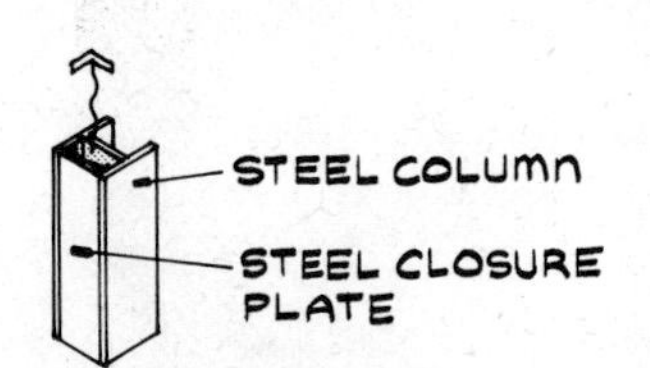

LIQUID-FILLED W-SHAPE COLUMN

The support systems for structural members (e.g., simple, fixed) influence resistance to fire under design dead and live load conditions. For example, a restrained steel beam with fixed supports (as shown in the margin) can resist collapse until its temperature reaches well above 1000°F because it cannot easily expand. An unrestrained steel beam with simple supports (also shown in the margin) can lose its strength at approximately 1000°F because it is able to expand longitudinally. Steel heated to 1000°F expands about $4\frac{1}{2}$ inches/50 ft of length.

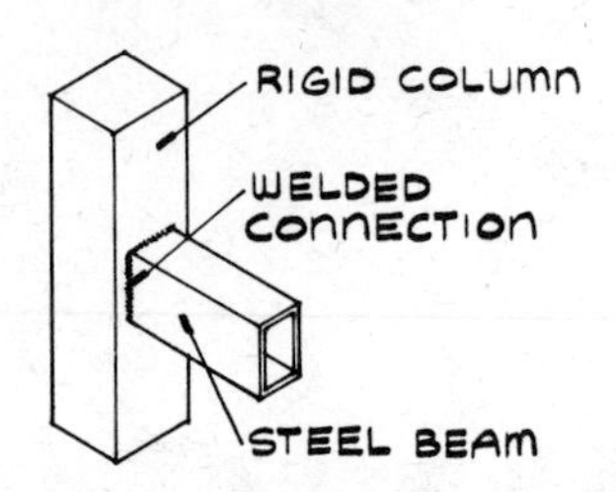

FIXED-END SUPPORTED BEAM

For further information on fire-resistance rating tests, including complete construction details, consult the Underwriters Laboratories publication, "Fire Resistance Index"; U.S. National Bureau of Standards reports; and other sources referenced in this chapter. It is important to know the complete test details. For example, the test may have been stopped before the endpoint criteria were reached. In addition, it is essential that fire-resistance rated constructions be installed in buildings in strict accordance with the design of the test specimen and the procedures used to erect it in the test furnace.

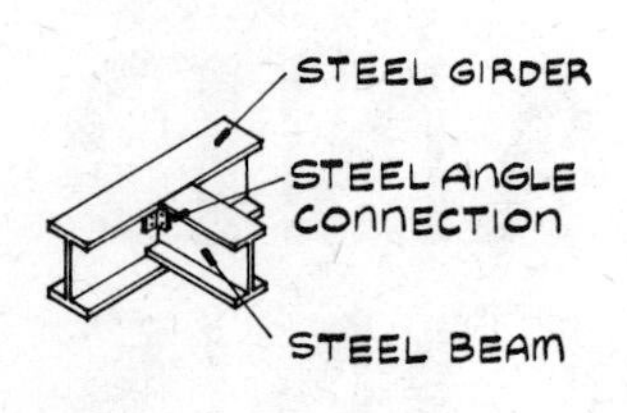

SIMPLY SUPPORTED BEAM

BUILDING MATERIALS AND CONSTRUCTIONS—STEEL BEAM SURFACE TEMPERATURES

The test data shown below are from standard time-temperature fire exposures on steel, floor-beam assemblies. The curves for top and bottom flange surface temperatures indicate that beam temperatures can differ by more than 600°F. In addition, due to the mass of the steel beams, interiors can be considerably cooler than the surface temperatures. Fire tests on steel beams are usually terminated when the average temperature reaches 1100°F at a cross section or 1300°F at a single point.

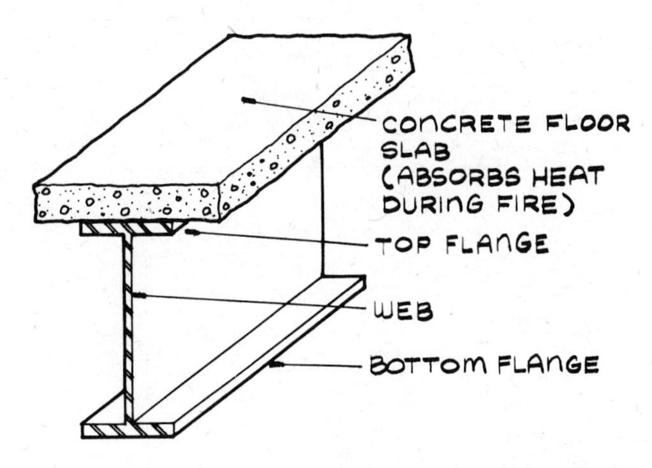

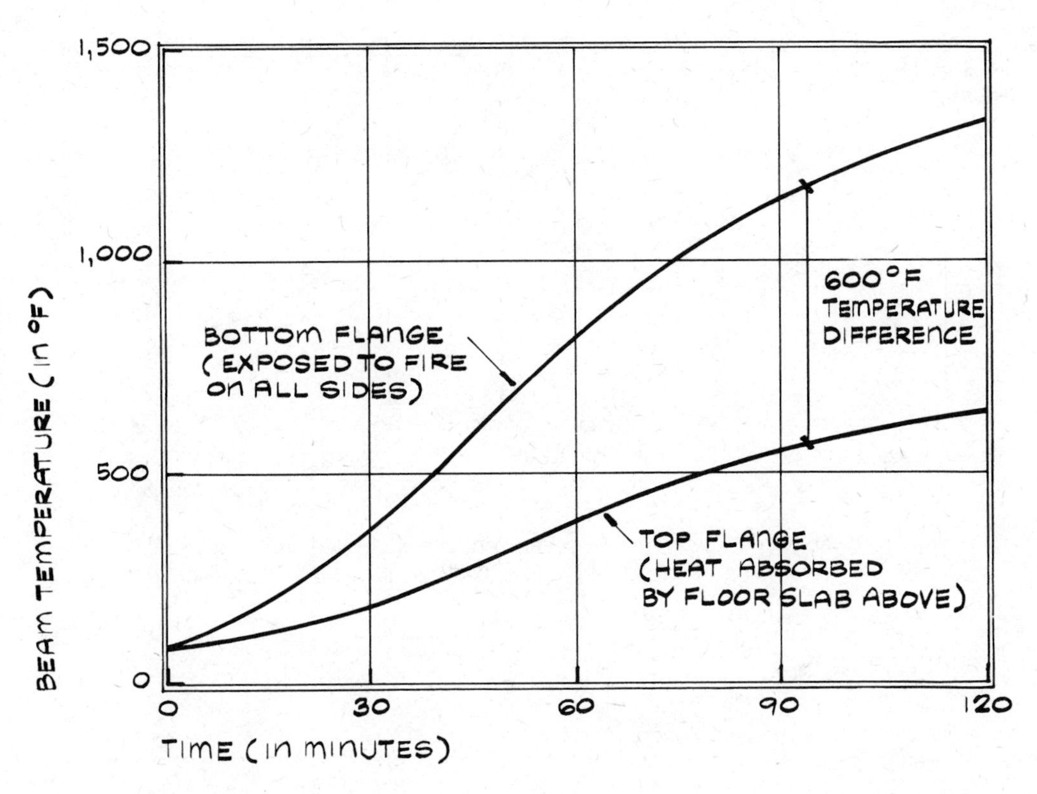

REFERENCE

Bletzacker, R. W. et al., "Fire-Resistance of Construction Assemblies," Engineering Experiment Station, Ohio State University, December 1969.

BUILDING MATERIALS AND CONSTRUCTIONS—TENSILE STRENGTH OF STEEL AND ALUMINUM AT ELEVATED TEMPERATURES

Above 400 to 500°F, the ultimate tensile strength of steel (tensile stresses pull the particles of steel apart as shown by the sketch) begins to decrease. With continuing increase in temperature, the strength drops rapidly as shown by the curves below representing the average performance of a range of unprotected A7, A36, and A440 structural steels. Also shown on the graph is the elevated temperature effect on the ultimate tensile strength of 6061-T6 wrought aluminum alloy. The data represent the lowest strength during $\frac{1}{2}$ to 10 hours exposure under no load conditions. Tensile strength is given in kips per square inch (ksi). One kip equals 1000 pounds.

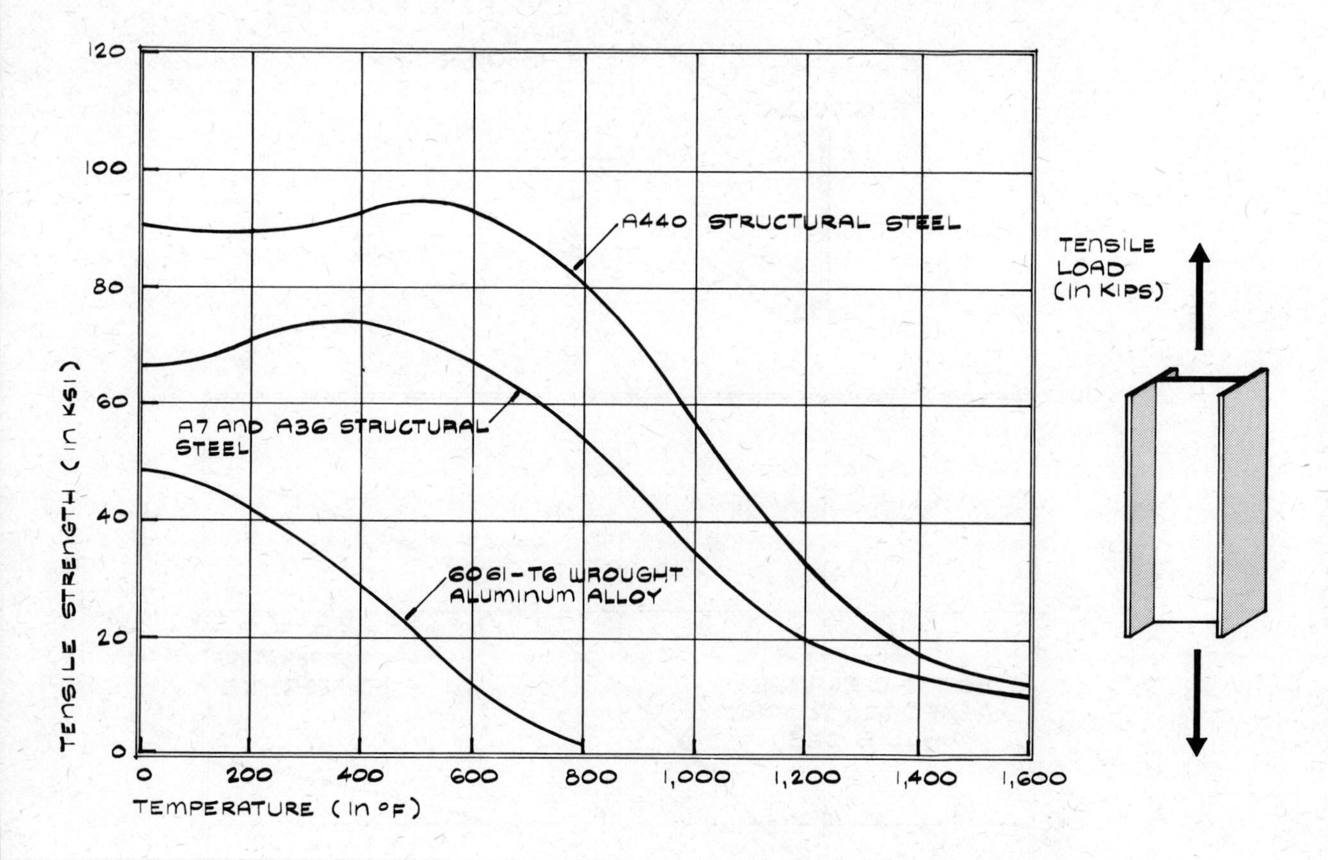

Note: The above data are from test specimens that were heated uniformly along their length and cross section. In actual building fires, steel beams are rarely heated uniformly.

REFERENCES

Fire Protection Through Modern Building Codes, New York: American Iron and Steel Institute (1971), pp. 153 and 155.

Metallic Materials and Elements for Aerospace Vehicles, Washington, D.C.: U.S. Department of Defense, Mil-Hdbk-5.

BUILDING MATERIALS AND CONSTRUCTIONS—COMPRESSIVE STRENGTH OF STEEL AT ELEVATED TEMPERATURES

Above 200 to 400°F, the compressive yield strength of steel (compressive stresses push the particles of steel together as shown by the sketch) begins to decrease rapidly. At 800 to 900°F, buckling (the bending outward of columns under compressive load) occurs. Compressive strength is given in kips per square inch (ksi). One kip equals 1000 pounds.

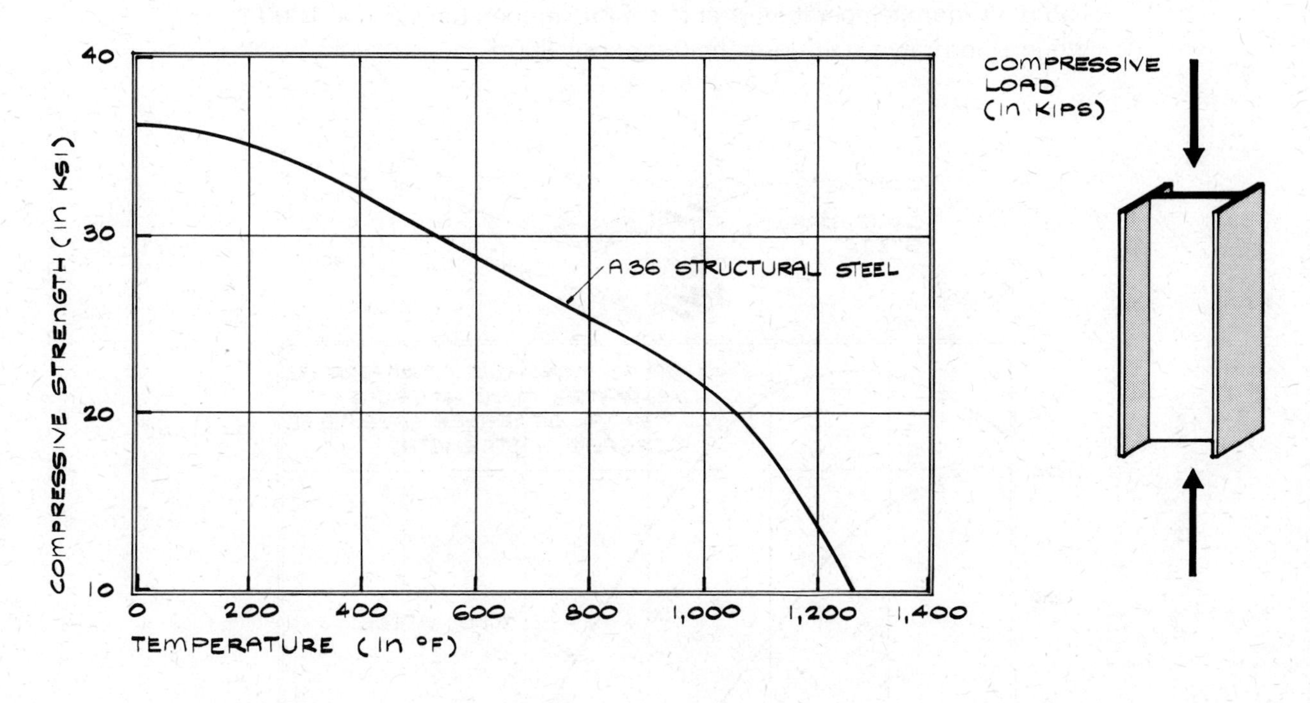

REFERENCE

Stanzak, W. W. and T. T. Lie, "Fire Resistance of Unprotected Steel Columns," *Journal of the Structural Division,* ASCE, Vol. 99, No. ST5, May 1973.

BUILDING MATERIALS AND CONSTRUCTIONS—TENSILE STRENGTH OF COLD-DRAWN STEEL AT ELEVATED TEMPERATURES

Above 500°F, the ultimate tensile strength of cold-drawn prestressing steel (i.e., cross-sectional area has been mechanically reduced by dies without heating) begins to diminish considerably as shown by the graph below. Also shown on the graph is the effect of elevated temperature on mild steel (i.e., iron having a very low carbon content). Mild steel loses half of its room temperature strength at fire temperatures above 1000°F, while cold-drawn steel loses half at about 800°F.

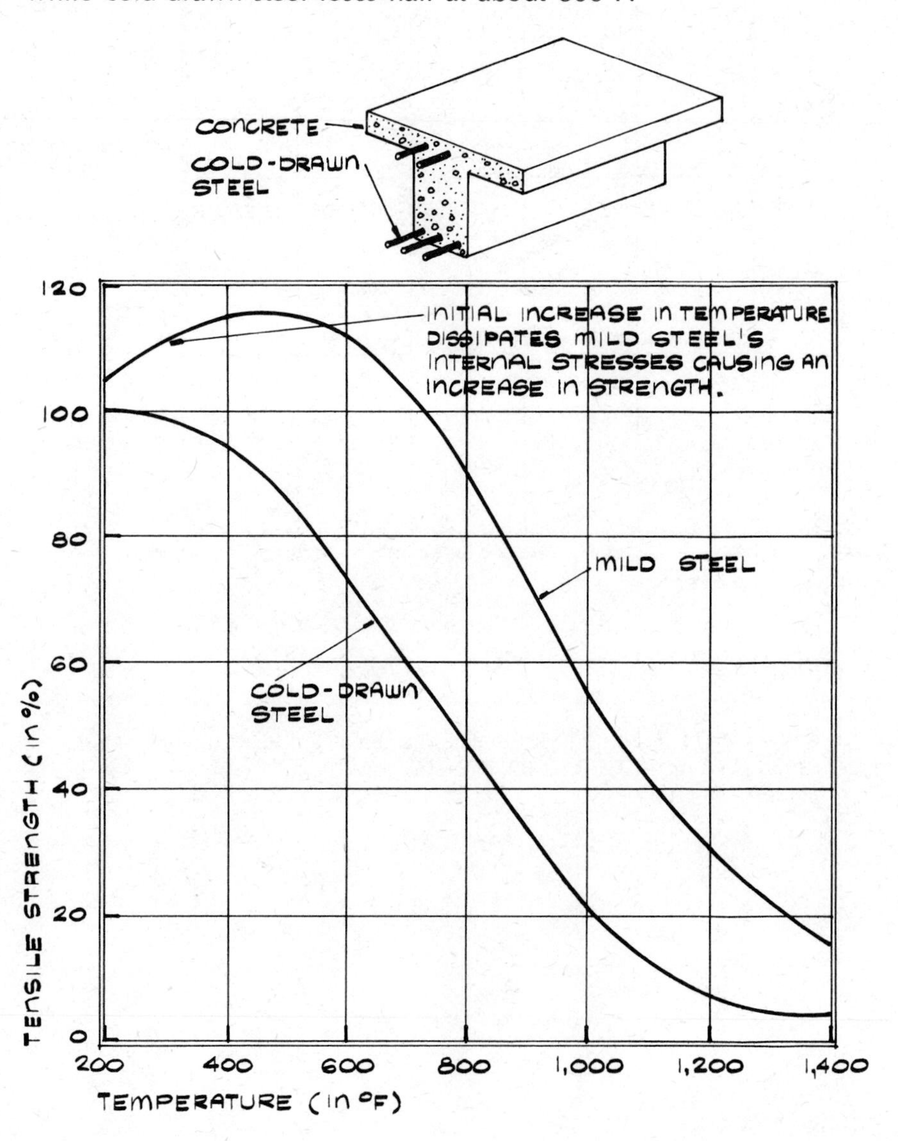

REFERENCE

Nelson, H. E. and A. Dailey, "Building Integrity," in *Reconvened International Conference on Firesafety in High-Rise Buildings,* Washington, D. C.: General Services Administration, October 1971.

BUILDING MATERIALS AND CONSTRUCTIONS—COMPRESSIVE STRENGTH OF CONCRETE AT ELEVATED TEMPERATURES

Above 600 to 700°F, the compressive strength of concrete begins to diminish considerably as shown by the graph below for heated concrete test cylinders. Conventional building structural member sizes, however, usually have sufficient mass to replenish surface moisture dissipated by fire and to absorb surface heat. For comprehensive information on the strength of concrete constructions at elevated temperatures, see A. H. Gustaferro, "Fire Resistance," in M. Fintel (ed.), *Handbook of Concrete Engineering,* New York: Van Nostrand Reinhold (1974), pp. 212–228.

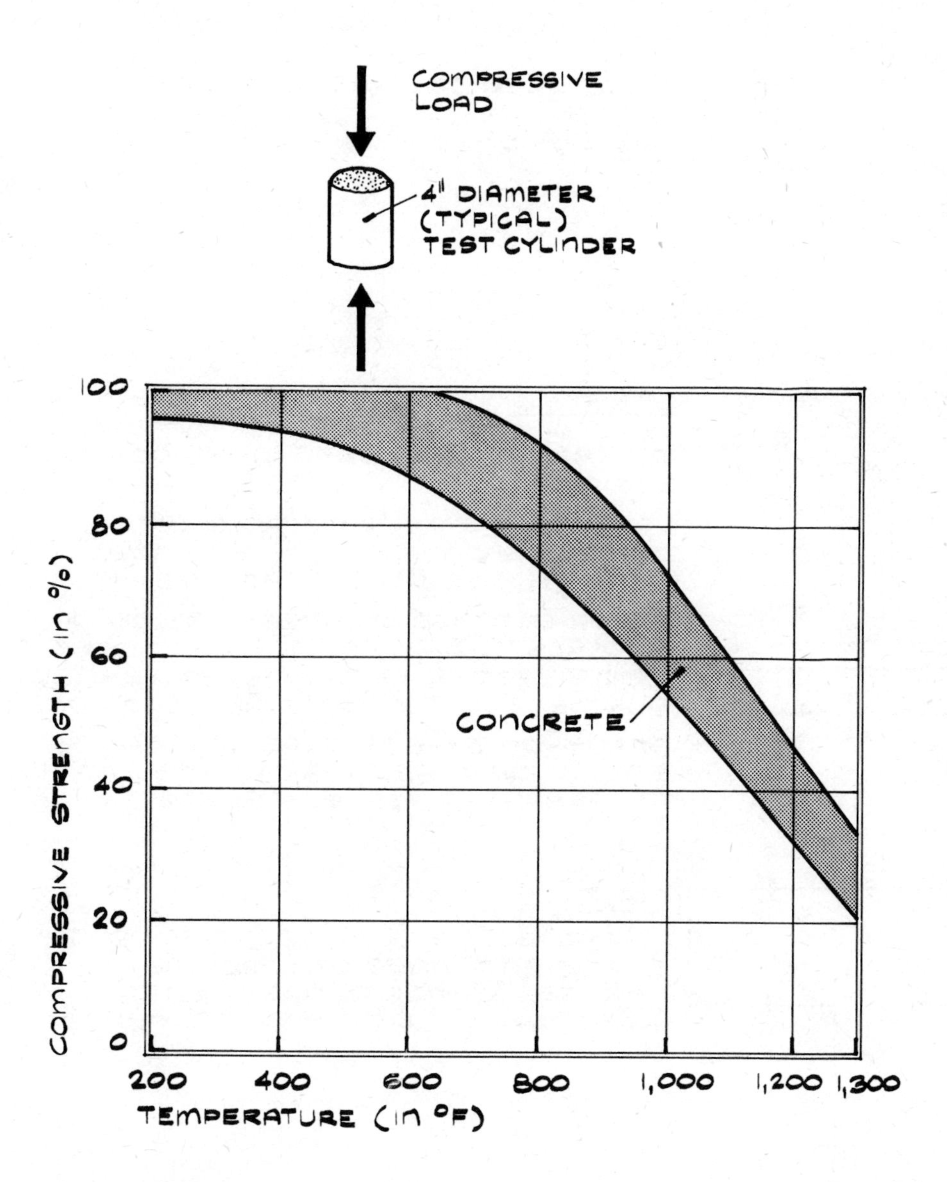

REFERENCE

Marchant, E. W. (ed.), *Fire and Buildings,* New York: Barnes & Noble (1973).

BUILDING MATERIALS AND CONSTRUCTIONS—STEEL COLUMNS

Shown below are example fire protection techniques for steel columns. For listings of various fire-rated structural members, see the latest edition of "Fire Resistance Design Data Manual," Gypsum Association; "Fire Resistance Index," Underwriters Laboratories; or "Fire Resistance Ratings," American Insurance Association.

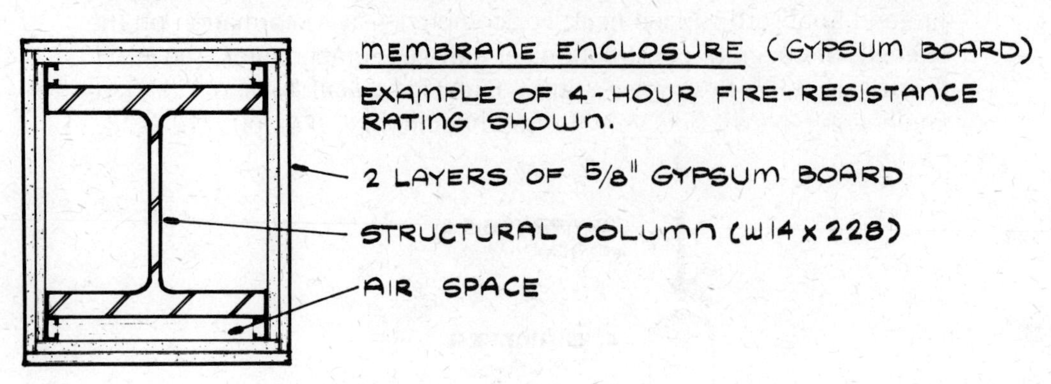

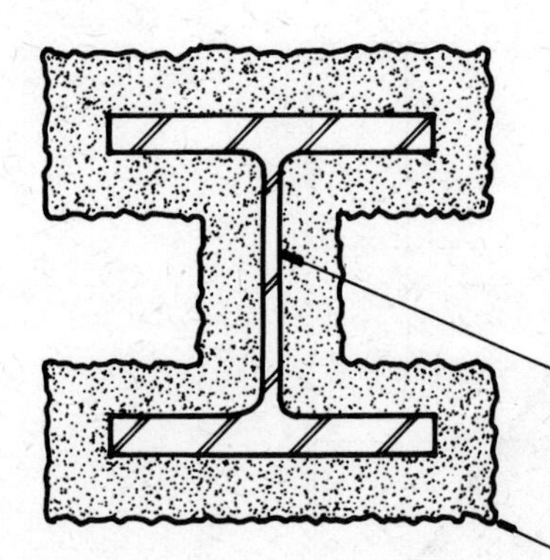

SPRAY-ON (GYPSUM PLASTER WITH VERMICULITE OR PERLITE AGGREGATE AND MINERAL WOOL FIBERS WITH INORGANIC BINDERS)

2 TO 4-HOUR FIRE-RESISTANCE RATING, DEPENDING ON ACHIEVED THICKNESS, DENSITY, AND OTHER FACTORS.

STEEL SURFACE MUST BE FREE OF DIRT AND RUST (E.G., APPLY PRIME COAT OF PAINT BEFORE SPRAYING, PREFERABLY DURING MILD-TEMPERATURE CONDITIONS)

SPRAY-ON GYPSUM PLASTER (CAN BE EASILY DAMAGED DURING INSTALLATION OF NEARBY PIPES, DUCTS, ETC.)

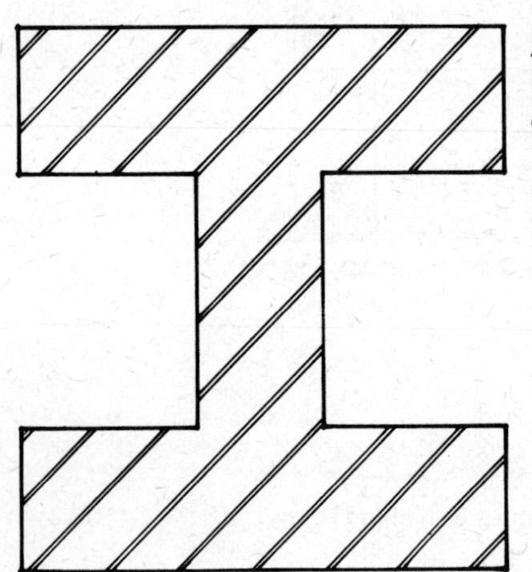

MASS

UNPROTECTED, BUILT-UP COLUMNS OF SUFFICIENT MASS CAN PROVIDE A FIRE-RESISTANCE RATING OF UP TO 1 HOUR (E.G., 30" x 26" COLUMN SHOWN).

BUILDING MATERIALS AND CONSTRUCTIONS—EFFECT OF STEEL COLUMN WEIGHT ON FIRE RESISTANCE

Shown below are the effects of column weight on fire resistance for steel columns with membrane or spray-on protection. Higher fire-resistance ratings can be achieved by the heavier columns because they absorb more heat. The membrane enclosures will provide higher fire-resistance ratings than the spray-on contours as the enclosure reduces the perimeter of steel column surface exposed to fire by forming a deep air space around the web.

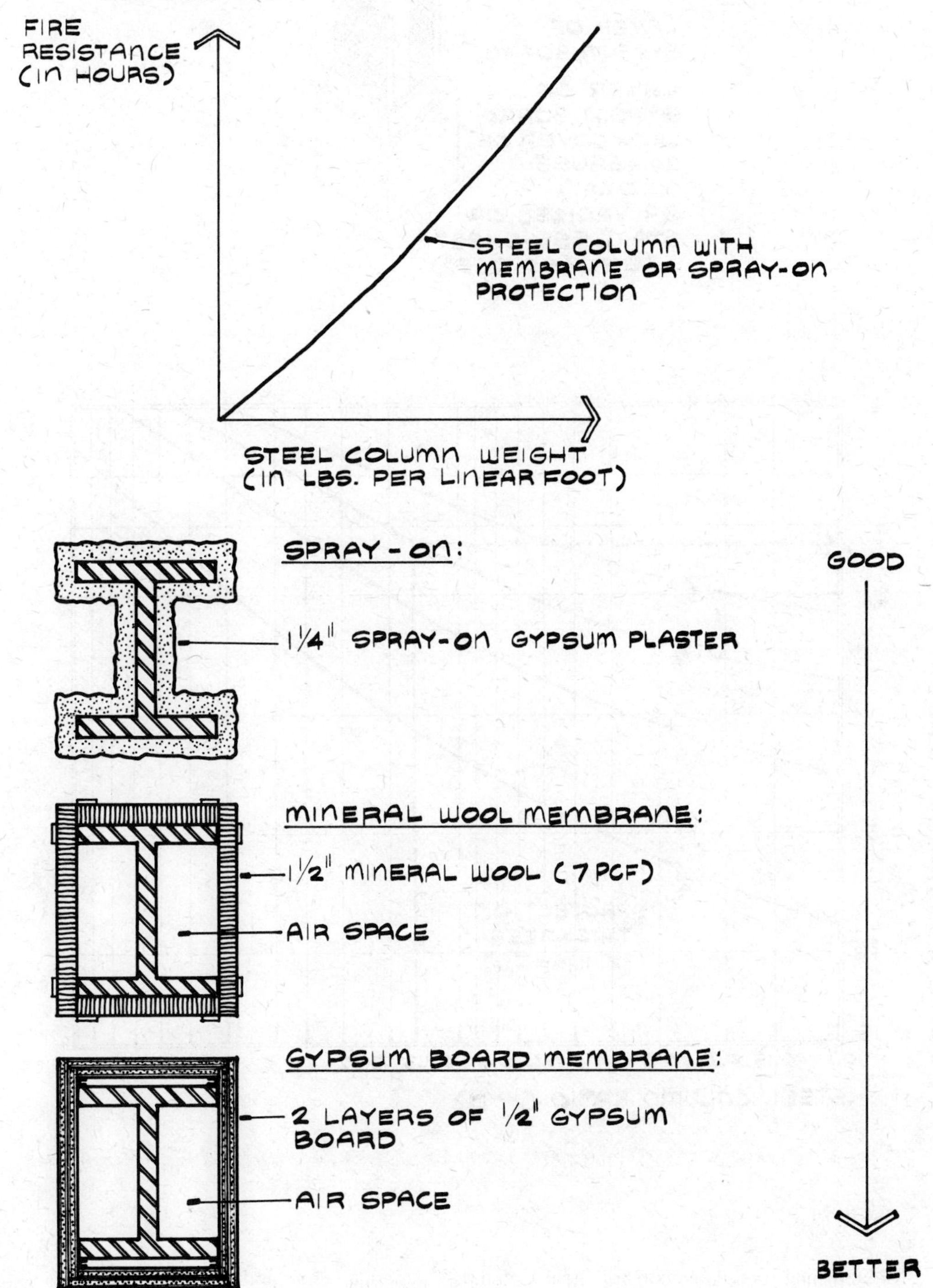

BUILDING MATERIALS AND CONSTRUCTIONS—FIRE RESISTANCE FOR STEEL COLUMNS PROTECTED BY GYPSUM BOARD

The graph below shows the thickness of gypsum board enclosure protection required to achieve a given fire-resistance rating in hours. For 2 hour ratings or less, the steel cover may be placed under the outer layer of gypsum board.

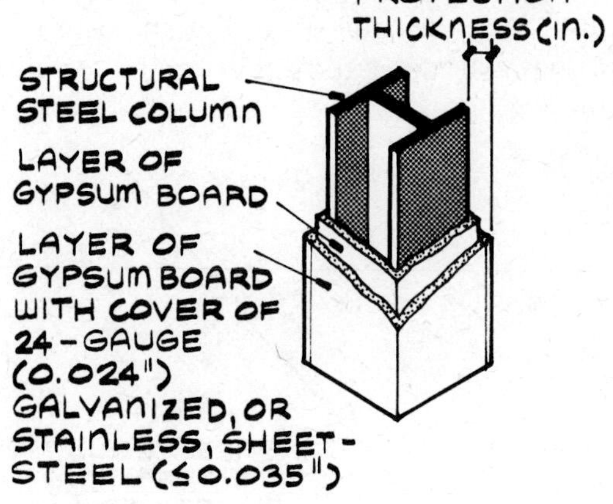

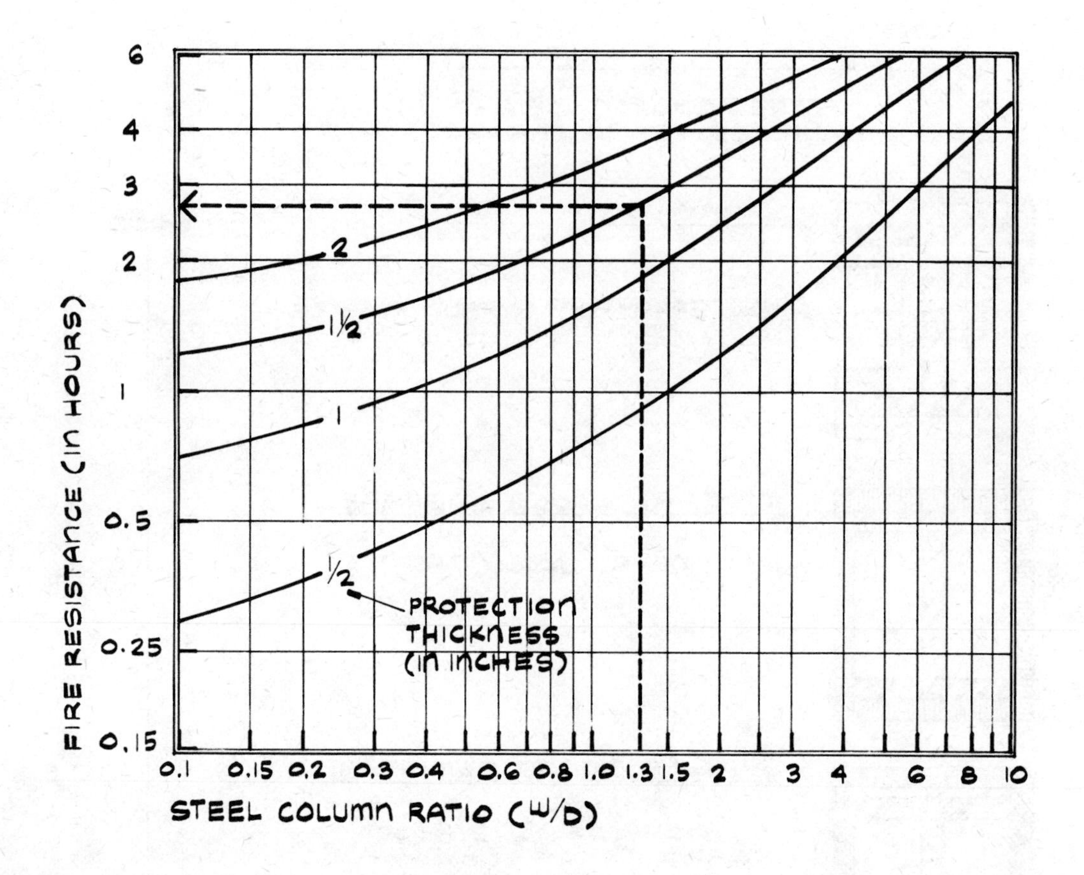

SOURCE

"Designing Fire Protection for Steel Columns," American Iron and Steel Institute, Washington, D.C., October 1975.

BUILDING MATERIALS AND CONSTRUCTIONS—FIRE RESISTANCE FOR STEEL COLUMNS PROTECTED BY GYPSUM BOARD (Continued)

Example—Use of Graph

Given: *W*10 X 49 steel column.

Procedure to find required gypsum board enclosure thickness for 2 hour fire rating:

1. Determine W/D ratio, where W = weight of steel in pounds per foot and *D* = heated perimeter of steel (i.e., backside of protection) in inches, *W/D* = 49/(4 X 10) = **1.3**

2. Enter graph at *W/D* = 1.3 and read opposite $1\frac{1}{2}$ in. curve to a fire resistance greater than 2 hours. Use gypsum board enclosure with a thickness of **$1\frac{1}{2}$ in.** or more.

thickness

BUILDING MATERIALS AND CONSTRUCTIONS—FIRE RESISTANCE FOR STEEL COLUMNS ENCASED IN CONCRETE

The graph below shows the protective cover in inches needed to achieve a given fire-resistance rating in hours for steel columns encased in concrete (sand and gravel aggregate). For example, an 8 in. column will need a 2 in. concrete cover to achieve a fire resistance of 4 hours (see dashed lines on graph). The larger columns require less cover for equivalent fire resistance as they have greater mass. However, a concrete cover of 2 in. or less can allow enough moisture penetration to cause corrosion of the structural steel.

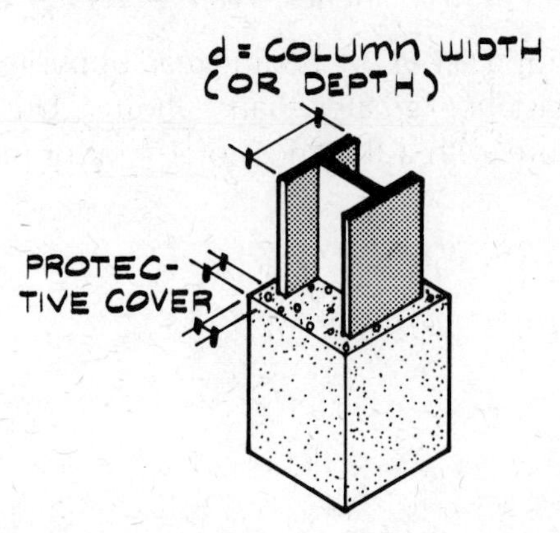

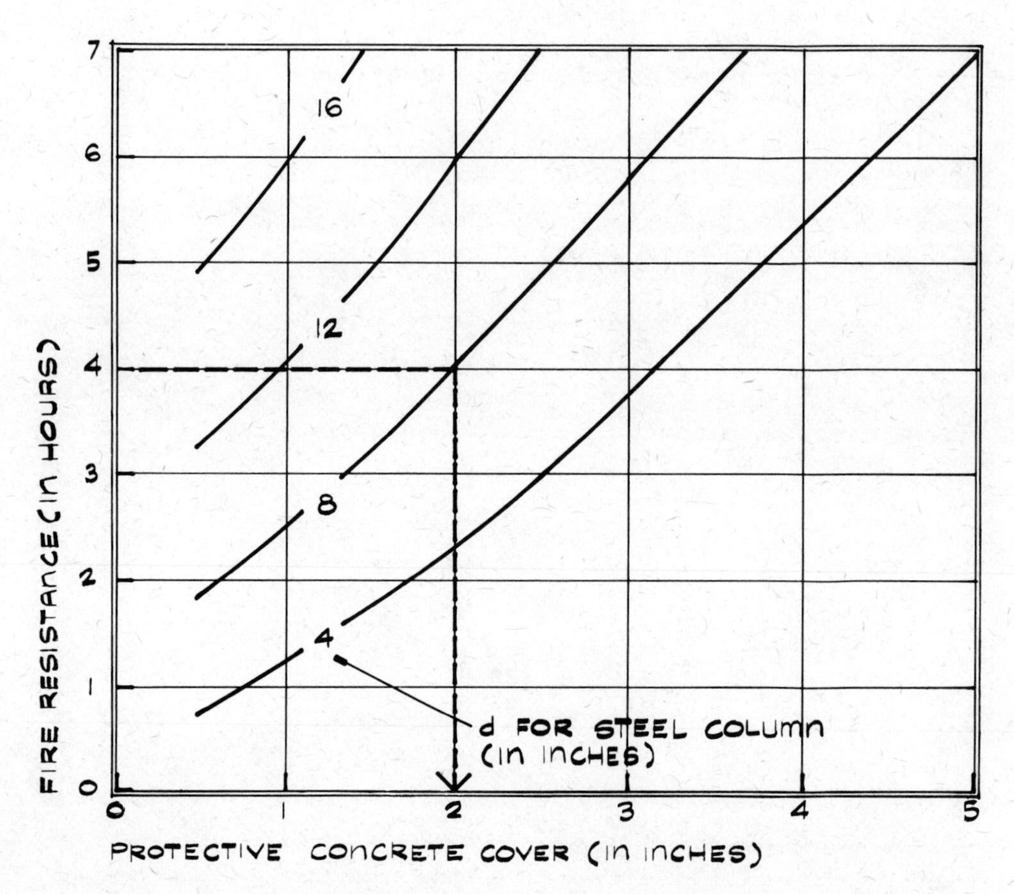

Note: For protective cover requirements for concrete of various aggregates, see G. P. McKinnon (ed.), *Fire Protection Handbook*, Boston, Mass.: National Fire Protection Association (1976), p. 6–59.

BUILDING MATERIALS AND CONSTRUCTIONS—STEEL BEAMS

Shown below are example fire protection techniques for steel beam floor-ceiling assemblies. For a comprehensive presentation of fire-resistance ratings of steel beams and columns, see M. Galbreath, and W. Stanzak, "Fire Endurance of Protected Steel Columns and Beams," Technical Paper No. 194, Division of Building Research, National Research Council of Canada, April 1965.

CONTINUOUS CEILING (GYPSUM BOARD)

EXAMPLE OF 2-HOUR FIRE-RESISTANCE RATING FOR UNRESTRAINED ASSEMBLY SHOWN.

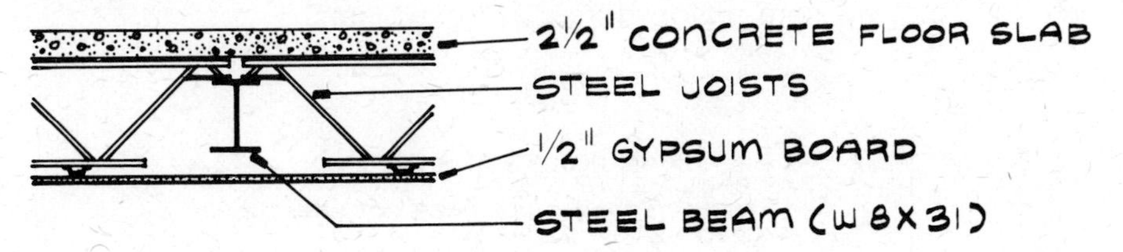

INDIVIDUAL ENCASEMENT (GYPSUM BOARD)

EXAMPLE OF 2-HOUR FIRE-RESISTANCE RATING SHOWN.

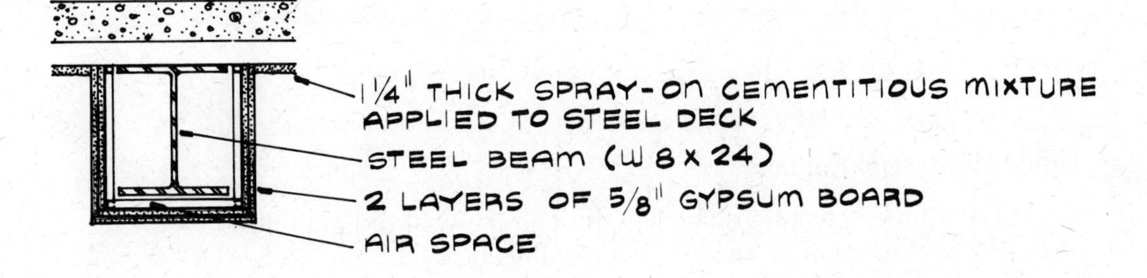

SPRAY-ON

1 TO 4-HOUR FIRE-RESISTANCE RATING, DEPENDING ON ACHIEVED THICKNESS, DENSITY, AND OTHER FACTORS.

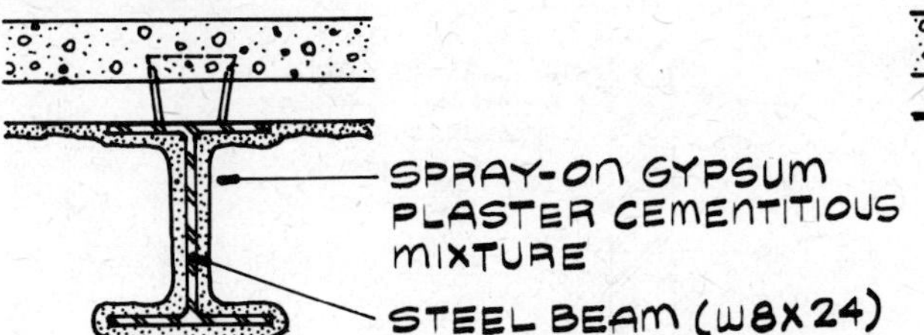

MASTIC COATING

EXAMPLE OF ½-HOUR RATING SHOWN.

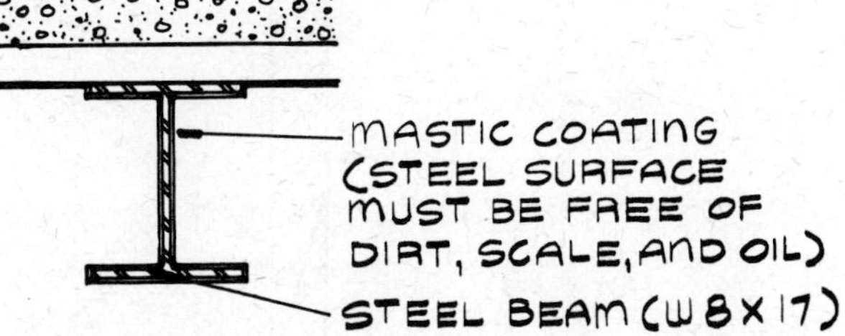

Note: When exposed to fire, intumescent mastic coatings release inert, nontoxic gases which cause the coating to expand into a multicellular layer of thermal insulation about 10 to 15 times its initial thickness. The gas-filled foam layer retards heat penetration and blocks off the oxygen supply. Coatings should be applied under well-ventilated conditions to avoid accumulation of flammable vapors.

Shown below are temperature gradients for a layer of gypsum (i.e., dihydrous calcium sulphate) and a liquid-filled steel column exposed to fire.

Gypsum

When gypsum is exposed to fire, chemically combined water within the gypsum is slowly released as steam (called calcination), retarding heat flow. Gypsum board cores are generally classified as "regular" or "type X." Type X has a specially formulated core using unexpanded vermiculite and siliceous clays to provide thermal stability. Glass-fibers also are used to reinforce the calcined core region during fire exposures.

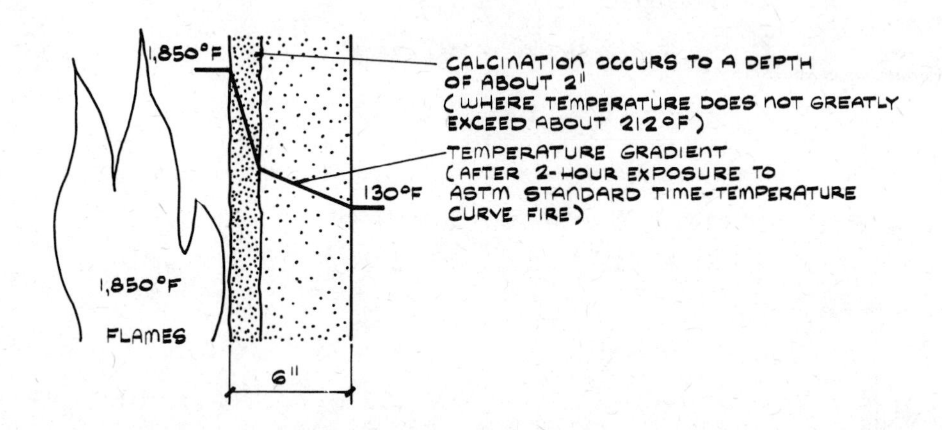

Liquid-Filled Steel Column

The maximum temperature difference through the wall of liquid-filled steel structural members is about 70°F/in. of thickness (cf., "Fire Protection Through Modern Building Codes," American Iron and Steel Institute, 1971, p. 170). Only columns or structural members with sufficient slope (e.g., cross-bracing) should be designed as liquid-filled members. For horizontal members, the venting of steam during fire exposures is difficult.

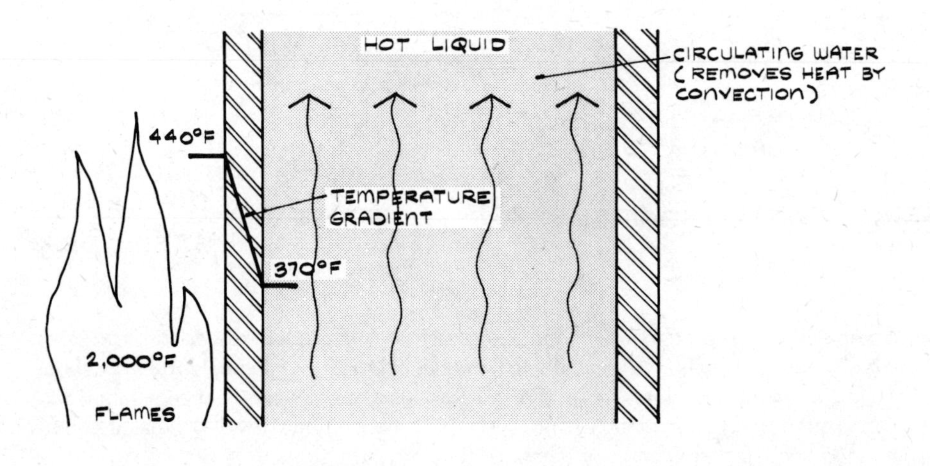

BUILDING MATERIALS AND CONSTRUCTIONS—LIQUID-FILLED COLUMN SYSTEMS

Shown below is a piping schematic for a zone within a liquid-filled exterior column fire protection system. These columns can be in the form of rectangular (or box) sections, tubing, or wide flange sections with a flat plate enclosing one side of the web to form a box. During a fire the liquid, circulating by convection between the fire floor columns and the storage tank, absorbs heat. Generated steam can be vented to the atmosphere through the top of the storage tank. Buildings having liquid-filled column fire protection include the U.S. Steel Building, Pittsburgh, Pa.; the Michelson Building, Newport Beach, Calif.; and the American Security Insurance Building, Atlanta, Ga.

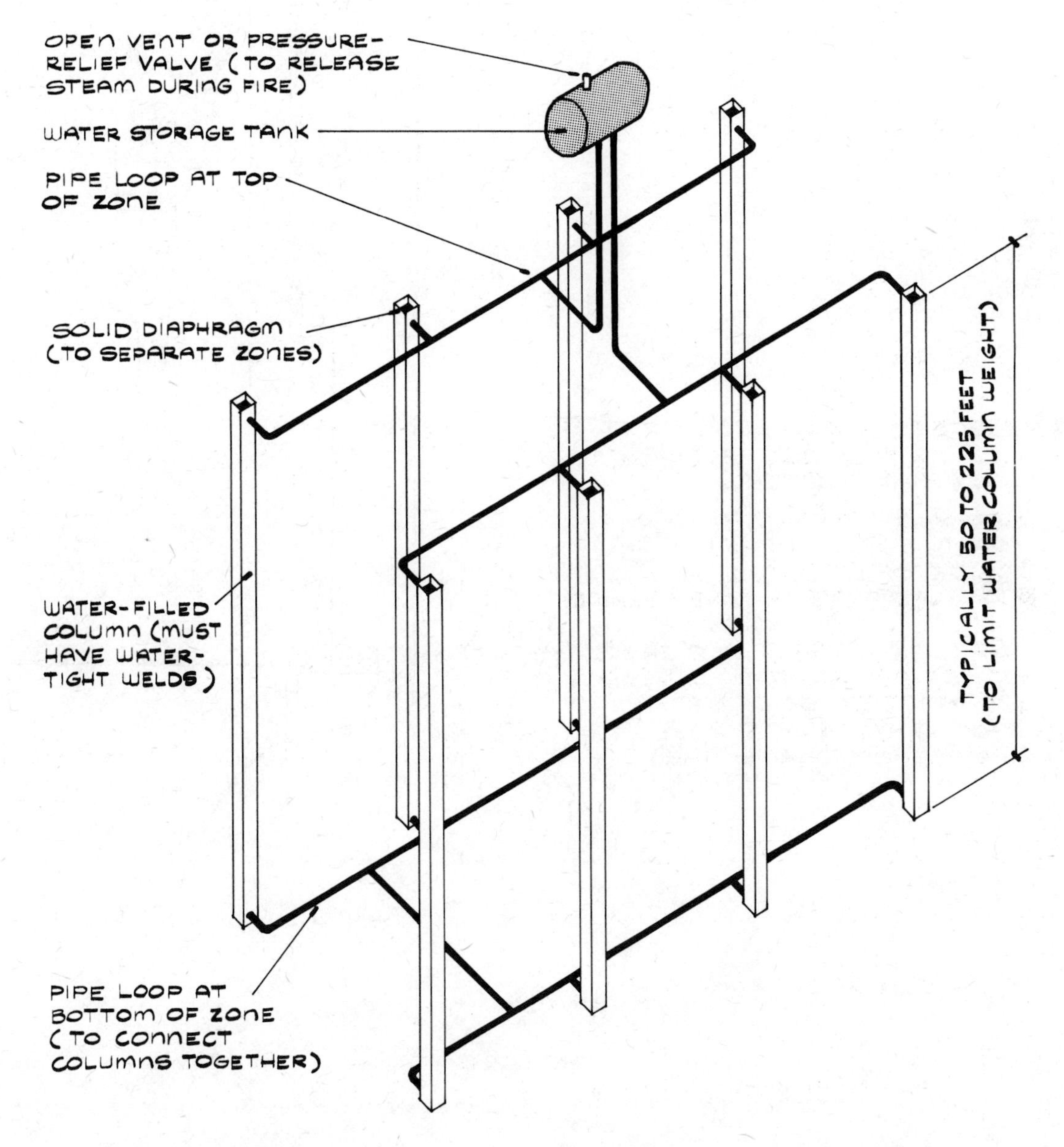

Note: In cold climate regions, antifreeze (e.g., potassium carbonate) must be added to the liquid in exterior column systems to maintain protection at low temperatures and to prevent ice damage. To prevent corrosion, rust inhibitors such as potassium nitrate can be used.

BUILDING MATERIALS AND CONSTRUCTIONS—FLAME-SHIELDS

The effects of fire from inside buildings on exterior structural members can be reduced by distance (e.g., locate members away from exterior openings) or by flame-shields as shown below. Steel cladding can deflect heat and flames from burning buildings away from exposed steel webs which can be covered by spray-on insulation on the inside surface.

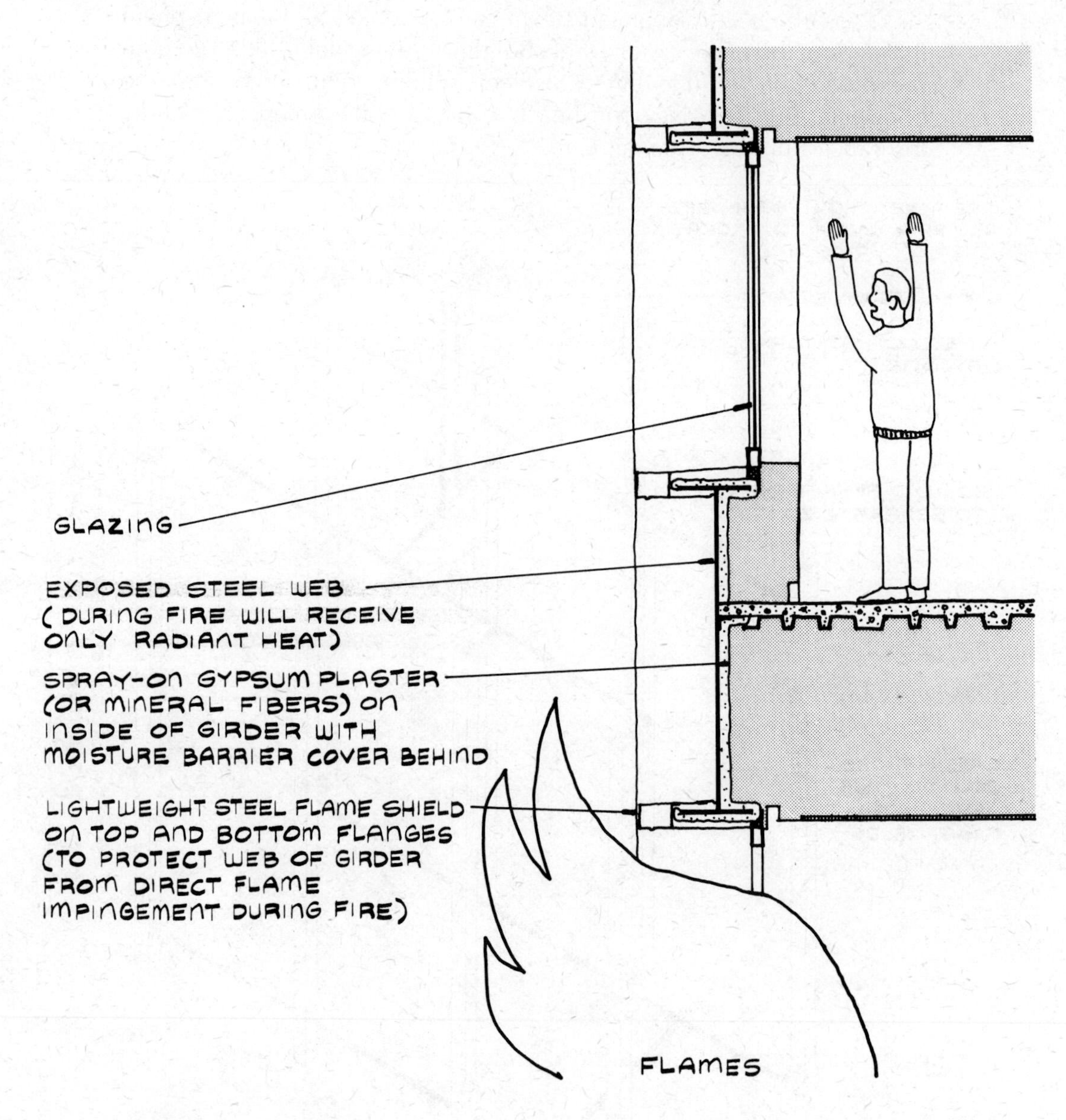

REFERENCE

Seigel, L. G., "Designing for Fire Safety with Exposed Structural Steel," *Fire Technology*, Vol. 6, No. 4, November 1970.

FIRE RATINGS FOR STRUCTURAL MEMBERS

Fire-resistance ratings in hours for building structural systems are given in the tables below. The data represent the average rating from the four model codes in the United States (Basic Building Code, National Building Code, Standard Building Code, and Uniform Building Code).

FIRE RATINGS FOR VARIOUS CODE CLASSIFICATIONS (IN HOURS)

Structural Element	*Highest*	*Second Highest*	*Third Highest*
Structural frame	4	3	2
Floors (and beams)	3	2	1
Roofs (and beams)	2	1½	1

FIRE RATINGS FOR VARIOUS OCCUPANCIES (IN HOURS)

Structural Element	*Office*	*Apartment*	*Hospital*
Structural frame	3	3	4
Floors (and beams)	2	2	3
Roofs (and beams)	1½	1½	2
Vertical openings (e.g., enclosed stairways)	2	2	2
Corridor (and occupant separation)	1	1	1

Model Building Codes

1. Basic Building Code (Building Officials & Code Administrators International, Inc., 1313 East 60th Street, Chicago, Ill. 60637).
2. National Building Code (American Insurance Association, 85 John Street, New York, N. Y. 10038).
3. Standard Building Code (Southern Building Code Congress, 3617 Eighth Avenue, South, Birmingham, Ala. 35222).
4. Uniform Building Code (International Conference of Building Officials, 5360 South Workman Mill Road, Whittier, Calif. 90601).

BUILDING MATERIALS AND CONSTRUCTIONS—TEST FURNACES

Testing laboratories design their test furnaces to achieve the standard time-temperature curve described by ASTM E 119. This curve is considered to represent fire development in an average room with a moderate level of ventilation and well-insulated enclosing surfaces. Shown below are schematic sketches of a wall furnace similar to the facility at the University of California at Berkeley and a floor furnace similar to the facility at the National Bureau of Standards, Gaithersburg, Md. Test specimens (beams, columns, floors, partitions, etc.) are constructed for testing in the manner they will be installed in finished buildings (e.g., loadbearing elements must be loaded during the test). However, test results should be interpreted with care as only one relatively small building element has been tested, not the entire structural system. The behavior of this element as part of a complete structural system may be quite different. Nevertheless, fire-resistance tests provide a useful basis of comparison for performance evaluations of individual building elements.

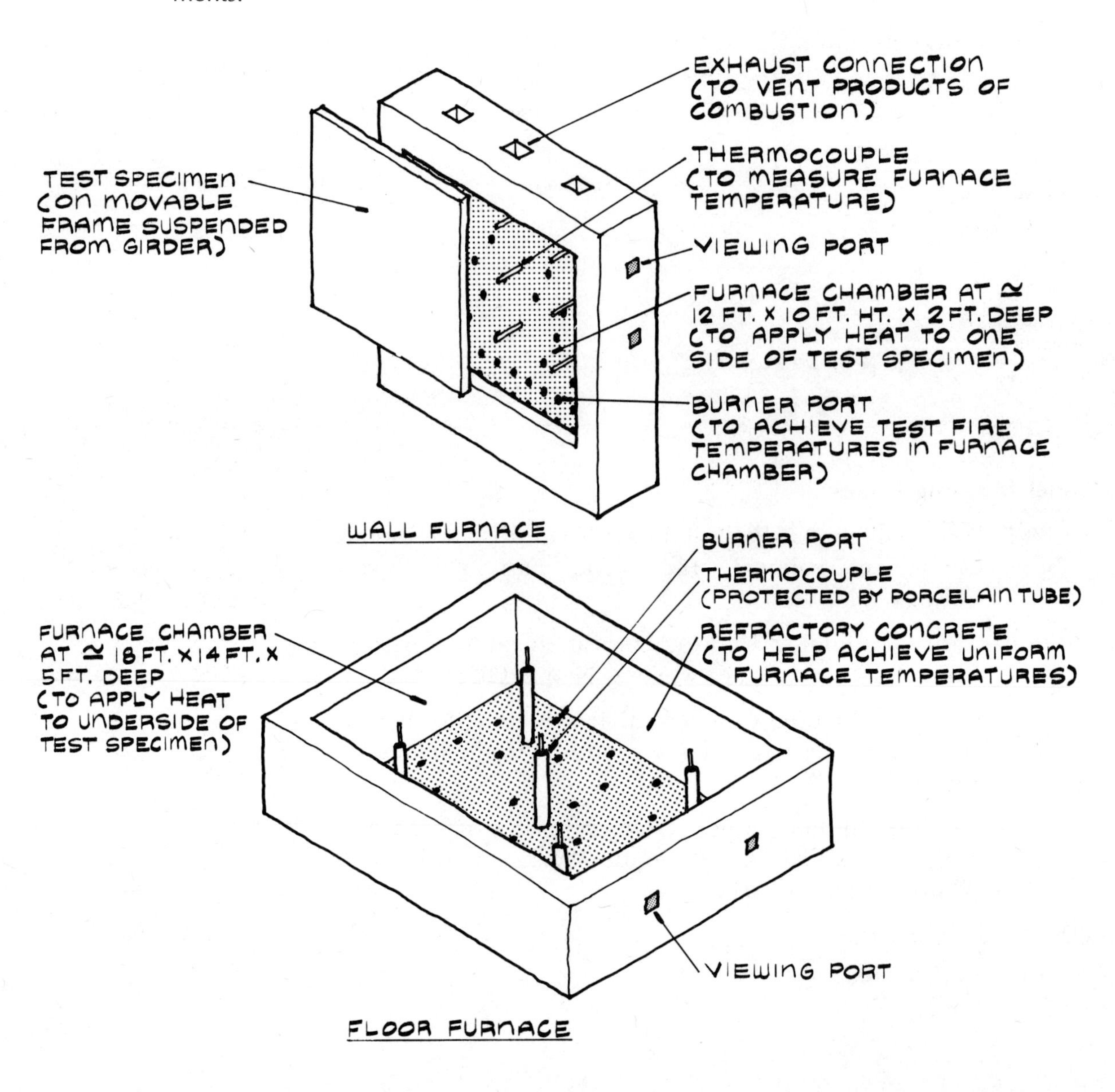

BUILDING MATERIALS AND CONSTRUCTIONS—FIRE-RATED WALL DETAILS

Fire-rated constructions should be installed in accordance with the design and procedures used to erect specimens in the test furnace. Shown below is an example fire-rated wall construction that is penetrated by a mechanical air duct.

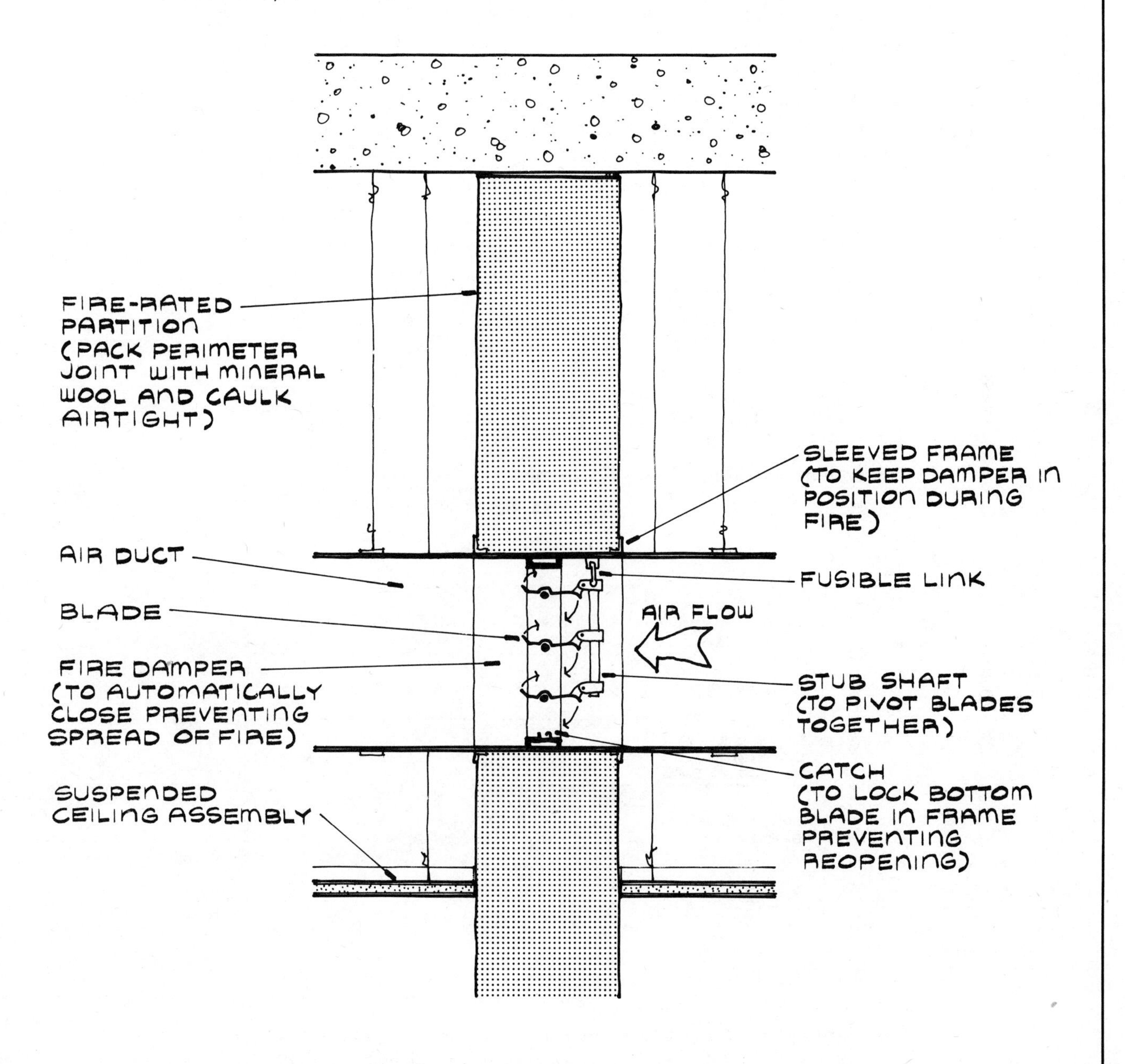

BUILDING MATERIALS AND CONSTRUCTIONS—FIRE RESISTANCE FOR CONCRETE MASONRY BLOCK WALLS

The fire resistance for hollow concrete masonry unit constructions depends on the type of aggregate and the thickness of the solid material. "Equivalent thickness" for hollow units is defined as the solid thickness that equals the concrete in a hollow unit. For example, an 8 in. hollow concrete block (calcareous gravel aggregate) is 55% solid. Its equivalent thickness is 0.55 X 7.625 (actual thickness) = 4.2 in. (See sketches below.) The approximate fire resistance will be 2 hours. (Enter table at calcareous gravel row and read opposite equivalent thickness of 4.2 in. to a fire-resistance value of 2 hours at the top of the table.) The fire resistance of concrete masonry unit constructions can be increased by filling the core voids (or cells) with an insulating material or with concrete, grout, and so on.

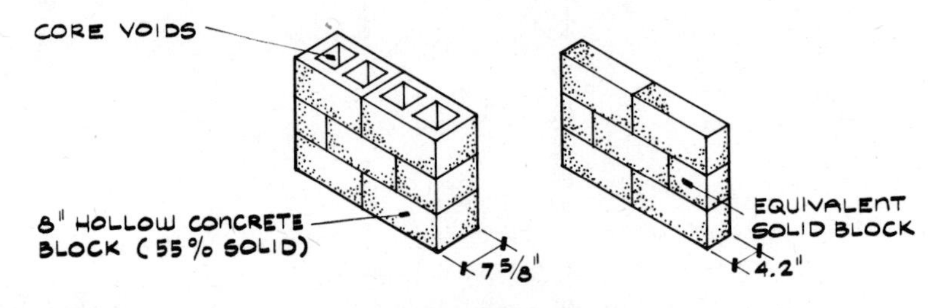

Equivalent Thickness Table (in Inches)

Type of Aggregate	Fire Resistance (in Hours): 4	3	2	1½	1
Expanded slag or pumice	4.7	4.0	3.2	2.7	2.1
Expanded clay or shale	5.7	4.8	3.8	3.3	2.6
Limestone, cinders, or air-cooled slag	5.9	5.0	4.0	3.4	2.7
Calcareous gravel	6.2	5.3	4.2	3.6	2.8
Siliceous gravel	6.7	5.7	4.5	3.8	3.0

TEST REFERENCE

"Methods of Testing Concrete Masonry Units," ASTM Method C 140. American Society for Testing and Materials (ASTM), 1916 Race Street, Philadelphia, Pa. 19103.

SOURCES

"Fire Resistance Ratings," New York, American Insurance Association, 1964 (amendments issued periodically).

"Fire Safety with Concrete Masonry," Arlington, Va., National Concrete Masonry Association, TEK 35, 1971.

BUILDING MATERIALS AND CONSTRUCTIONS—FIRE RESISTANCE FOR BRICK WALLS

Shown below are fire-resistance ratings in hours for various brick load-bearing walls.

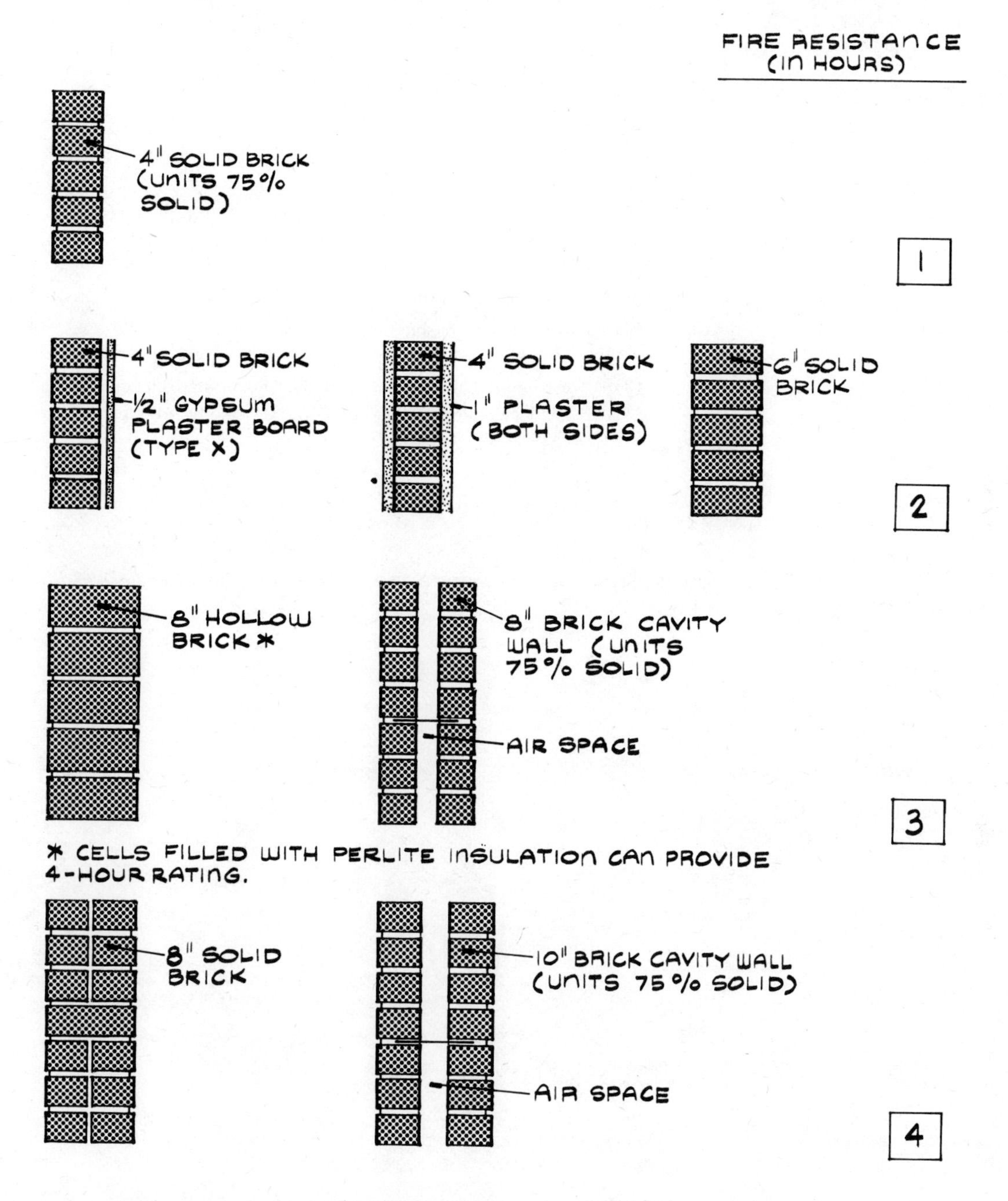

Note: Where combustible members (e.g., wood joists, beams, girders) are framed-in to 2 hour and greater wall constructions, subtract 1 hour from the rating shown above. There should be at least 4 in. of brick between the ends of the combustible members and the opposite face of the wall.

REFERENCE

Gross, J. G. and H. C. Plummer, "Principles of Clay Masonry Construction," McLean, Va., Structural Clay Products Institute, 1970.

STRUCTURAL CLAY TILE PARTITIONS

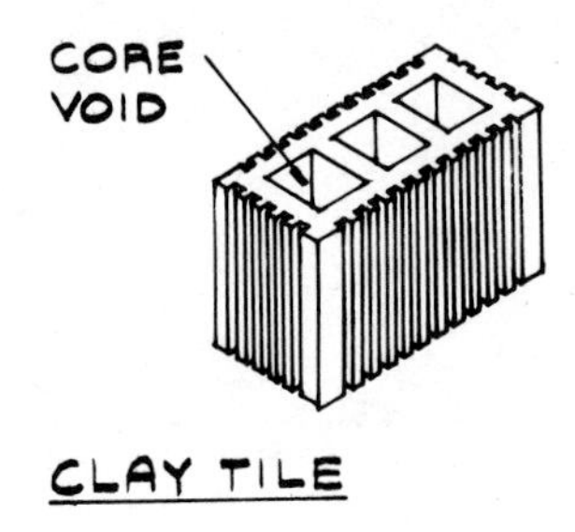

Fire resistance in hours is given for various structural clay facing-tile partitions. All plastered partitions are plastered on the unexposed side.

Thickness (in.)	*Description*	*Fire Resistance (hours)*
6	Glazed or unglazed facing tile, cored not in excess of 25% (i.e., 75% solid tile), 2 units in wall thickness.	3
6	2 in. glazed or unglazed facing tile, cored not in excess of 25%, and 4 in. structural tile cored not in excess of 40%, plastered on unglazed side with $\frac{3}{4}$ in. gypsum sand plaster.	3
4	Glazed or unglazed facing tile, cored not in excess of 25%, plastered on unglazed side with $\frac{3}{4}$ in. vermiculite plaster.	2
4	Glazed or unglazed facing tile, cored not in excess of 30%, plastered on unglazed side with $\frac{3}{4}$ in. gypsum sand plaster.	1
4	Glazed or unglazed facing tile, solid shell horizontal cell units not less than $\frac{3}{4}$ in. thick, plastered on unglazed side with $\frac{3}{4}$ in. gypsum sand plaster.	1

SOURCE

Callender, J. H. (ed.), *Time-Saver Standards For Architectural Design Data,* New York: McGraw-Hill (1974), p. 312.

REINFORCED-CONCRETE CONSTRUCTIONS

Fire-resistance ratings for the reinforced-concrete constructions in the table should be considered approximate as the quality of materials used, method of fabrication, specimen size, applied load, and edge restraint conditions may vary between the tested specimen and the actual structural element used in a building.

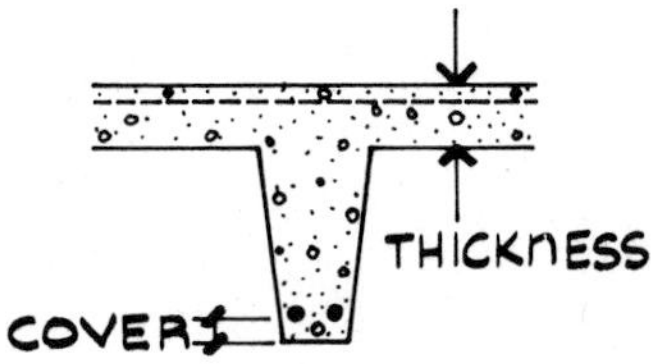

Description	*Fire Resistance (hours)*
Slab floor or ceiling	
(1) 4½ in. slab: expanded slag aggregate, with ¾ in. protection reinforcement (concrete cover shields reinforcing steel from heat).	4
(2) 6 in. slab: air-cooled slag aggregate, with 1 in. protection reinforcement.	4
(3) 6 in. slab: electrical raceways and junction boxes, with ¾ in. protection reinforcement.	3
(4) 4 ¾ in. slab: traprock or siliceous gravel aggregate, with ¾ in. protection reinforcement.	2
(5) 2 ½ in. slab: pan and joist construction, with ¾ in. cementitious-mixture cover applied to bottom of pan sections.	1½
(6) 3 in. slab: limestone aggregate, with ¾ in. protection reinforcement.	1
(7) 2½ in. slab: construction no. 5 without cementitious-mixture cover.	½
*Beams, girders, and trusses**	
(8) 1½ in. concrete: coarse aggregate; air-cooled slag, expanded slag, crushed limestone, calcareous gravel, siliceous gravel, or traprock.	4
(9) 1 in. concrete: same as no. 8.	1
*Columns**	
(10) 2½ in. concrete: coarse aggregate; granite, sandstone, or siliceous gravel (16 in. or larger round or square column).	4
(11) 1½ in. concrete: coarse aggregate; limestone, calcareous gravel, traprock, or blast-furnace slag (12 in. or larger round or square column).	4
(12) 1½ in. concrete: coarse aggregate; granite, sandstone, or siliceous gravel (16 in. or larger round or square column).	3

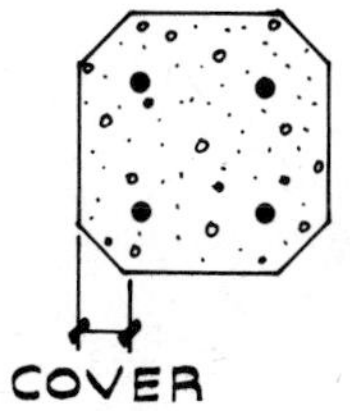

THICKNESS
COVER
CONCRETE FLOOR SLAB

COVER
CONCRETE COLUMN

*Thickness given is concrete cover over steel reinforcement.

REINFORCED-CONCRETE CONSTRUCTIONS (Continued)

Description	*Fire Resistance (hours)*
Walls	
(13) 6 $\frac{1}{2}$ in. thick with 1 in. protection reinforcement.	4
(14) 5 in. thick with $\frac{3}{4}$ in. protection reinforcement, expanded shale or slag aggregate.	4
(15) 4$\frac{1}{2}$ in. thick with $\frac{3}{4}$ in. protection reinforcement, expanded shale or slag aggregate.	3
(16) 5 $\frac{1}{2}$ in. thick with 1 in. protection reinforcement.	3
(17) 4 in. thick with $\frac{3}{4}$ in. protection reinforcement, expanded shale or slag aggregate.	2
(18) 3 in. thick with $\frac{3}{4}$ in. protection reinforcement.	1

Note: For a comprehensive presentation of fire-resistance ratings of concrete assemblies, see M. Galbreath, "Fire Endurance of Concrete Assemblies," Technical Paper No. 235, Division of Building Research, National Research Council of Canada, November 1966.

SOURCES

Bono, J. A., "Fire Protection for Structural Components and Assemblies," in R. Jensen (ed.), *Fire Protection for the Design Professional*, Boston, Mass.: Cahners Books (1975).

Reese, R. C., *CRSI Design Handbook*, Chicago, Ill.: Concrete Reinforcing Steel Institute (1973).

BUILDING MATERIALS AND CONSTRUCTIONS—FIRE RESISTANCE FOR CONCRETE FLOORS

The graph below shows typical fire-resistance ratings in hours for reinforced-concrete floors of sand and gravel aggregate and lightweight aggregate. Lightweight aggregate concrete (usual density less than 115 pcf) has better fire resistance than normal weight concrete due to its higher moisture content and higher thermal resistance to heat flow. The achieved fire resistance of concrete beams and slabs can be greatly improved where the edge restraint to lateral thermal expansion is between fixed and simple support conditions. For ribbed constructions, use the thickness shown on the sketch when this dimension is less than about one-fourth the spacing between ribs. At closer rib spacings, however, a greater equivalent thickness can be used.

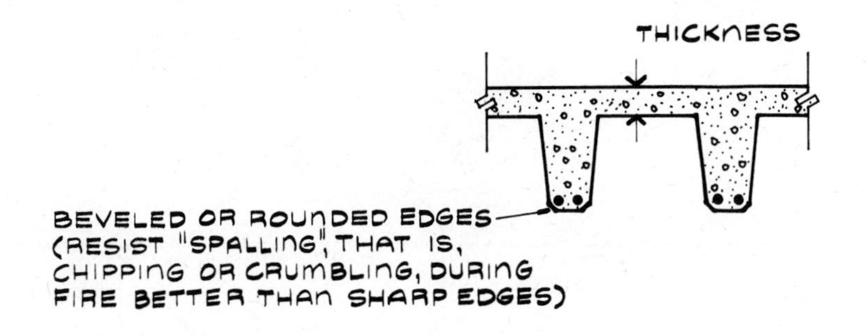

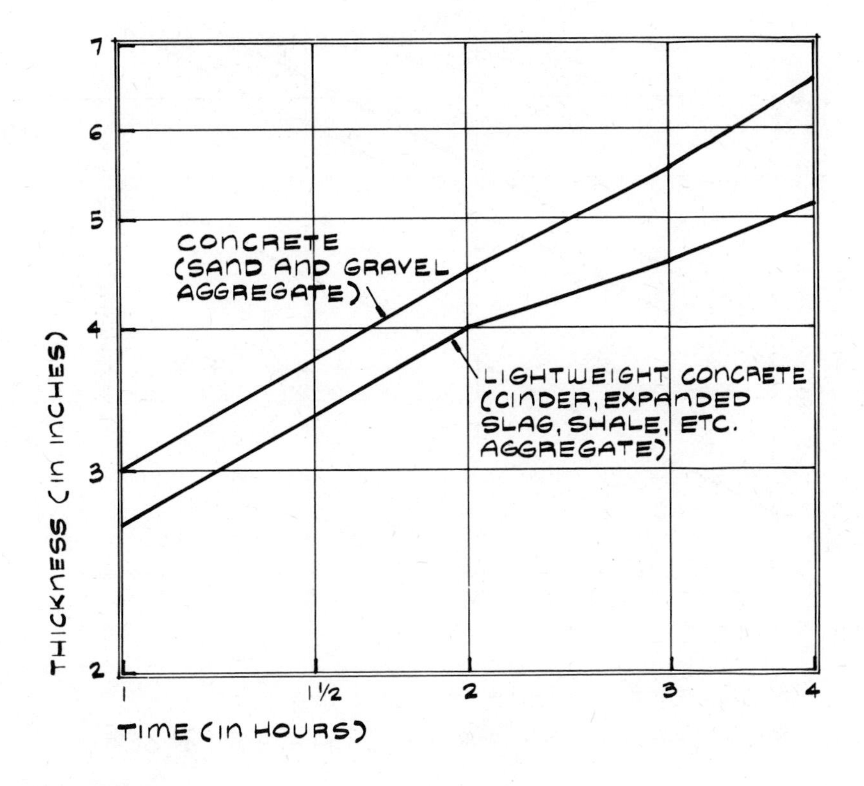

REFERENCE

McKinnon, G. P. (ed.), *Fire Protection Handbook,* Boston, Mass.: National Fire Protection Association (1976), p. 6–62.

BUILDING MATERIALS AND CONSTRUCTIONS—FIRE RESISTANCE FOR CONCRETE WALLS

The graph below shows typical fire-resistance ratings in hours for reinforced-concrete walls of various aggregates. The fire resistance of concrete depends on its thickness, aggregate (expanded shale, calcareous gravel, siliceous gravel), moisture and air content, cover over the reinforcing steel, and nature of the structural loads during fire exposure.

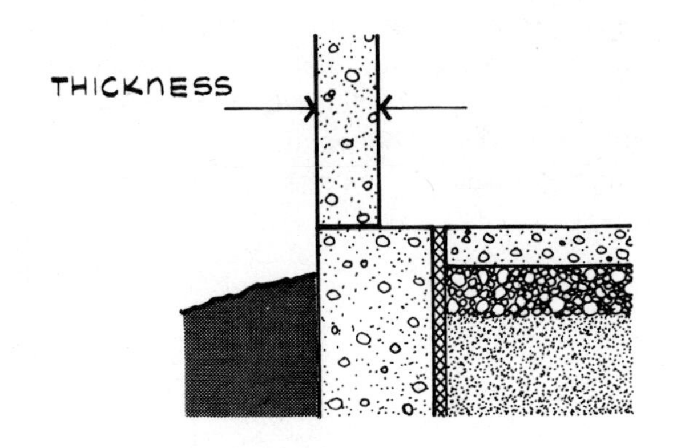

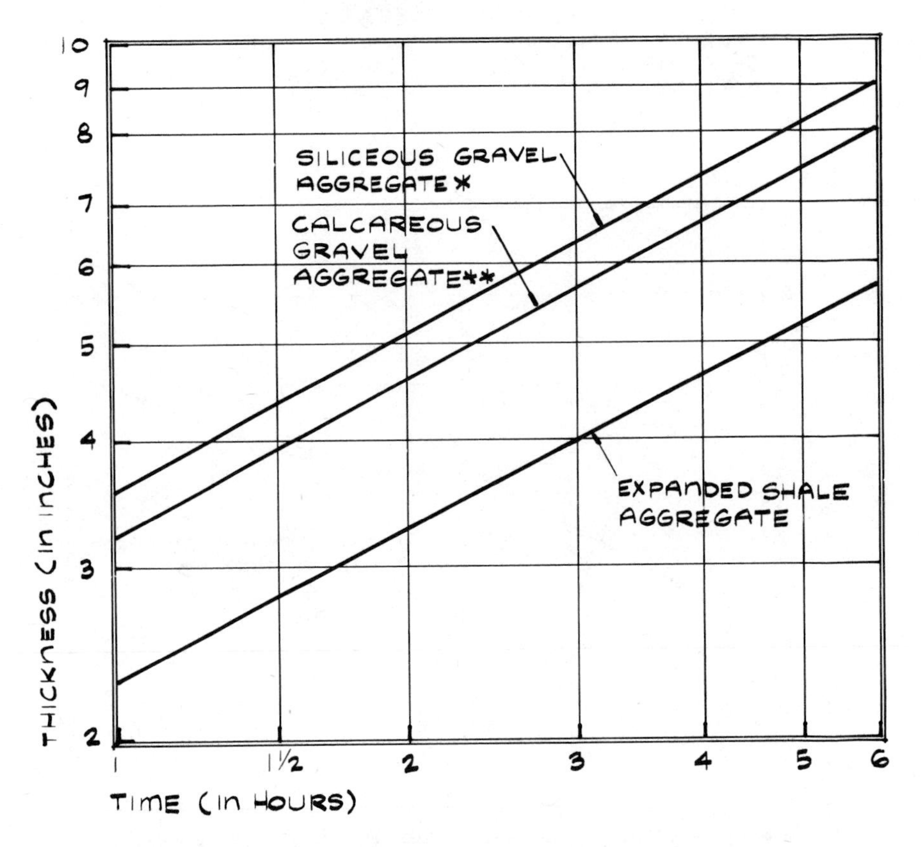

* GRANITE, SANDSTONE, QUARTZ

** LIMESTONE, TRAP ROCK, SLAG, BURNT CLAY

REFERENCE

McKinnon, G. P. (ed.), *Fire Protection Handbook,* Boston, Mass.: National Fire Protection Association (1976), p. 6–63.

BUILDING MATERIALS AND CONSTRUCTIONS—FIRE RESISTANCE FOR GYPSUM PARTITIONS

The graph below shows typical fire-resistance ratings in hours for gypsum plaster on metal lath and gypsum board partitions of various thicknesses presented as the sum of both partition facings. For actual constructions, achieved fire-resistance ratings also depend on the applied load, size and shape of studs, method of gypsum board attachment (i.e., spacing and type of fasteners), quality of construction, and on duplicating any special features that were used in the tested assembly.

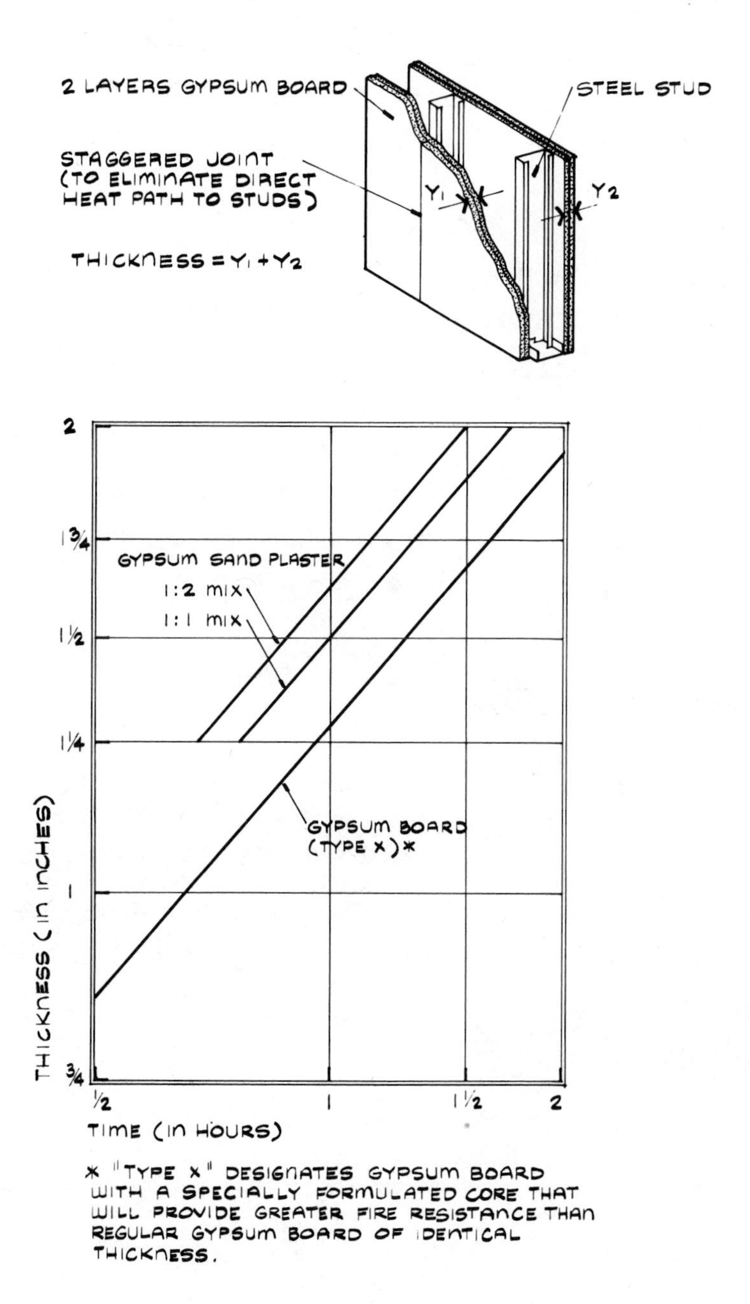

* "TYPE X" DESIGNATES GYPSUM BOARD WITH A SPECIALLY FORMULATED CORE THAT WILL PROVIDE GREATER FIRE RESISTANCE THAN REGULAR GYPSUM BOARD OF IDENTICAL THICKNESS.

REFERENCE

McKinnon, G. P. (ed.), *Fire Protection Handbook,* Boston, Mass.: National Fire Protection Association (1976), p. 6–62.

BUILDING MATERIALS AND CONSTRUCTIONS—WALL ASSEMBLIES

Shown below are example loadbearing party wall assemblies having a fire-resistance rating of 1 hour. Fire-resistance tests of loadbearing constructions should be conducted with live and dead design loads applied to the test specimen. To achieve the tested fire resistance, assemblies must be constructed in strict conformance to the tested assembly, including all components and observing any size limitations.

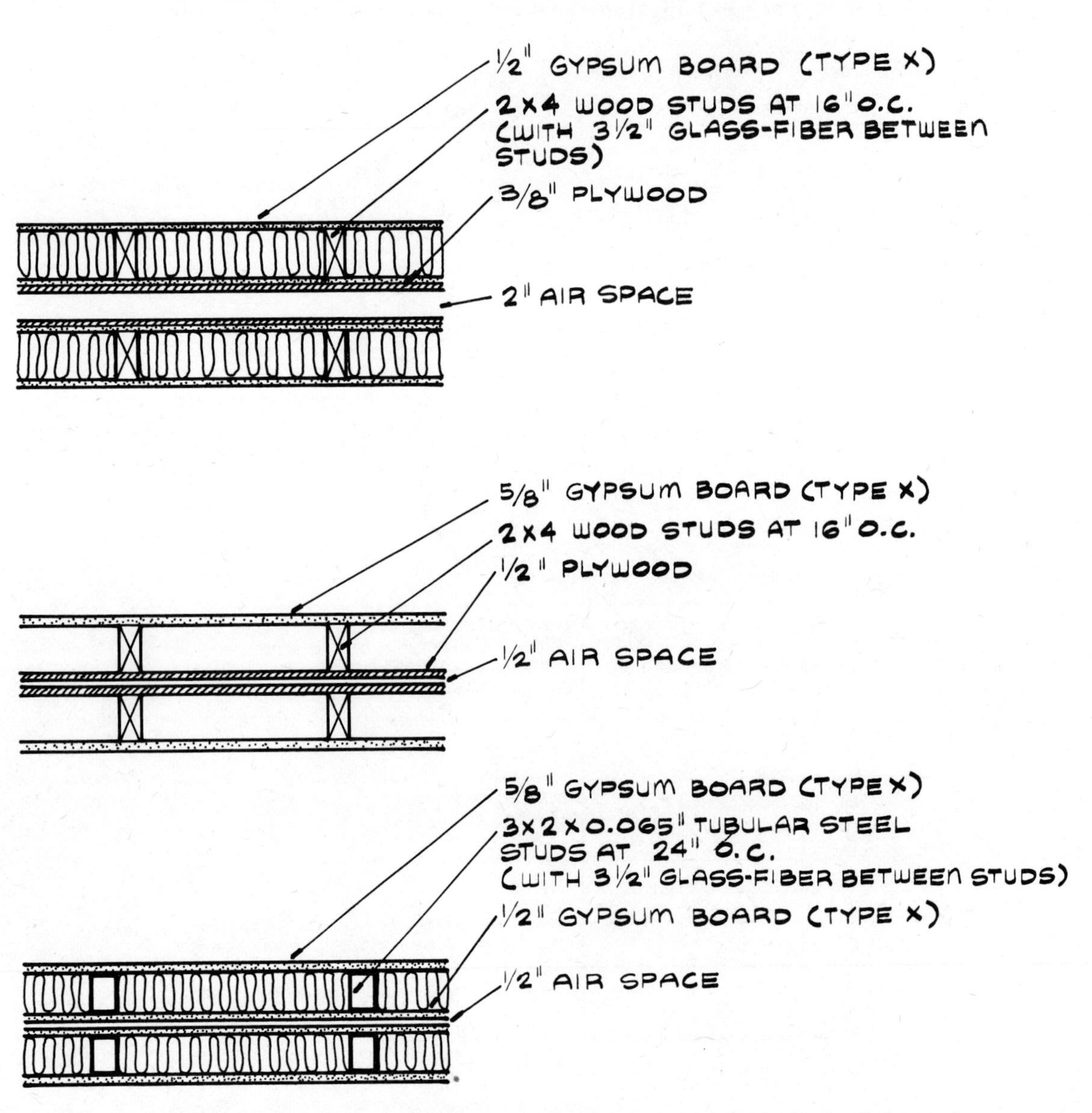

Note: For a comprehensive presentation of fire-resistance ratings of wall and other assemblies, see "A Compendium of Fire Testing" from "Operation Breakthrough Feedback" report series, U.S. Department of Housing and Urban Development, Vol. 5, 1976.

BUILDING MATERIALS AND CONSTRUCTIONS—FLOOR-CEILING ASSEMBLIES

The fire-resistance rating of a structural element does not necessarily mean it is noncombustible. For example, shown below are combustible floor-ceiling assemblies having a fire-resistance rating of 1 hour. The rating indicates that fire will not penetrate through the assemblies within a 1 hour fire exposure to the ASTM standard time-temperature exposure, but fire may penetrate into the assemblies. Open spaces above suspended ceilings can become unwanted paths for the spread of smoke and hot gases. When exposed to ASTM time-temperature curve conditions, bare wood floors with exposed joists typically fail in less than 15 minutes.

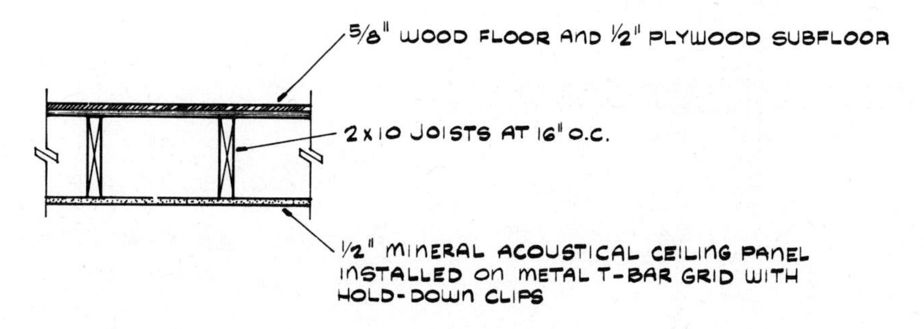

Note: Ceiling panel units, removed for maintenance or other reasons, must be mechanically refastened to insure fire integrity of the ceiling assembly.

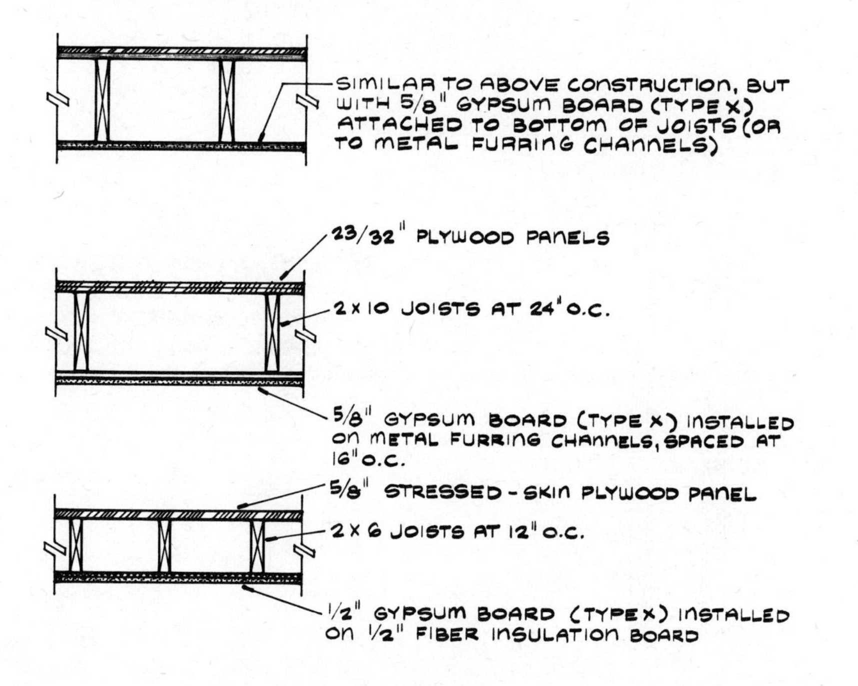

Note: Fire-rated assemblies must be constructed in buildings with the same fire integrity as the assembly erected in the test furnace.

BUILDING MATERIALS AND CONSTRUCTIONS—FIRE-STOP DETAILS (WALLS)

Openings above fire-rated walls and partitions should be "fire-stopped" to form a tight seal with the floor slab above. Example techniques for achieving continuous vertical, fire-resistant barriers are shown below.

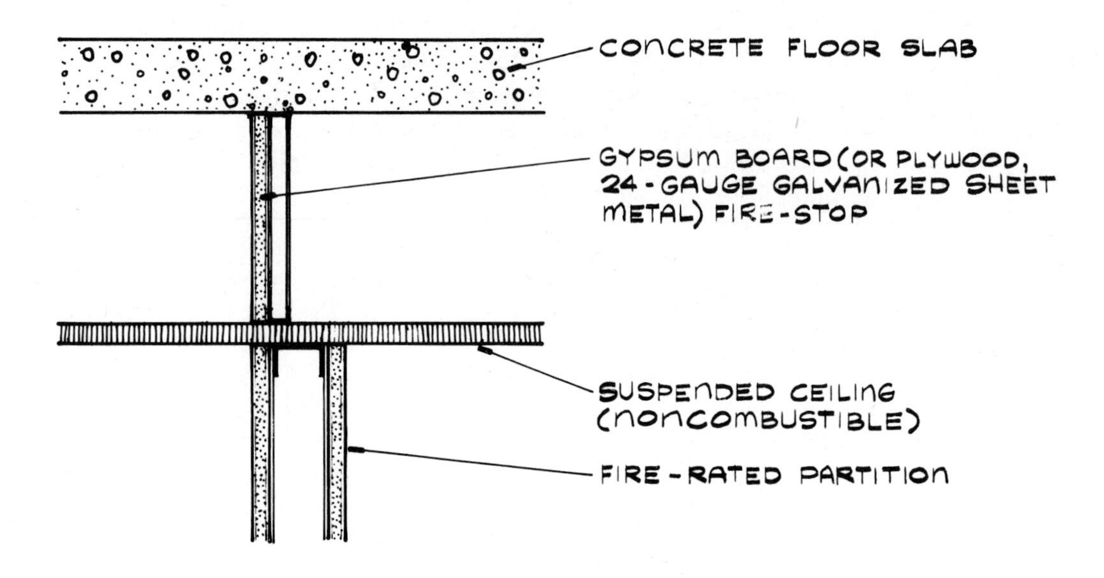

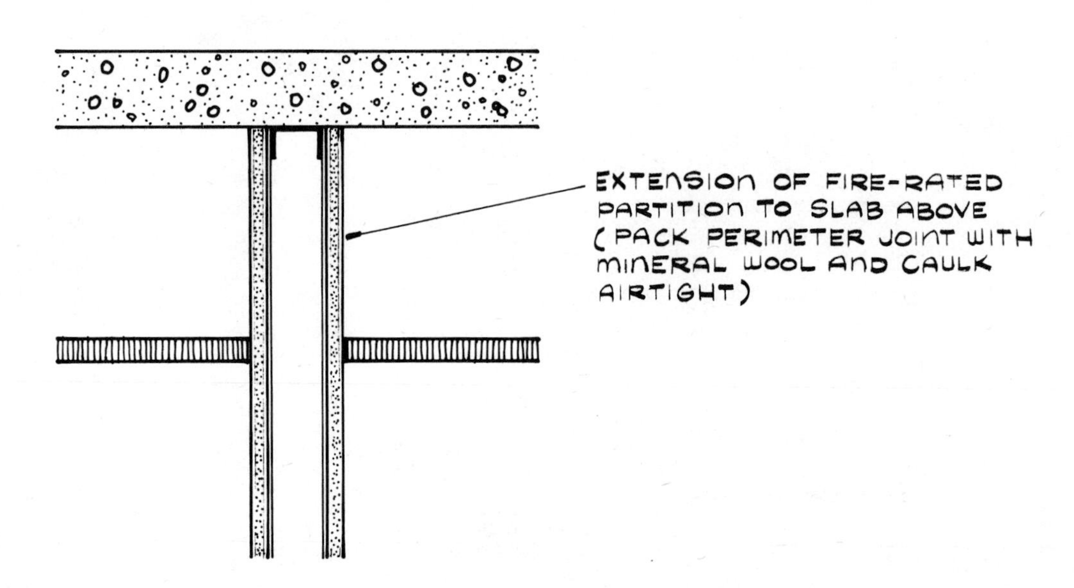

Note: It is good practice to subdivide unoccupied attic spaces into small areas with fire-stops. Similarly, "cocklofts" (i.e., shallow attic spaces between upper-story ceilings and roofs in residential buildings) should be fire-stopped to prevent fire and smoke penetration into adjacent areas.

BUILDING MATERIALS AND CONSTRUCTIONS—FIRE-STOP DETAILS (WALL CAVITIES)

Fire-stops in walls and partitions are essential to prevent the spread of fire and smoke. The open cavity spaces between vertical studs can act as concealed flues for fires originating behind wall surfaces.

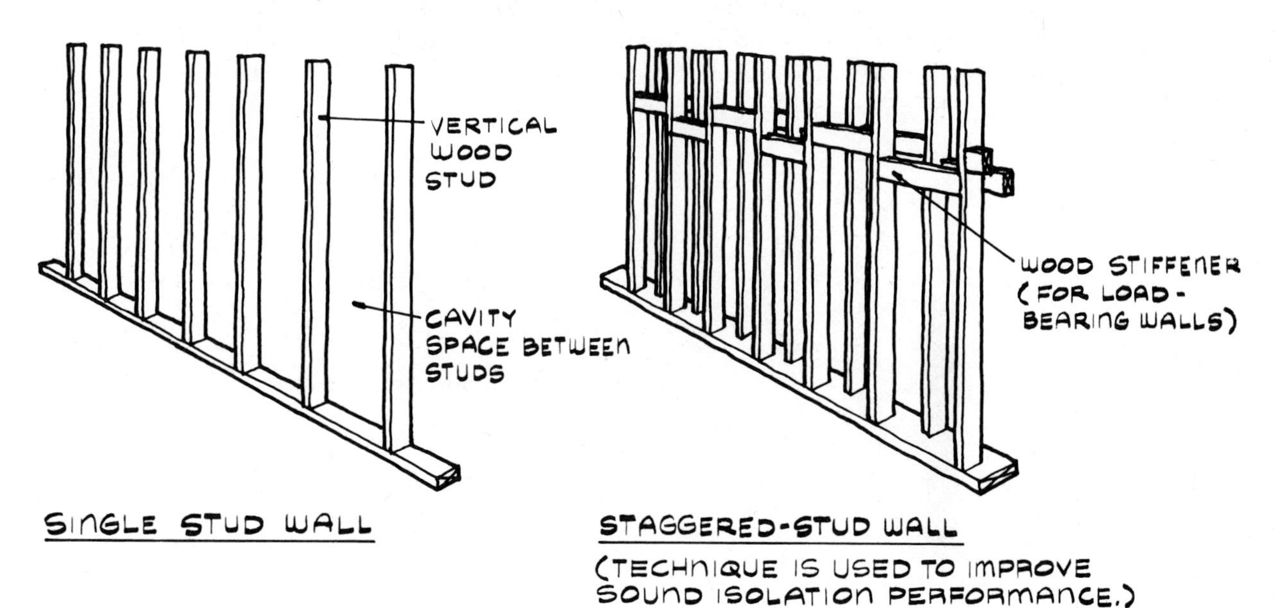

SINGLE STUD WALL

STAGGERED-STUD WALL

(TECHNIQUE IS USED TO IMPROVE SOUND ISOLATION PERFORMANCE.)

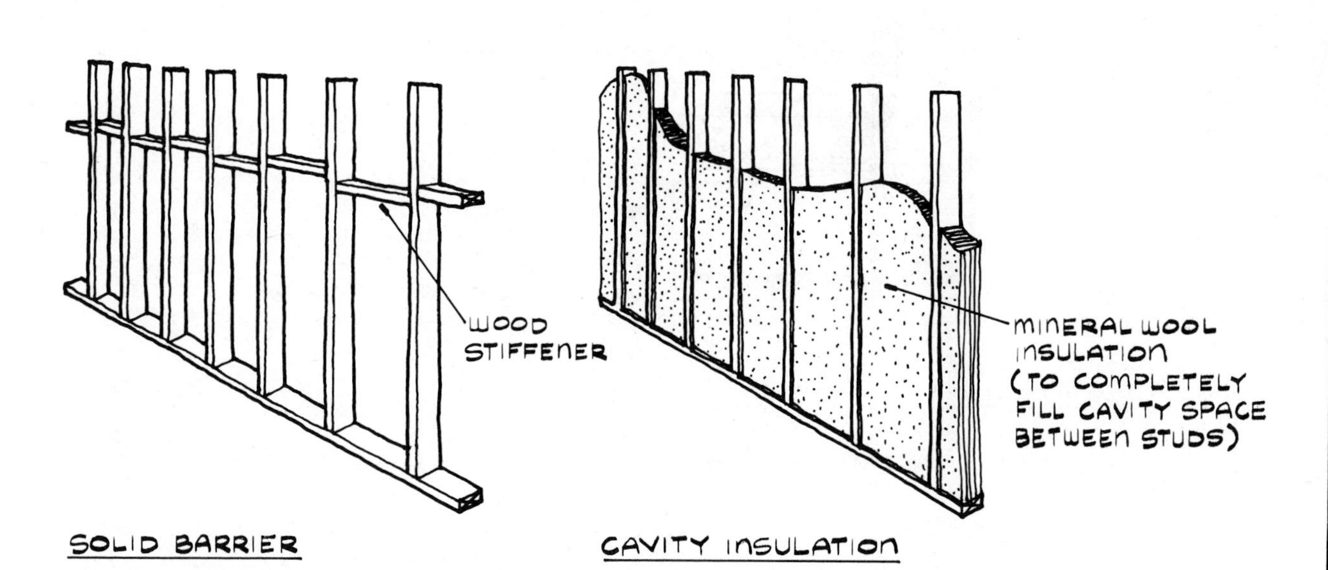

SOLID BARRIER

(CONTINUOUS ROW OF HORIZONTAL FURRING STIFFENERS ACTS AS A BARRIER TO VERTICAL SPREAD OF SMOKE AND FIRE.)

CAVITY INSULATION

(MINERAL WOOL, WITH HIGH MELTING POINT, BLOCKS SPREAD OF SMOKE AND FIRE.)

BUILDING MATERIALS AND CONSTRUCTIONS—FIRE-STOP DETAILS (FLOOR-CEILINGS)

Openings between curtain walls and floors, penetrations through floors for building services, and so on should be fire-stopped to prevent the spread of fire, smoke, and gases. Example fire-stop details using mineral wool (or rock wool) are shown below. Fire-stop materials that melt or disintegrate in a fire should not be used. The fire-stop material should withstand the ASTM standard time-temperature curve for a duration equal to the rating of the assembly being penetrated.

Metal Curtain Wall

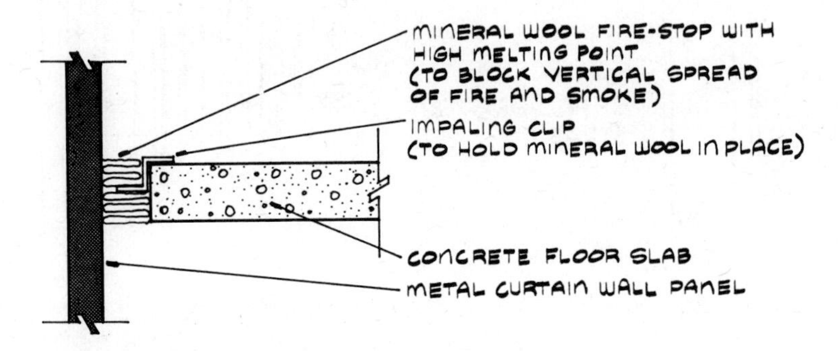

Glass Curtain Wall

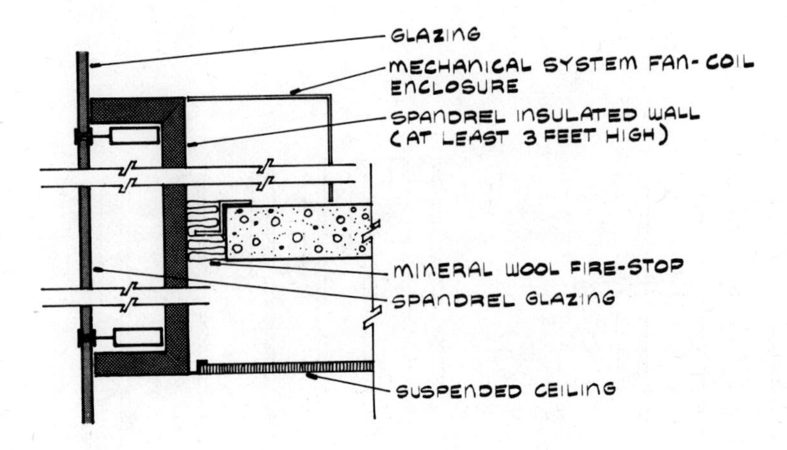

REFERENCE

Yosick, D. A., "High-Melt Fire Stops Help Avoid Towering Disasters," *Form & Function*, Issue 4, 1974.

BUILDING MATERIALS AND CONSTRUCTIONS—FLOOR PENETRATIONS

Example metal pipe and electrical conduit penetrations through floor slabs are shown below. Where pipe penetrates through three or more floors, it should be enclosed in a shaft. Protection for field-drilled holes for electrical services are shown on the following page.

Poor

Open penetrations provide paths for vertical fire spread.

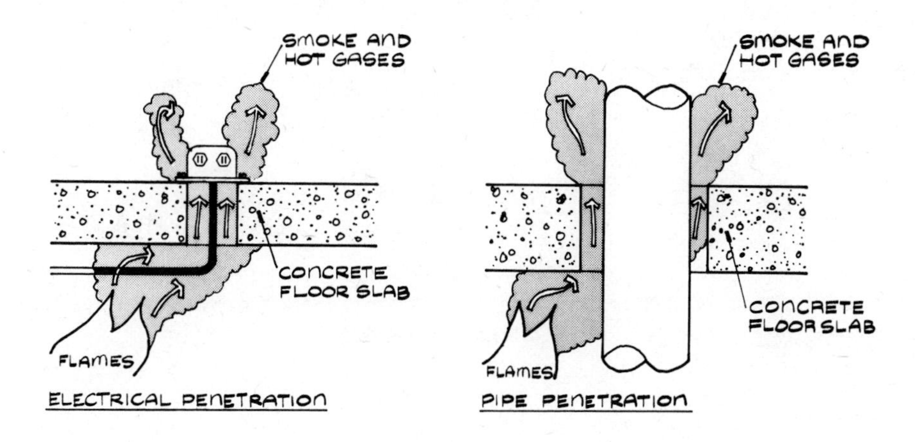

Better

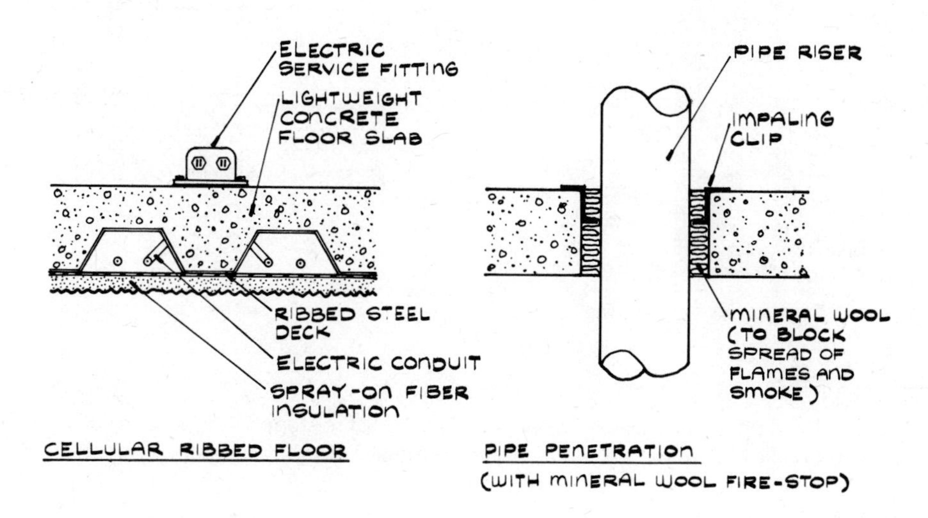

BUILDING MATERIALS AND CONSTRUCTIONS—POKE-THROUGH ELECTRICAL FITTINGS

Fittings for electrical power, telephone service, and the like installed through drilled holes in the floors of completed buildings are called "poke-through" assemblies. To prevent the vertical spread of fire, the underside of poke-through assemblies can be protected by (1) spray-on insulation undercoating, (2) intumescent mastic coating on the conduit, or (3) heat shields. Unprotected, poke-through assemblies can lower the fire resistance of floors to only a few minutes!

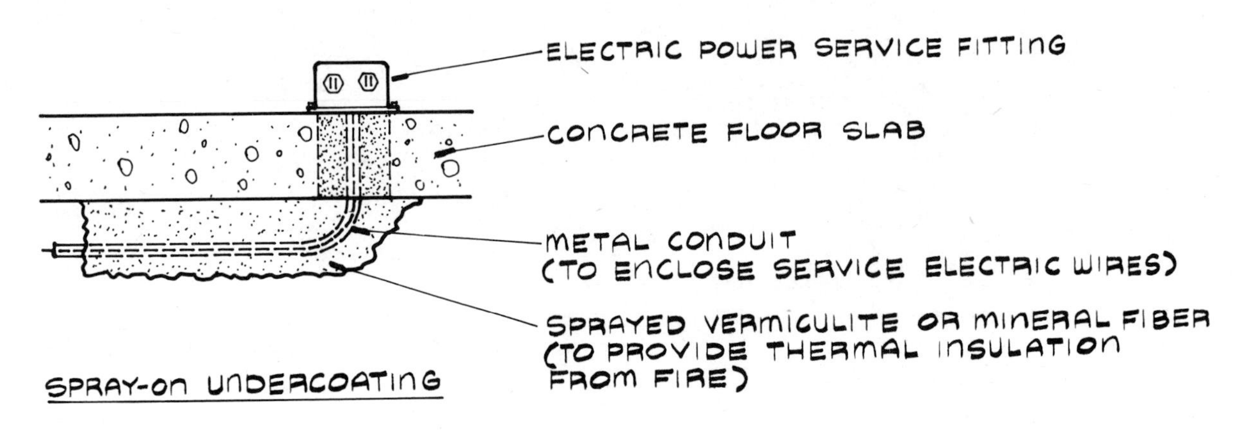

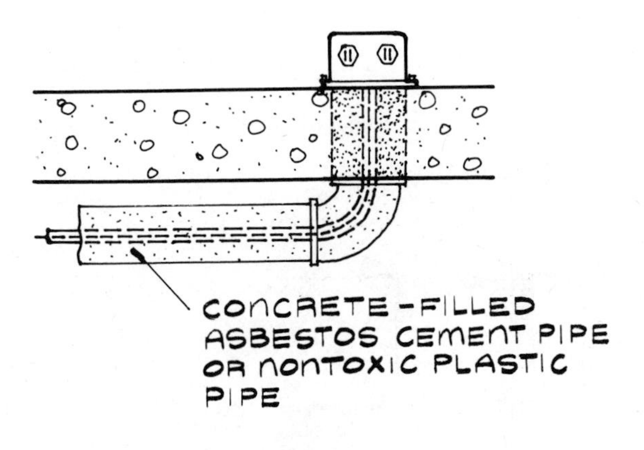

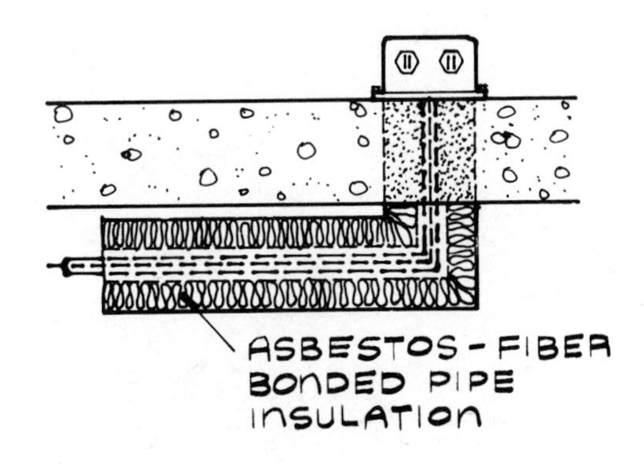

HEAT SHIELDS

Note: Where poke-through openings are not properly fire-stopped, suspended ceilings can be installed to upgrade the overall fire resistance of the floor-ceiling assembly.

REFERENCE

Abrams, M. S. and A. H. Gustaferro, "Fire Tests of Poke-Through Assemblies," *Fire Journal,* May 1971. (See also PCA Bulletin RD008.01B, available from Portland Cement Association, Old Orchard Road, Skokie, Ill. 60076.)

BUILDING MATERIALS AND CONSTRUCTIONS—PLENUM PENETRATION OF WALLS

Shown below in order of increasing fire protection are various ceiling constructions. Where floor-ceiling plenum spaces used for mechanical services penetrate fire-rated walls, the ceiling construction itself should have compatible fire resistance. Fire-resistance ratings for the example constructions vary from ½ to 2 hours for 30 in. × 32 in. samples exposed to the standard ASTM time-temperature curve in a small electric furnace.

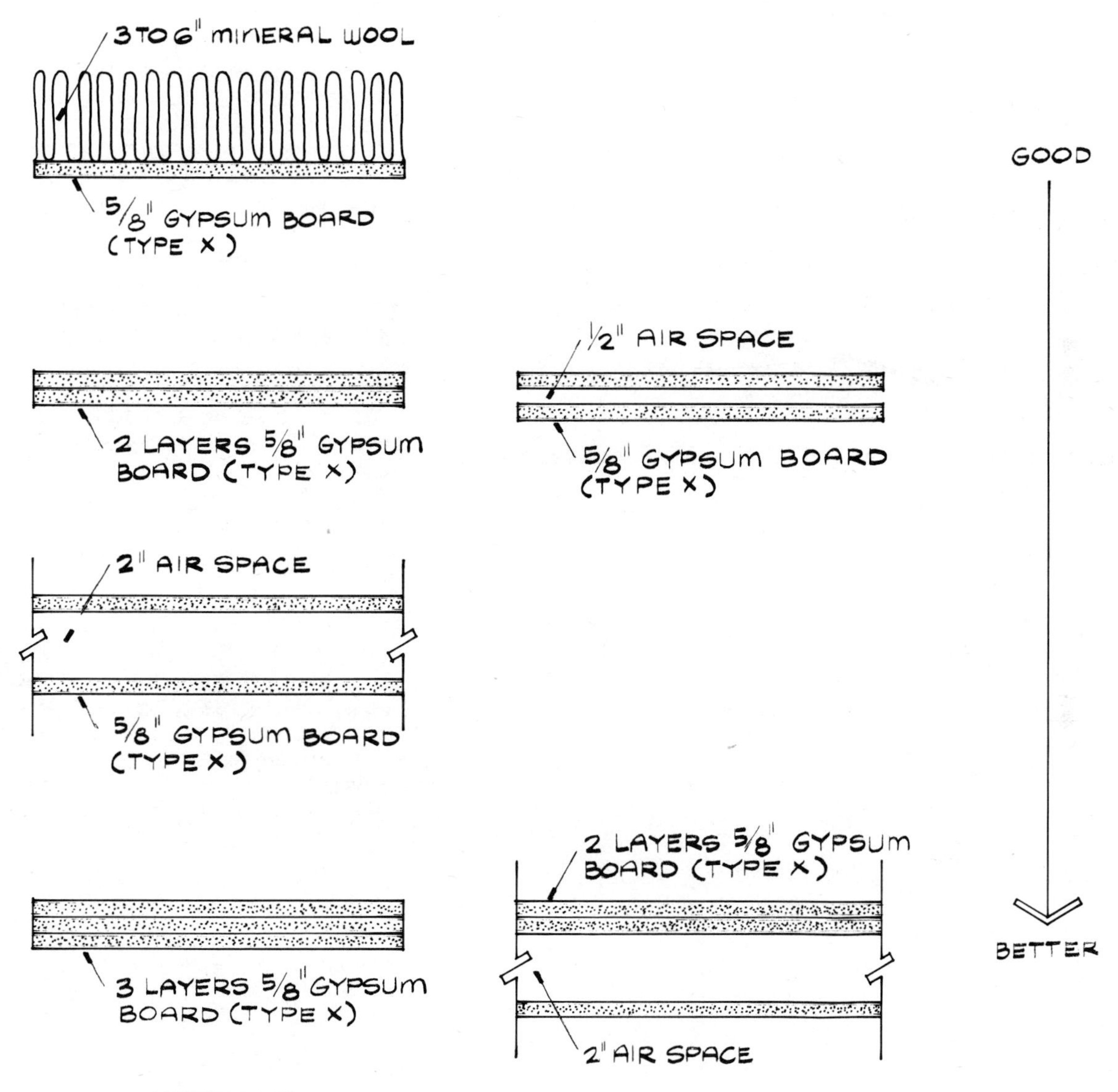

REFERENCE

Konicek, L. and M. Galbreath, "Fire Protection of Horizontal Service Spaces," Division of Building Research, National Research Council of Canada, BR Note No. 108, February 1975.

BUILDING MATERIALS AND CONSTRUCTIONS—MECHANICAL SYSTEM PENETRATION OF A CEILING

Where openings (e.g., air duct, electrical outlet) do not exceed about 0.7% of the total ceiling area (not floor slab area), there may be little effect on the ceiling assembly's fire resistance. However, for ceilings with larger open areas, dampers should be used. Dampers consist of a single blade (as shown below) or of connected multiple blades. They can be operated directly by a heat-sensitive device such as a fusible link or remotely by a fire detection device.

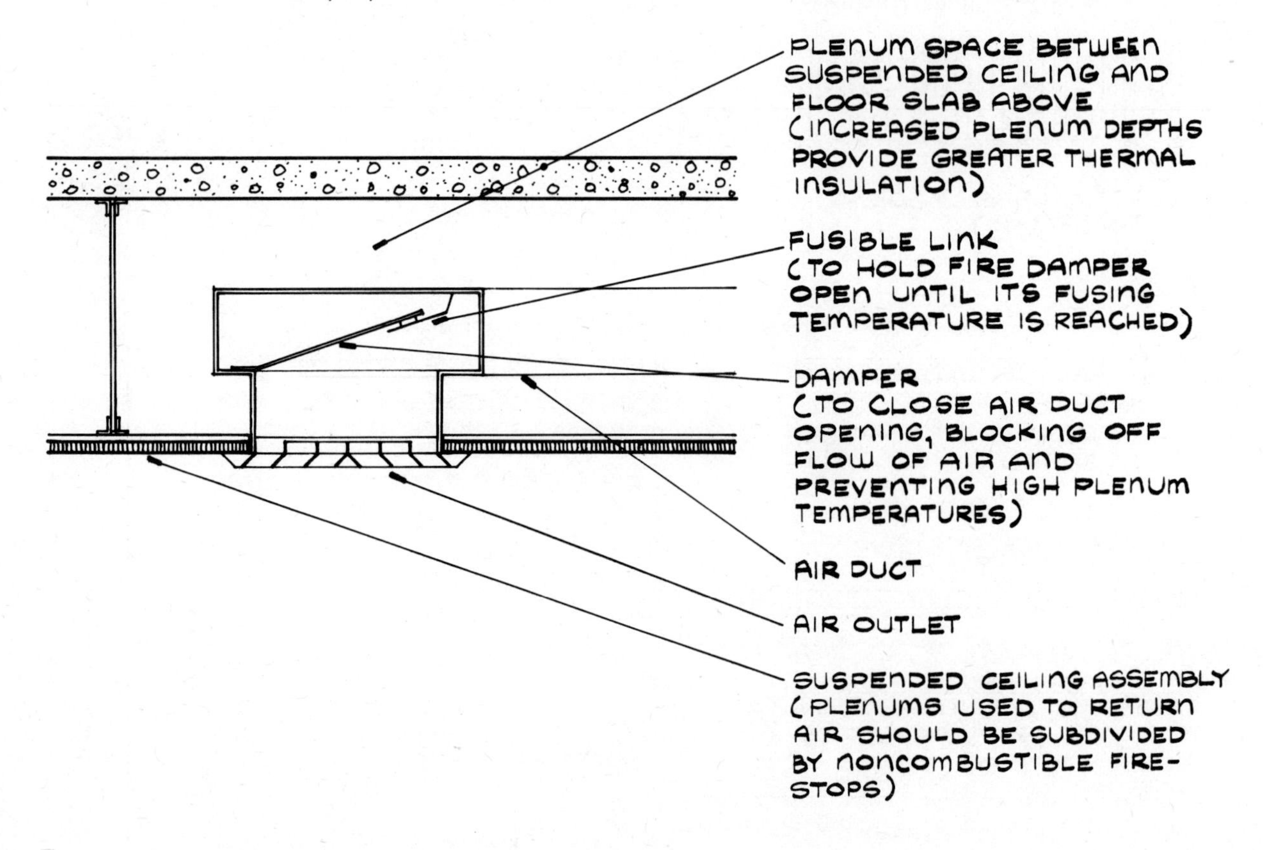

BUILDING MATERIALS AND CONSTRUCTIONS—CEILING AIR DUCT OUTLETS

Shown below are damper, enclosure, and internal lining methods to retard heat transmission through air duct outlets. For a detailed description of fire tests on various ceiling outlet protection systems, see "Alternate Methods of Fire Protection for Ceiling Outlets of Galvanized Steel Ducts," American Iron and Steel Institute Research Report, 1972.

Single Blade Damper

When closed, dampers retard heat penetration at ceiling outlets. However, dampers are obstructions to airflow causing pressure losses and noise.

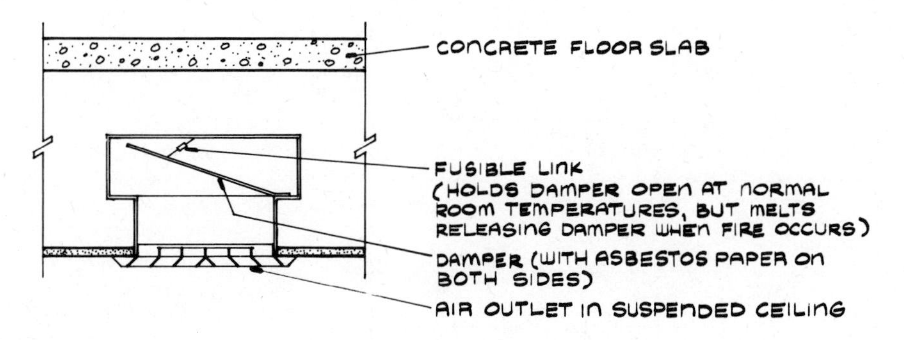

Duct Enclosure

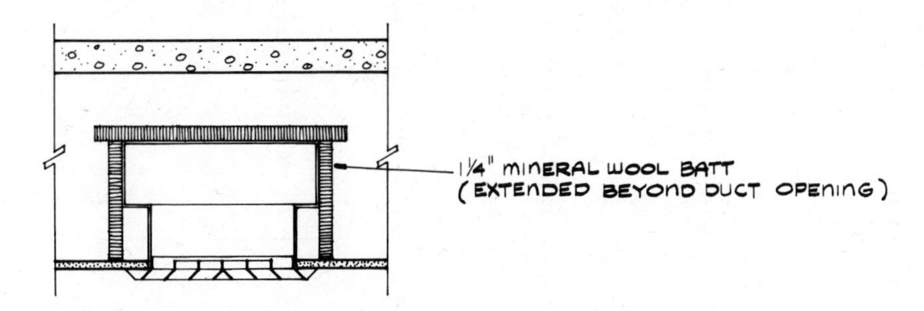

Internal Lining

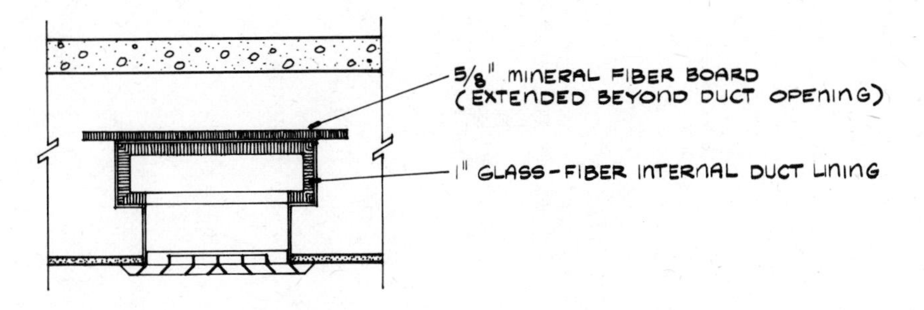

BUILDING MATERIALS AND CONSTRUCTIONS—LIGHTING FIXTURES

Various fire protection details for lighting fixtures in fire-rated ceiling assemblies are shown below. Full box (shown in side and end elevation views), tent, and flat cover enclosures are depicted. Covering materials include mineral-fiber board, mineral-wool blanket, and mineral-wool batt.

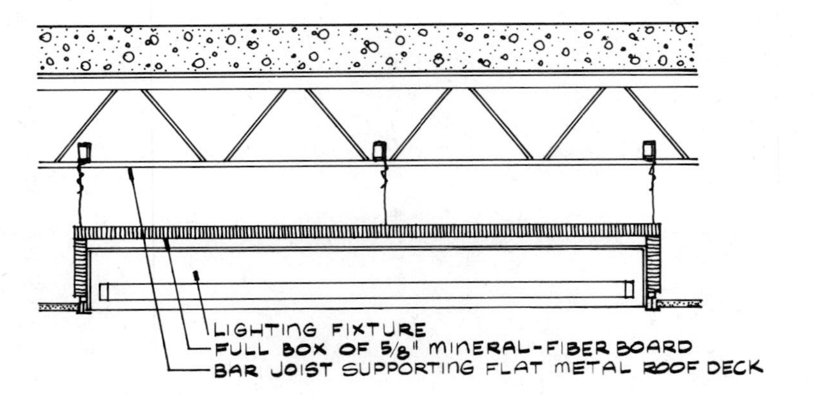

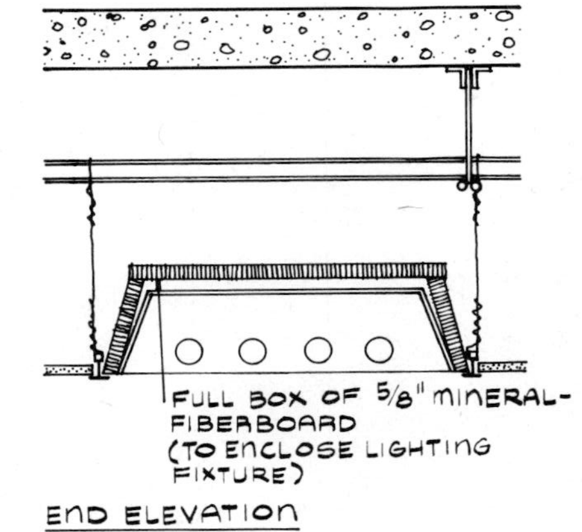

Example Lighting Fixture Protection Systems

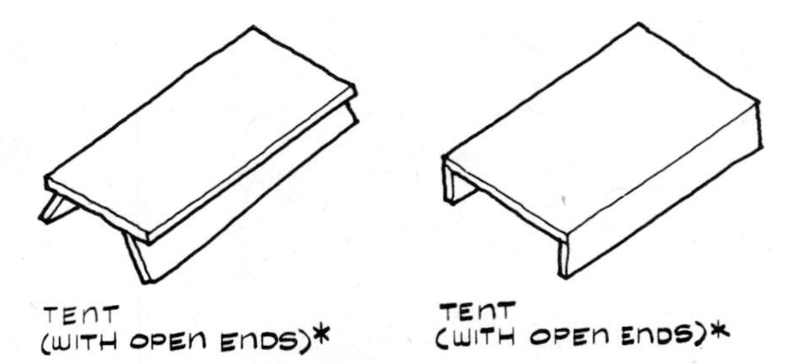

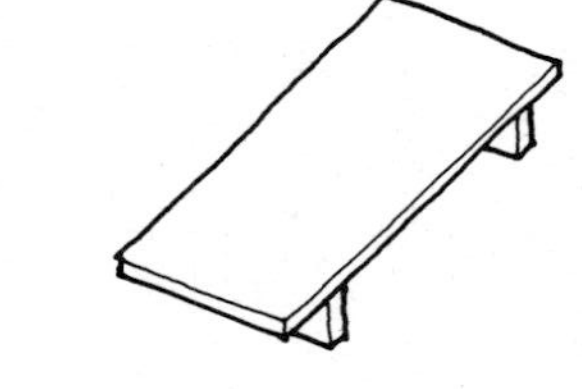

* TO RELEASE HEAT FROM LIGHTS

FIRE TEST FOR DOORS

In the ASTM E 152 fire test for door assemblies, the door and frame to be tested are installed in a wall construction. The door assembly is then subjected to a fire exposure according to the ASTM standard time-temperature curve. The door must remain in place during the test, must not excessively separate from the hinge or latch side of the frame, and must withstand the thermal shock from a standard water hose stream applied to the heated door surface (i.e., water must not penetrate the door which must remain in its frame). In addition, surface temperatures on the unexposed side of the door assembly are recorded during the first 30 minutes of the test.

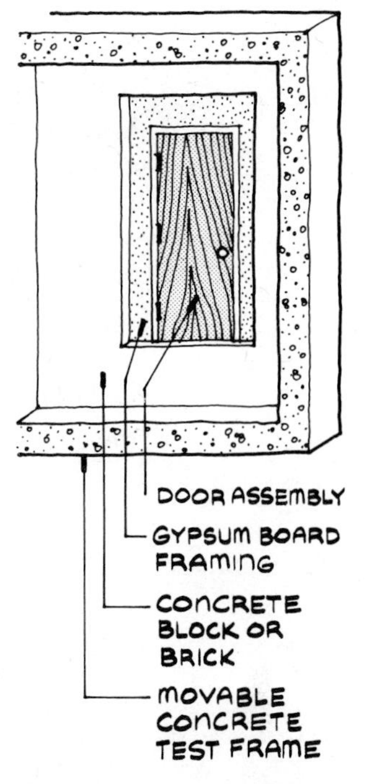

DOOR FIRE TEST ASSEMBLY

Fire tests for walls differ somewhat from the fire test procedures for doors described above. For walls, surface temperatures on the unexposed side are recorded throughout the wall test as temperature rise is an important endpoint criterion. In addition, cotton waste placed on the unexposed side of the wall surface must not ignite. Finally, a duplicate wall construction is subjected to another ASTM standard time-temperature test fire exposure of one-half the length of time for which the wall is to be rated, but not more than 1 hour (e.g., $\frac{1}{2}$ hour for 1 hour rating; 1 hour for 2, 3, or 4 hour ratings). After this exposure, a standard water hose stream is applied to the heated wall surface. No through openings must develop due to the thermal shock from the water hose stream.

REFERENCES

"Fire Tests of Door Assemblies," ASTM Standard Method E 152.

Degenkolb, J. G., "The Twenty-Minute Door," *Doors and Hardware,* July 1975.

BUILDING MATERIALS AND CONSTRUCTIONS—HEAT TRANSMISSION THROUGH DOORS

The curves below show example surface temperature increases on the laboratory side of $1\frac{3}{4}$ in. thick metal and wood doors subjected to the ASTM standard time-temperature curve on the furnace side. The ASTM standard time-temperature curve from 0 to 5 minutes is also plotted on the graph for comparison. The fire performance of doors depends on the behavior of the door, frame, and hardware at elevated temperatures and also on the interaction of these door assembly components. Surface temperatures for the hollow-metal door with honeycomb core are extremely high during most of the 1 hour period shown due to the efficient heat conductance of metal. Surface temperatures are considerably lower, however, for hollow-metal doors having thermal-insulating cores and solid-core wood doors.

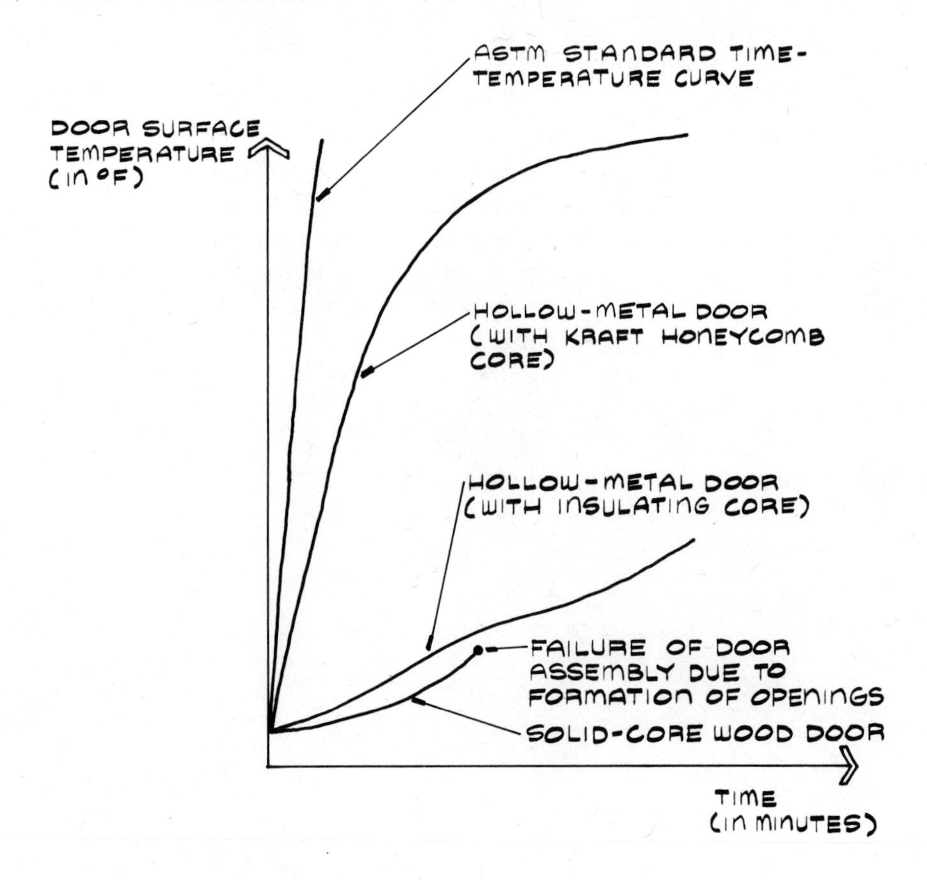

REQUIREMENTS FOR DOORS

If an opening occurs in a fire-rated wall or partition, the opening must have a door which, when closed, will retard the spread of fire for a period of time approximately equal to the fire resistance of the wall in which it is installed. Codes usually permit lower fire-resistance ratings for doors than that of the wall because doors take up a small percentage of the overall surface area and combustible furnishings generally are not placed close to doors. To achieve equivalent fire resistance, however, sprinklers can be installed at door openings (see sketch in the margin). Water spray nozzles can be used to provide a film of flowing water over fire exposed doors and to inhibit burning adjacent to doors.

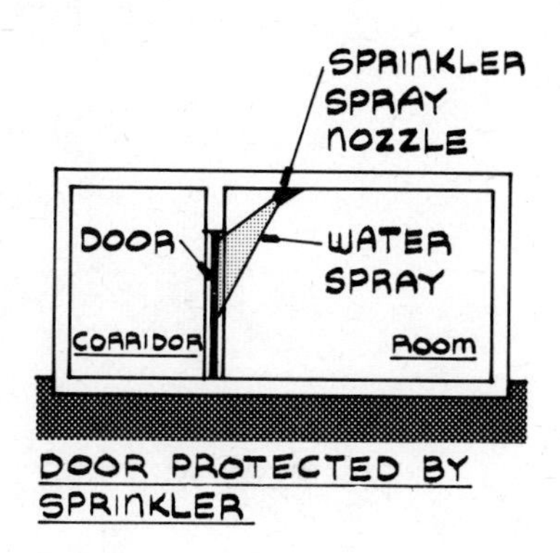

DOOR PROTECTED BY SPRINKLER

The table below lists example fire-resistance requirements for fire doors (cf., "Fire Protection Through Modern Building Codes," American Iron and Steel Institute, 1971, p. 73). Consult local building codes for the prevailing requirements at a given door location.

Door Location	*Fire-Resistance Rating (hours)*	*Wired Glass Maximum Dimension (in.)*	*Wired Glass Maximum Area (sq. in.)*
Walls separating buildings, fire walls	2 or 3	Not permitted	Not permitted
Enclosures for vertical movement (e.g., stairways, elevators) or horizontal movement (e.g., horizontal exits, smoke barriers)	1 or 1½	12	100
Corridors, room partitions	¾*	54	1296
Exterior walls subject to "severe" fire exposure from outside	1½	Not permitted	Not permitted
Exterior walls subject to "moderate" fire exposure from outside	¾	54	720

*Most codes permit a 20 minute fire-resistance rating (e.g., 1¾ in. solid-core wood door with appropriate frame and hardware) at these locations.

BUILDING MATERIALS AND CONSTRUCTIONS—FIRE RESISTANCE FOR DOORS

Shown below are maximum fire-resistance ratings for typical 1¾ in. thick metal and wood doors.

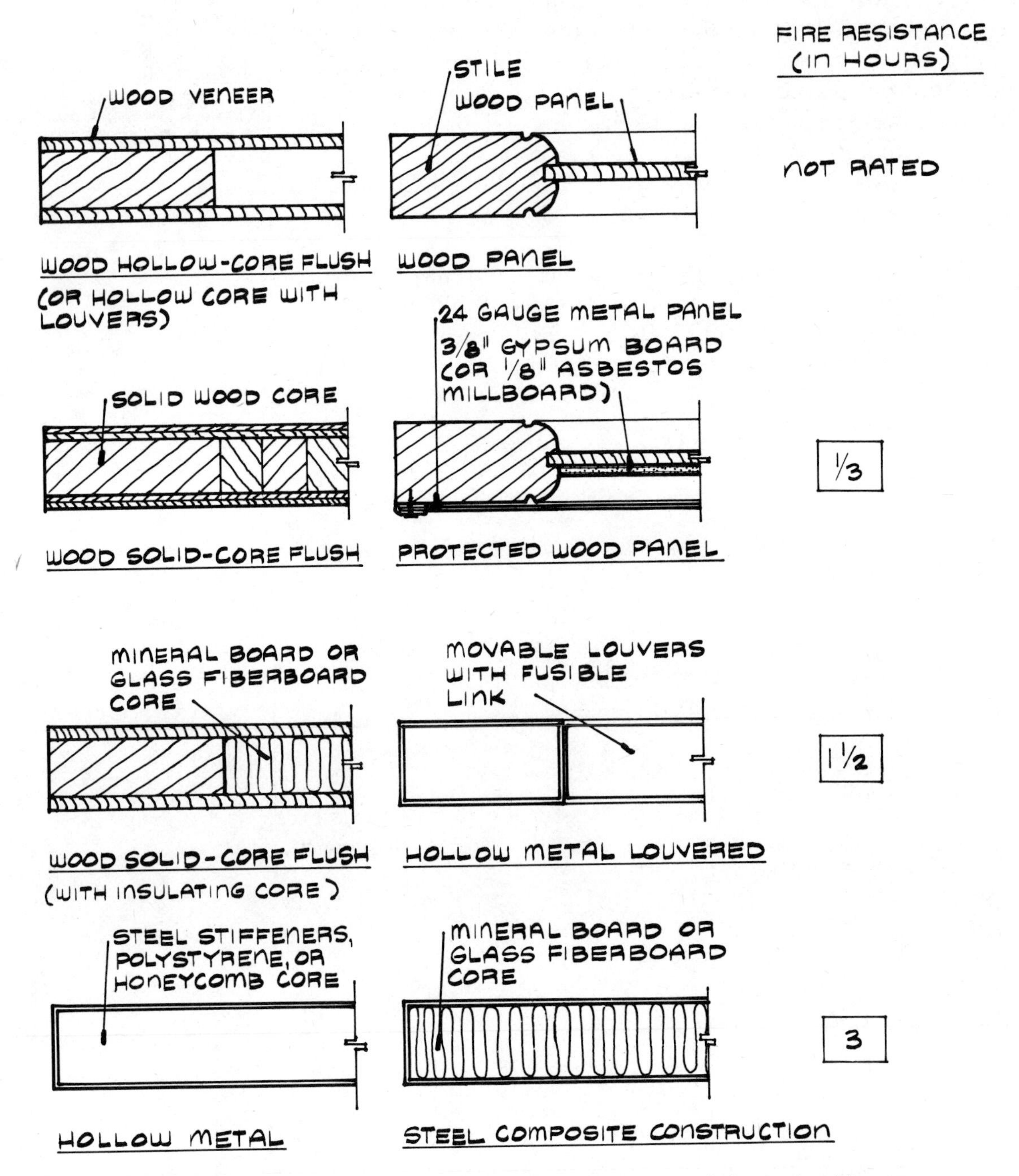

Note: Before selecting doors to provide a barrier to fire spread, the results of fire tests should be carefully examined. For example, doors and frames that are labeled (e.g., by Underwriters Laboratories, Factory Mutual, or other recognized testing laboratories) indicate the fire-resistance rating and temperature rise during 30 minute exposure to the ASTM standard time-temperature curve. Where no reference is made to temperature rise on the label, the door surface temperature on the unexposed side exceeded 650°F in 30 minutes.

BUILDING MATERIALS AND CONSTRUCTIONS—FIRE WALLS

A fire wall is a fire-resistant barrier that normally extends through the roof to prevent the spread of fire from one side to the other and to provide a shield for fire fighters. However, fire walls often do not extend through fire-resistant roof decks. In industrial applications, self-supporting fire walls have traditionally been constructed of thick brickwork. Fire walls normally require 2 to 4 hour fire-resistance ratings, depending on the anticipated fire severity. Typical constructions are reinforced concrete, brick, concrete block, and gypsum plaster. Example fire wall details for apartment buildings are shown below. To prevent the short circuit of fire around a fire wall, there should be no openings adjacent to its ends. The overhang of gable and mansard roofs also must be fire-stopped to prevent the short circuit of fire through the eaves.

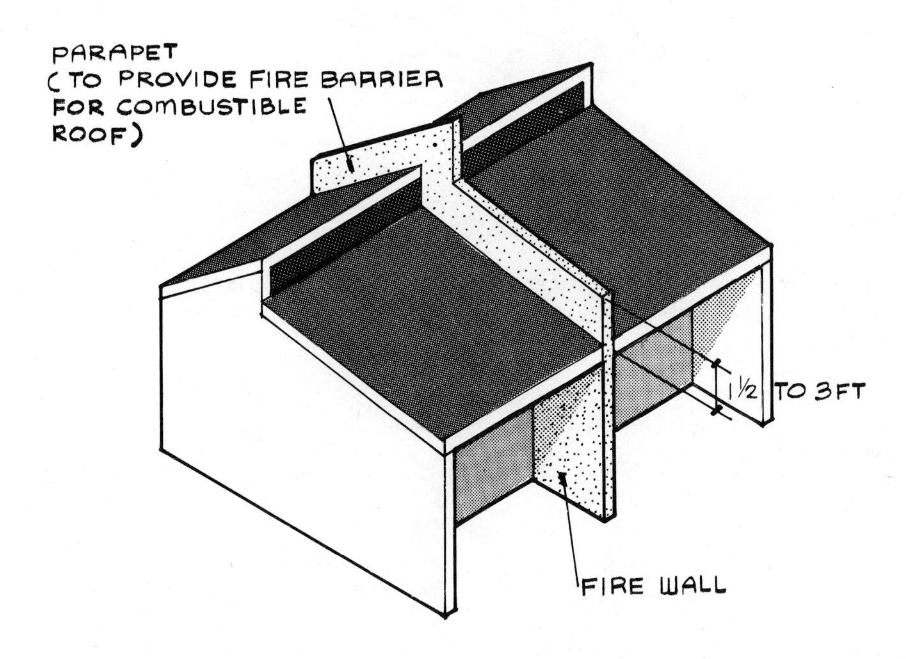

BUILDING MATERIALS AND CONSTRUCTIONS—FIRE WALLS (Continued)

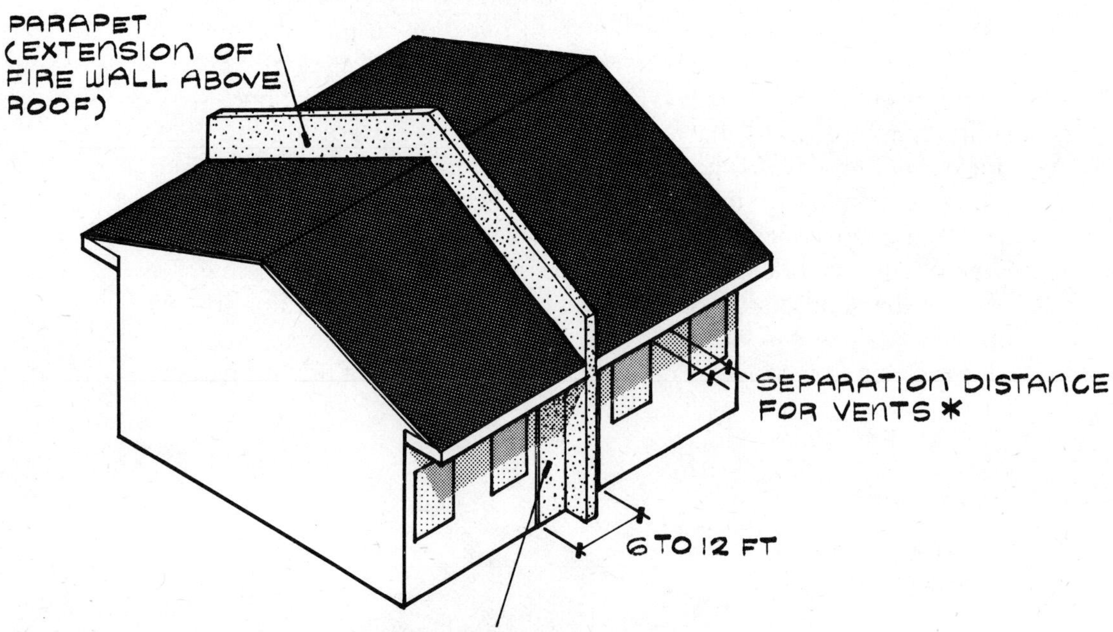

* VENTS LOCATED IN THE SOFFIT OF ROOF OVERHANG SHOULD BE AT LEAST 3 FEET AWAY FROM VERTICAL PROJECTIONS OF ADJACENT WINDOW AND DOOR EDGES.

Note: In apartments, condominiums, and so on, fire walls should extend beyond exterior walls to separate combustible balcony constructions.

BUILDING MATERIALS AND CONSTRUCTIONS—CHIMNEYS AND VENTS

Chimneys and vents should extend far enough above buildings so that wind from any direction will not cause a pressure against the upward exhaust flow of smoke and gases. Roof shape, adjacent buildings, wind conditions, and other factors must be considered when determining the appropriate projection of chimneys and vents. Typical clearances for chimneys and vents that exhaust gases and smoke are shown below. Refer to codes for required chimney wall thickness, connection details, and other installation guidelines. The example layouts shown below are based on chimney locations less than 10 ft from ridge line, wall, or parapet.

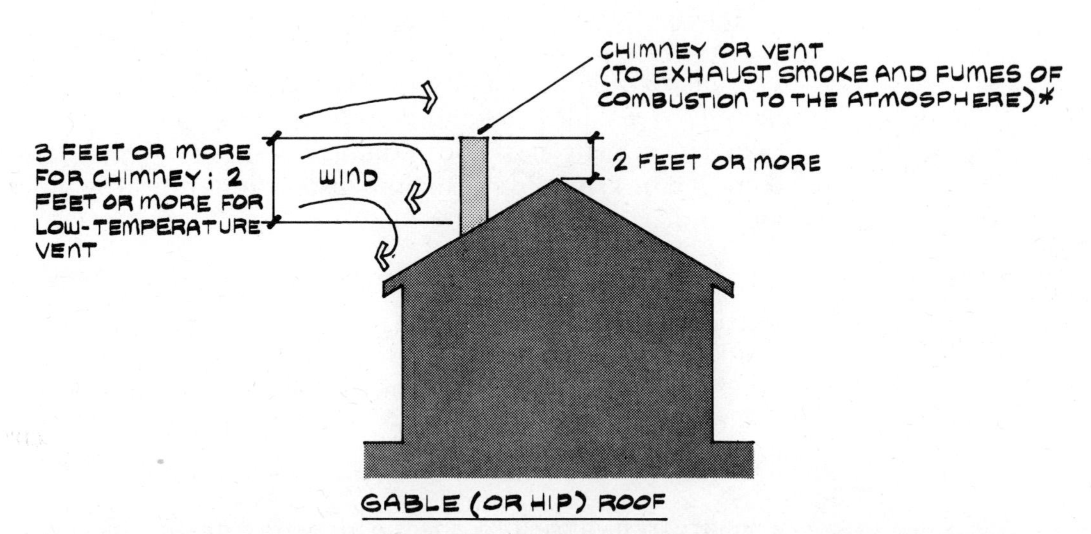

GABLE (OR HIP) ROOF

* NOTE: FOR CHIMNEY (OR VENT) LOCATED 10 FEET OR MORE AWAY FROM RIDGE LINE, EXTEND CHIMNEY AT LEAST 2 FEET ABOVE ANY ROOF SURFACE WITHIN 10 FEET HORIZONTAL DISTANCE FROM CHIMNEY.

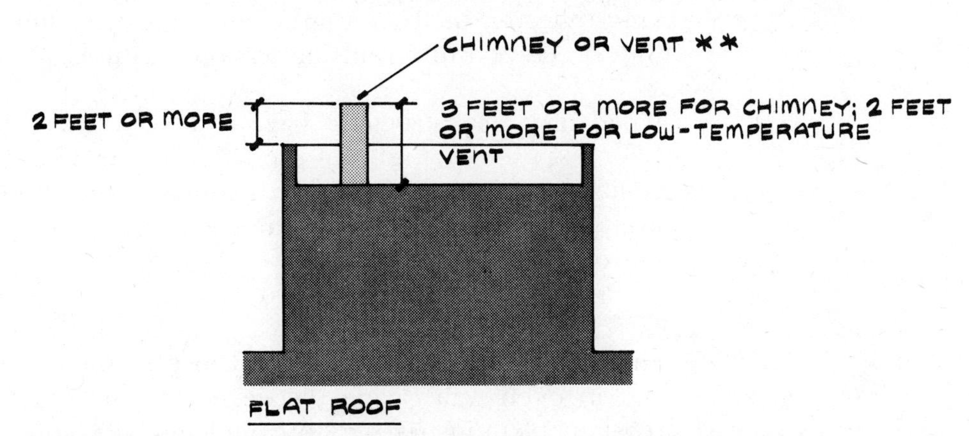

FLAT ROOF

** NOTE: CHIMNEY (OR VENT) NEED NOT EXTEND ABOVE PARAPET (OR WALL) LOCATED MORE THAN 10 FEET AWAY.

REFERENCE

McKinnon, G. P. (ed.), *Fire Protection Handbook,* Boston, Mass.: National Fire Protection Association (1976), p. 7–59.

BUILDING MATERIALS AND CONSTRUCTIONS CHECKLIST

When mild or hot-rolled steel is heated to temperatures greater than about 1000°F, which are common in severe fires, its strength diminishes rapidly. Cold-drawn steel tendons used to prestress concrete structural elements are also sensitive to high temperatures and will lose their prestressing strength at about 800°F.

Structural steel members can be fire-protected with membrane enclosures (e.g., mineral wool, gypsum board), spray-on materials (e.g., gypsum plaster with lightweight aggregate and mineral-wool fibers), concrete encasement, magnesium oxychloride inorganic coatings, and intumescent mastic coatings. Spray-on materials should be applied only to steel surfaces that are free of dirt, oil, and scale. To prevent corrosion, use a prime coat of paint on surfaces to be sprayed. The achieved thickness and density of the spray-on material determines its effectiveness. Be certain that spray-on materials are not damaged or removed during the installation of nearby pipes, ducts, and similar equipment.

Structural steel members located outside a building can be protected from fire by increasing the separation distance from the building to prevent direct flame impingement and to diminish heat build-up, by flame shields to help deflect heat and flames, and by liquid-filled columns to remove the heat from fire by circulating liquid in the columns.

The fire resistance of concrete depends on its thickness, aggregate, moisture and air content, and cover over the reinforcing steel. Lightweight aggregates such as vermiculite and perlite will provide greater fire resistance than conventional aggregates such as granite, sandstone, or quartz. Aggregate properties, however, may vary widely between different fabrication locations.

Wood frame constructions have relatively little structural strength when exposed to fire. However, thicker structural members (e.g., laminated arches, frames) can maintain their strength for a relatively long time. Wood is consumed by fire to produce charcoal at a rate of about $\frac{1}{40}$ in./minute. The charcoal in turn provides a protective coating for the unburned wood.

Fire can easily spread undetected in plenum spaces above ceilings and in open wall cavities between vertical studs. Use solid, fire-resistant plenum barriers to completely close off the openings or extend the fire-rated partition or wall to the underside of the floor slab.

BUILDING MATERIALS AND CONSTRUCTIONS CHECKLIST (Continued)

Poke-through penetrations in floor slabs can be an unwanted path for the vertical spread of fire and smoke. In addition, unprotected poke-through assemblies can lower the fire-resistance rating of floors to only a few minutes. Fire-stop the underside of poke-through assemblies with spray-on insulation undercoating, insulated heat shields, or install suspended ceiling underneath.

Poke-through openings for electrical services can be protected by (1) packing the opening with mineral-wool fibers, (2) metal frame units with elastomer insert blocks fabricated to fit snuggly around cable and conduit, and (3) silicone elastomer foams.

Lighting fixtures and air duct outlets should not be points of vulnerability to fire spread. Air outlet devices can be protected by dampers that close the opening during a fire and by insulating enclosures. Lighting fixtures can be protected by insulation enclosures of various shapes such as flat covers, tents, or boxes.

FIRE DETECTION AND SUPPRESSION

FIRE DETECTION AND SUPPRESSION—FIRE DETECTION AND ALARM WARNING

Early detection of fire and alarm warning are essential to prevent loss of life or harm to building occupants. As shown by the bar graph below, if the time needed to detect fire ignition and warn building occupants can be reduced, a corresponding vital margin of safety for evacuation can be provided. The alarm signal to building occupants should be well above the background noise. Audible alarm devices (e.g., bells, buzzers, chimes, horns) generally should produce a sound level in decibels* of at least 85 dBA at 10 ft away from the detector. In addition, early detection is necessary to ensure that the fire department will be rapidly notified or that automatic suppression will be initiated.

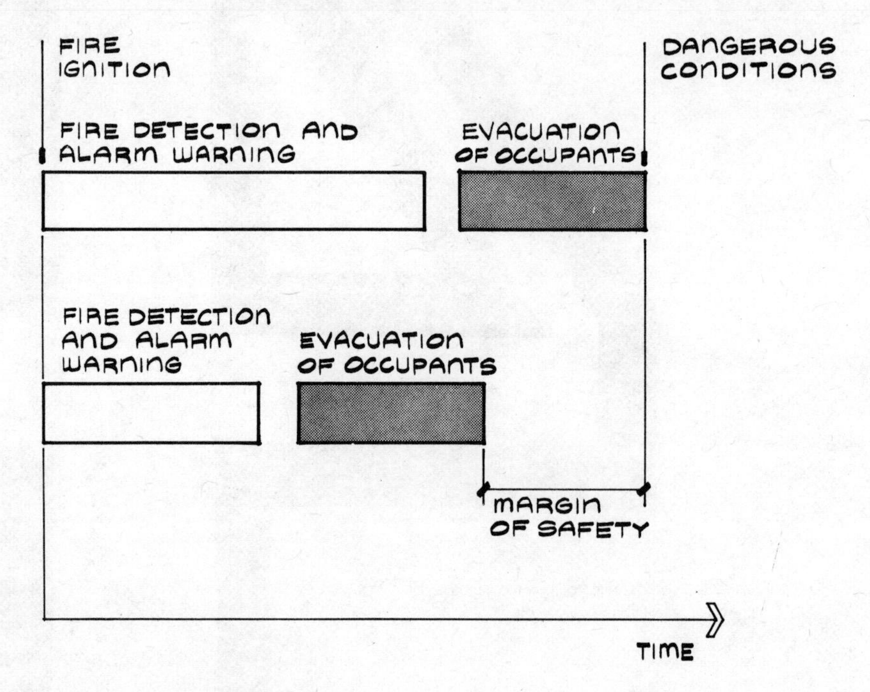

*Decibel is a dimensionless unit used to express the ratio of sound energies on a logarithmic scale. The reference energy usually is taken at the threshold of hearing. Human perception of sound depends on frequency as well as energy level. "dBA" refers to decibels measured by a sound-level meter with an electrical A-weighting network that tends to represent the frequency discrimination characteristics of the human ear. For a detailed presentation of sound and hearing, see M. D. Egan, *Concepts in Architectural Acoustics*, New York: McGraw-Hill (1972), pp. 3–22.

FIRE DETECTION AND SUPPRESSION—RESPONSE OF FIRE DETECTORS

Fire detectors have individual characteristics which should be carefully matched to the anticipated fire hazard (e.g., photoelectric for smoldering fires, ionization for flaming fires, infrared for flash fires). The time-temperature curves below represent growth to hazardous conditions for smoldering, flaming, and uncontrolled heat stages of fire. The types of detectors capable of providing rapid response to these fire conditions are indicated on the curves. Detectors should be well-maintained and be installed in buildings at positions where heat or smoke will collect.

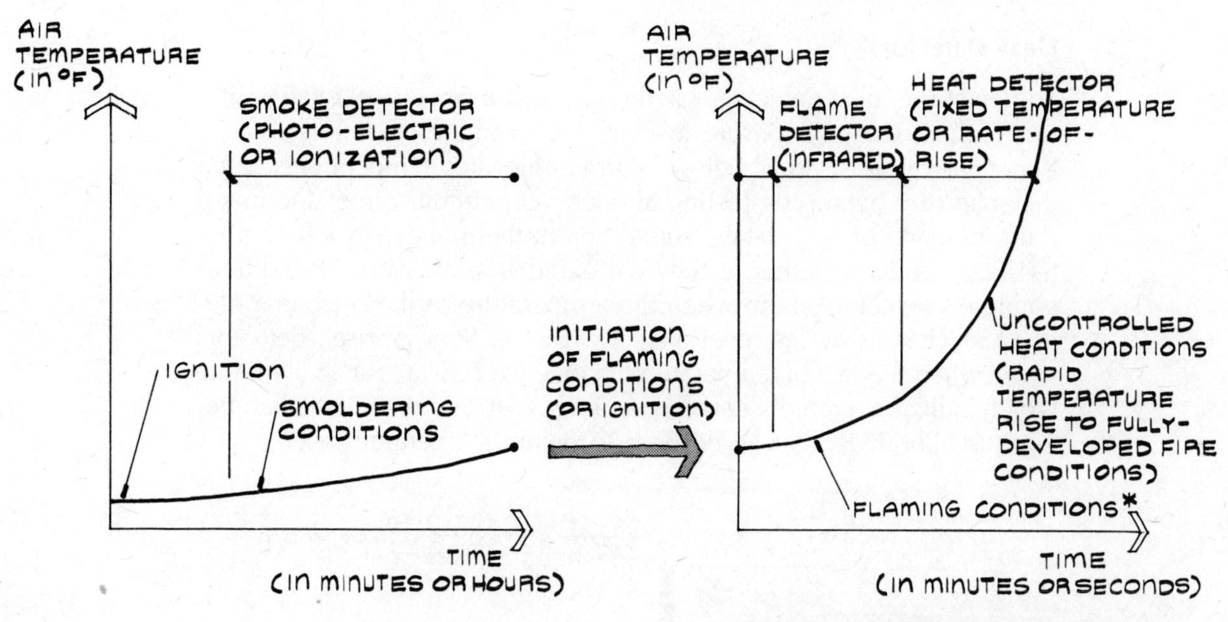

Note: For comprehensive presentations of fire detectors, see J. L. Bryan, *Fire Suppression and Detection Systems*, Beverly Hills, Calif.: Glencoe Press (1974) and R. L. P. Custer, and R. G. Bright, "Fire Detection: The State-of-the-Art," U.S. National Bureau of Standards, NBS TN 839, June 1974.

FIRE DETECTION AND SUPPRESSION—HEAT AND SMOKE DETECTORS

Shown below are an example heat detector which responds to the convected heat of a fire and a smoke detector which responds to both visible smoke and invisible products of combustion. Other types of detectors include ultraviolet and infrared units (detect flicker of flame), continuous line tubing (e.g., steel wire conductors separated by fusible thermoplastic that when melted completes an electrical circuit), water flow devices (detect water flow in pipes to open sprinkler heads), and laser beams (transmitted energy is attenuated by smoke).

Heat Detector

To detect fire, heat detectors use the physical or electrical change of a material caused by exposure to heat. For example, detectors using a bimetallic disc (as shown below) alarm at elevated temperatures when the snap disc bends completing an electrical circuit. Bimetallic materials are made of two metals, with different thermal expansion characteristics, bonded together so they will bend when heated. "Fixed temperature" detectors alarm when the temperature of the operating element reaches its design level (usually 135°F). "Rate-of-rise" detectors alarm when the rate of temperature change exceeds about 15°F/minute which indicates a rapidly developing fire. Expansion of air in a chamber with a calibrated vent can be used to detect temperature rise.

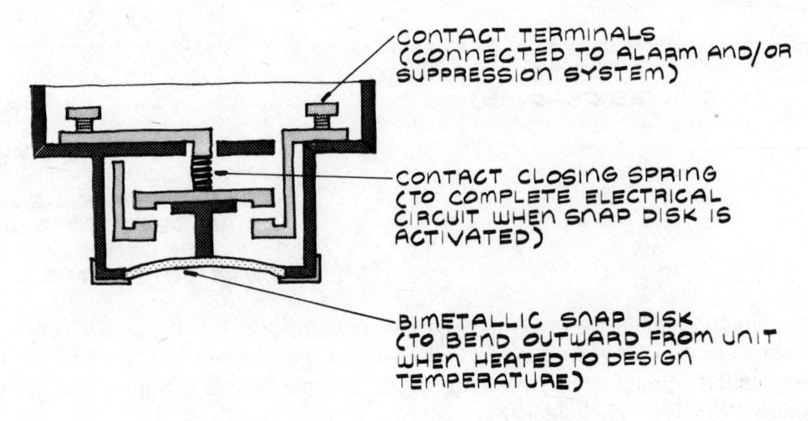

Note: Combination rate-of-rise units to detect rapidly developing fires and fixed-temperature units to detect slowly developing, smoldering fires also are available.

FIRE DETECTION AND SUPPRESSION—HEAT AND SMOKE DETECTORS (Continued)

Smoke Detector

To detect smoke, ionization detectors use the interruption of a small current between electrodes in the ionized sampling chamber (containing Americium 241 or Radium 226 source). In a dual-chamber detector, the reference chamber (exposed only to air temperature, pressure, and humidity) reduces false alarms caused by changing ambient conditions. Ionization detectors, with a single chamber, are also commercially available.

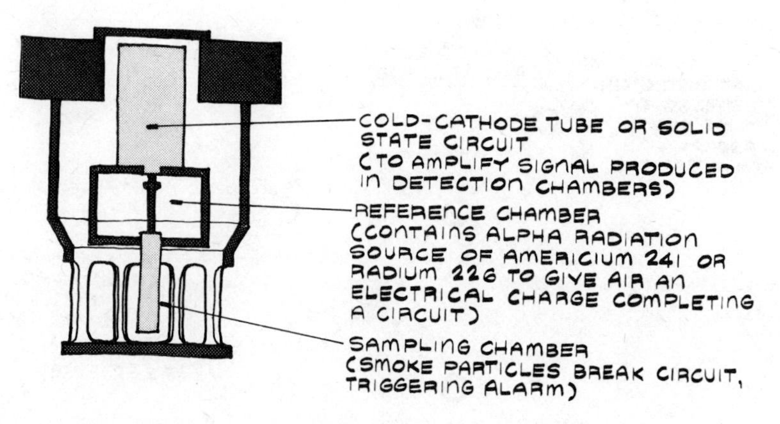

Note: In the photoelectric smoke detector, smoke causes a light beam from an incandescent lamp source to be scattered into the view of a photocell.

FIRE DETECTION AND SUPPRESSION—HEAT DETECTOR SPACING

Heat detector spacing ratings can be determined under test fire conditions in a 60 ft × 60 ft room described by the sketches below. Ratings for maximum permissible spacings are given for distances between detectors installed in typical spacing increments of 15, 20, 25, 30, 40, 50, or 60 ft. Following test fire ignition, the detector being tested must operate within about 2 minutes (plus or minus 10 seconds) before operation of the automatic sprinkler heads.

Detector Spacing Layout

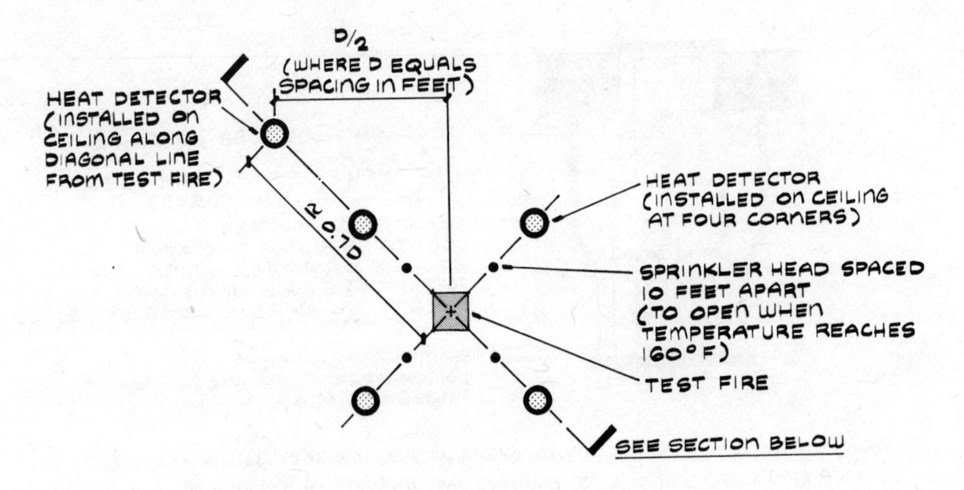

Section View

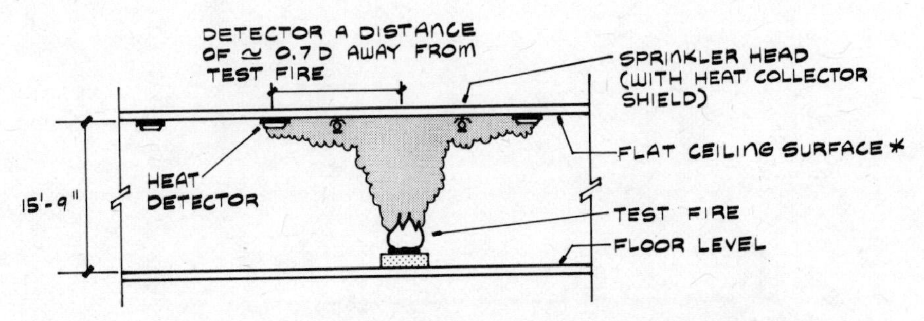

* A FLAT, UNOBSTRUCTED CEILING SURFACE IS A FAVORABLE CONDITION FOR DISTRIBUTION OF HEAT FROM A FIRE. FOR OTHER LESS FAVORABLE CEILING CONFIGURATIONS, DETECTOR SPACINGS MAY HAVE TO BE REDUCED.

Note: Smoke detectors can be evaluated in a small chamber containing an air circulation system and a smoke density meter to measure the light obscuration by smoke from a cotton wick test fire at the moment of alarm (cf., "Photoelectric Type Smoke Detectors for Fire Alarm Service," UL 168).

TEST REFERENCES

"Fire-Detection Thermostats," UL 521, 1974.

"Thermostats for Automatic Fire Detection," Factory Mutual Research Corporation (1151 Boston-Providence Turnpike, Norwood, Mass. 02062) Class No. 3210, 1973.

FIRE DETECTION AND SUPPRESSION—SMOKE DETECTOR LOCATIONS

Solar radiation can cause a hot air layer to form directly under roof surfaces. This thermal barrier can prevent the products of combustion from reaching smoke detectors. To avoid the effects of thermal barriers, detectors can be installed as shown by the sketches below. For example, a detector for a gable roof 30 ft high (*H*) should be placed at least 33 in. (*d*) below the ridge line. (See dashed lines on graph.)

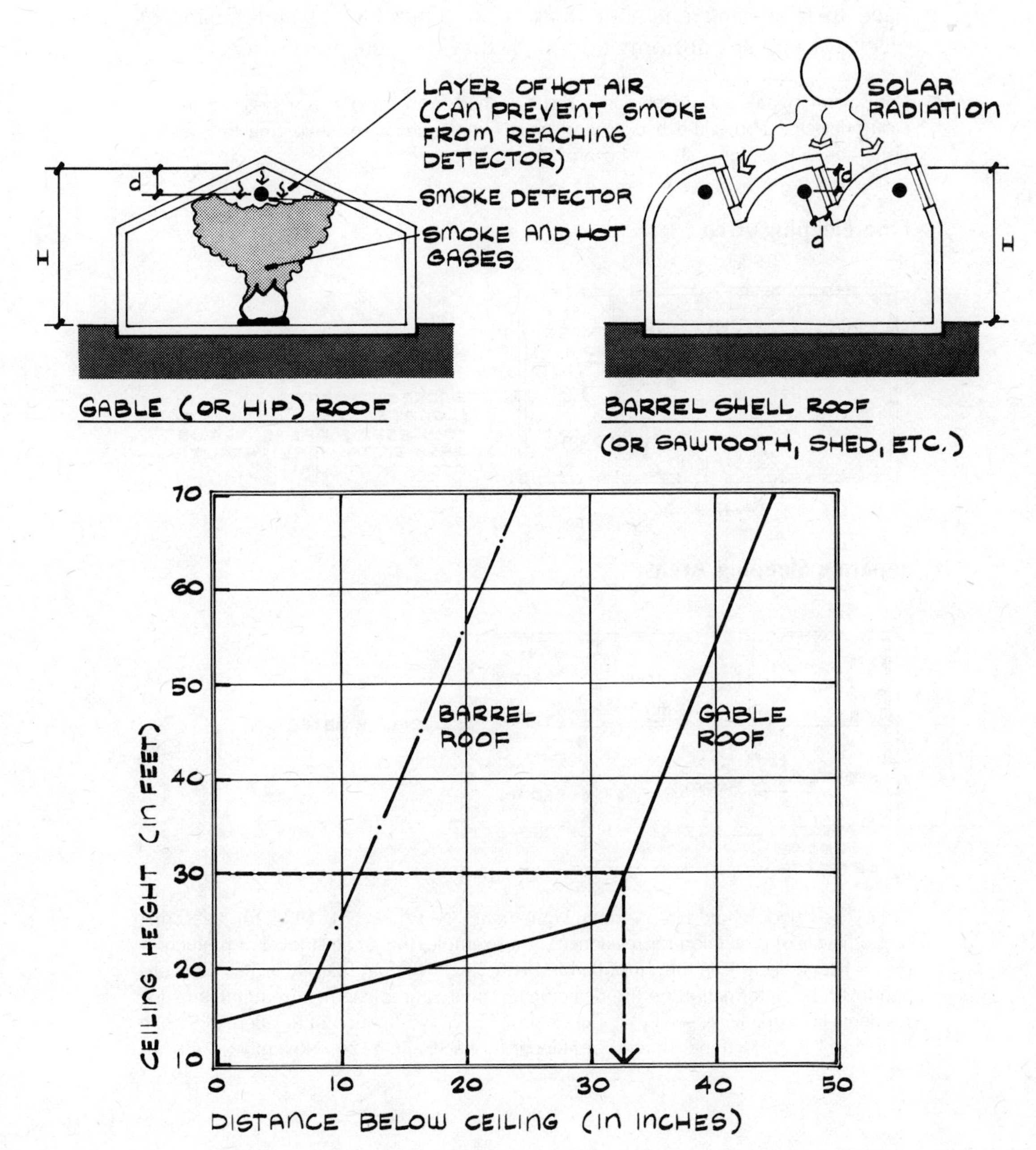

REFERENCE

Johnson, J. E., "Engineering Early Warning Fire Detection," *Fire Technology*, Vol. 5, No. 1, 1969.

FIRE DETECTION AND SUPPRESSION—FIRE DETECTOR LOCATIONS FOR RESIDENCES

Fire detection devices should sound an audible warning* at a designated level of smoke or heat to give building occupants sufficient time for evacuation to safety. In addition, detectors (called "unit" smoke detectors) can be used to activate dampers in air ducts, sprinklers, fire doors, and so on. Shown below are example "single-station" smoke detector layouts for single-story residences. Single-station detectors have their sensing chamber, alarm-sounding device, and means of electrical power transformation all within the detector enclosure.

---

*Fire alarm signals should be easily distinguished from other alarm systems and be easily detected above the background noise. For applications involving the deaf, a visual (or tactile) signal should be provided.

One Sleeping Area

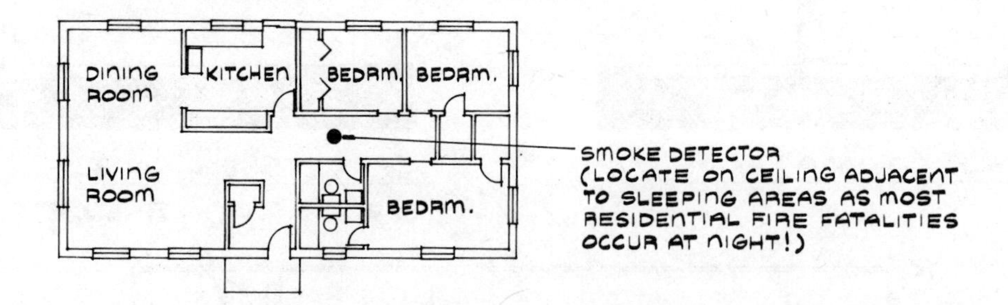

Separate Sleeping Areas

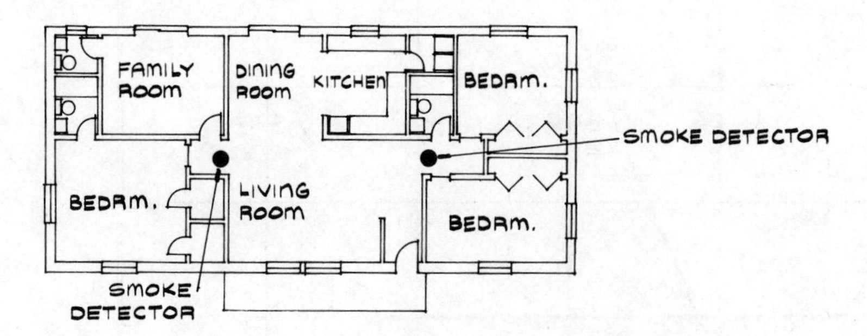

Note: See "Household Fire Warning Equipment," NFPA No. 74, 1974, for a description of levels of protection for residences. For example, the greatest level of protection means heat detectors are also installed in bedrooms, living and family rooms, kitchen, and foyer. For information on fire detector sensitivity and placement requirements for residential occupancies, see R. W. Bukowski, "Field Investigation of Residential Smoke Detectors," U.S. National Bureau of Standards, NBSIR 76-1126, November 1976.

CHECKLIST FOR EFFECTIVE USE OF FIRE DETECTORS IN RESIDENCES

The areas that should be protected by smoke detectors, in decreasing order of importance, are (1) every occupied floor and basement, (2) sleeping areas and basement near stairs, and (3) sleeping areas only.

Locate smoke detectors on ceilings near the center of rooms (or upper wall surfaces within about 6 to 12 in. of ceiling), corridors, or top of stairways away from obstructions such as beams or large lighting fixtures. Where corridors are longer than about 30 ft, consider using two or more detectors. Wall locations are preferred where thermal barriers will be created at ceilings by radiant panel ceiling heating systems.

Where ceiling beams, open joists, or other architectural features will interrupt the spread of smoke and heat, detectors should be located closer together than recommended by the manufacturer. Install detectors on the bottom surface of beams, joists, and the like.

When placing detectors, consider the room airflow patterns from mechanical air distribution systems. Stagnant areas can occur which prevent smoke from reaching detectors (e.g., in corners of a room). Avoid detector locations in or near the supply air stream from air registers or diffusers. The supply air stream can purge the detector of smoke and prevent fire detection. Do not locate detectors where smoke can be diluted by return airflow from other parts of the residence where there is no fire (e.g., near return air grilles located at the top of stairways).

Use ordinary temperature rated heat detectors (e.g., 135 to 165°F) in remote areas where serious fire conditions could develop before smoke could reach a detector. Example spaces normally separated from living areas include basement shop areas, unoccupied attics and storage areas, and attached garages.

Proper cleaning and maintenance of detectors is important to achieve the designed operating characteristics. For example, dirt, grease, and insects can collect on protective screens of smoke detectors and block the flow of smoke into the unit. Detectors should be tested periodically to assure that they are in good working condition.

FIRE DETECTION AND SUPPRESSION—DRY STANDPIPES

"Dry standpipes" are water pipes used to connect hoses that fire fighters use in buildings to ground level fire hydrants. They are designed to save time when placing fire hose streams in service. Fire hoses can be connected to dry standpipe outlets in the building to fight fires as shown below. Standpipes are often installed in stairways which means that exit doors will be ajar during fire-fighting operations. Dry standpipes also can be installed on the outside of buildings.

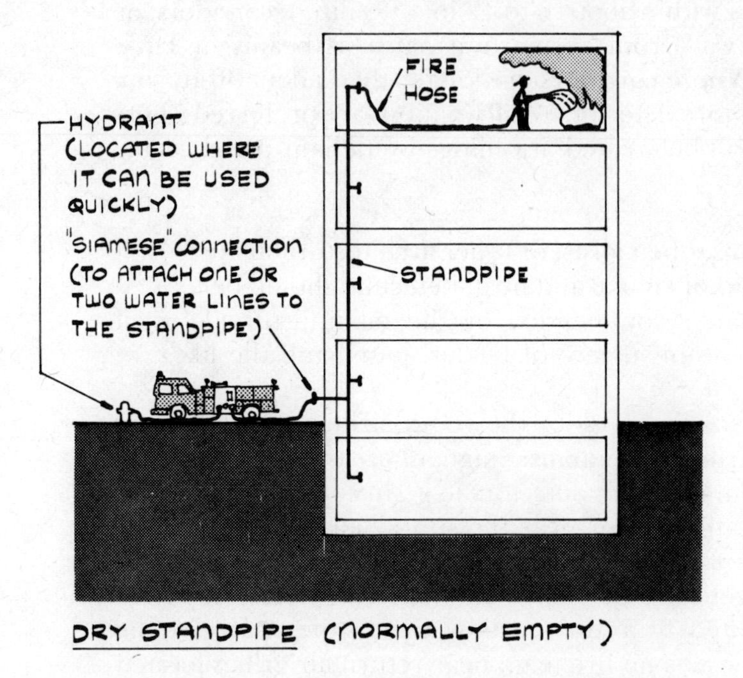

DRY STANDPIPE (NORMALLY EMPTY)

FIRE DETECTION AND SUPPRESSION—WET STANDPIPES

"Wet standpipes" are generally used in tall buildings and in industrial buildings. Building occupants can use the fire hose cabinets on each floor as shown below. Provision of a wet standpipe system, however, establishes the need for building occupants to be trained in the use of fire hoses. Experience is normally required to effectively use fire hoses because nozzle reaction forces can be considerable. Users also should be aware that linen hose will leak until the fabric swells and that kinks can completely block the flow of water. Standpipes should be located so hoses will be able to reach every part of the building. For example, codes usually base requirements for number of standpipes on a 100 ft length of extended, unlined fire hose (or 75 ft of lined hose). A dependable water source must serve the wet standpipe system. Water sources may be connections to public or private water mains, gravity tanks (e.g., at 50 to 150 ft elevations), pressure tanks (air pressure in tank provides energy for immediate discharge of water), natural water bodies, or in certain situations even swimming pools.

"Primed standpipes" are similar to wet standpipe systems. The pipes are water-filled, but do not have sufficient operating pressure until a fire pump starts or the fire fighters connect water lines to the system.

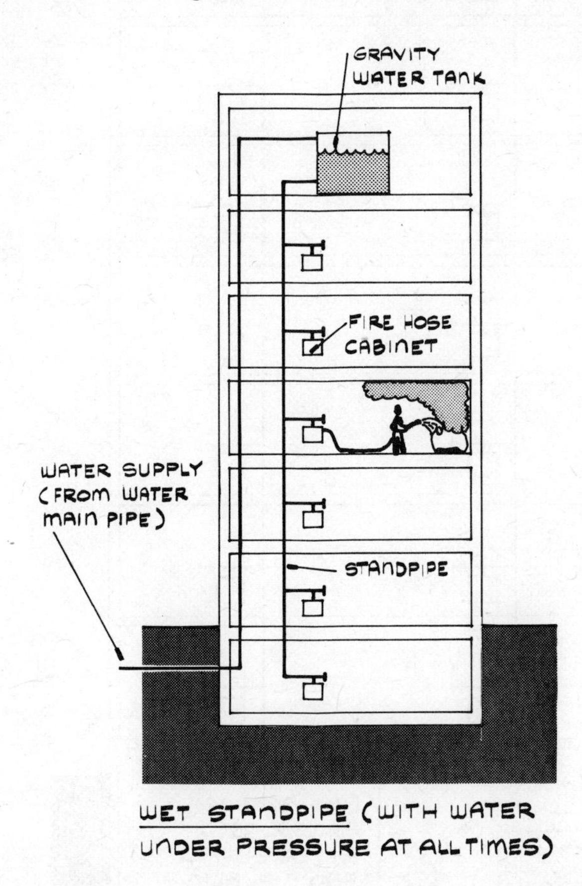

WET STANDPIPE (WITH WATER UNDER PRESSURE AT ALL TIMES)

Note: In residences, copper tubing (or protected plastic pipe) can be used to supply domestic system cold water to "hose stations" consisting of a hose bib with attached garden hose coiled on a rack. Locate hose stations at top of basement stairs and in upper-story closets.

FIRE DETECTION AND SUPPRESSION—COMBINATION STANDPIPE SYSTEM

Shown below is an example "combination" wet and dry standpipe fire protection system using a gravity tank. Wet standpipe systems are intended for use by building occupants on small fires. Most tall buildings should have automatic fire pumps that use a water supply from public or private mains, storage tanks, or reservoirs.

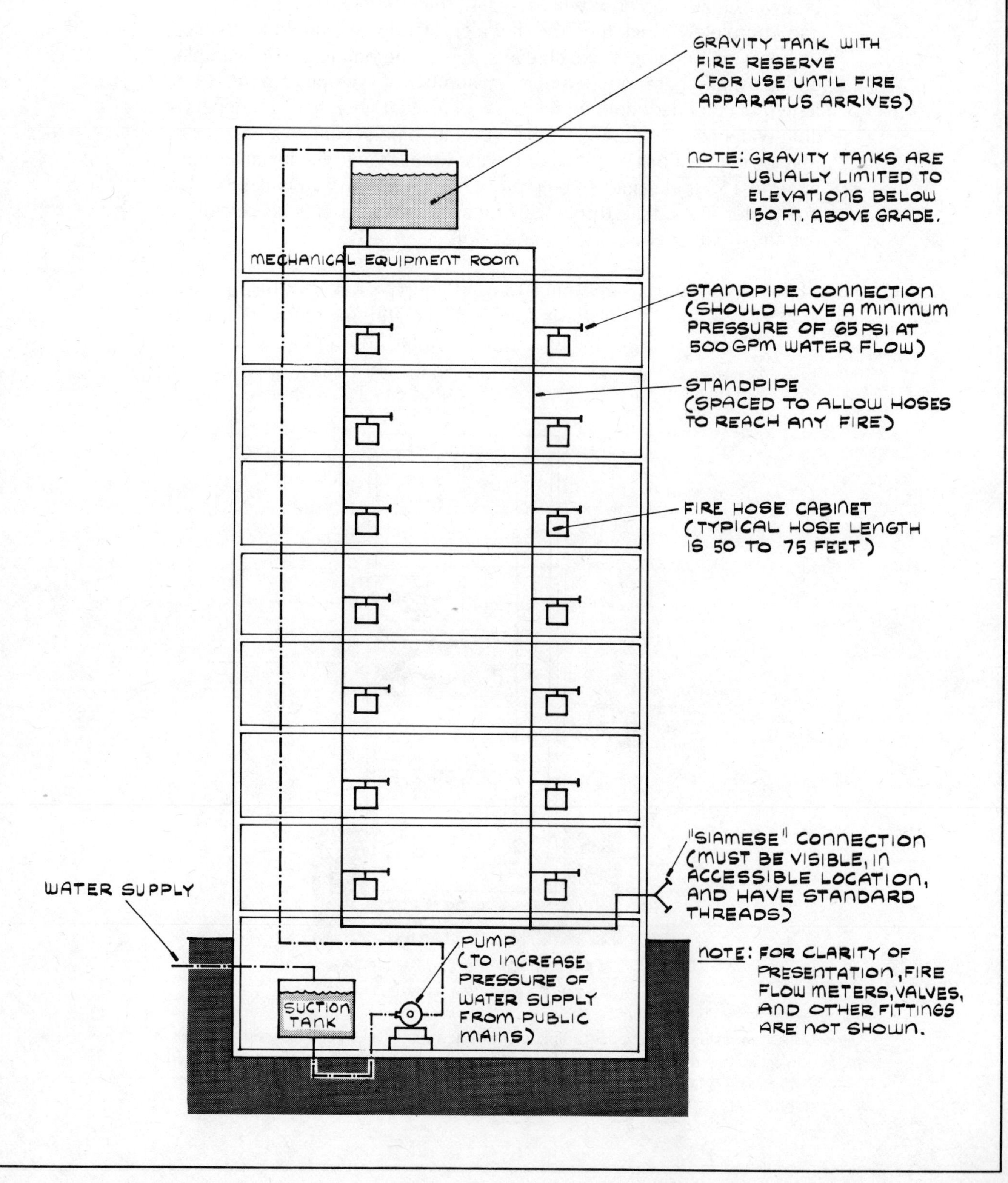

FIRE DETECTION AND SUPPRESSION—STANDPIPE-SPRINKLER SYSTEM

An example "joint-use" standpipe system for both fire hose and sprinkler heads is shown below. An adequate water supply for simultaneous operations must be provided, usually through a standpipe riser of at least 6 in. diameter. Take-offs for fire hose and sprinklers should have separate valves to allow independent control at each floor level.

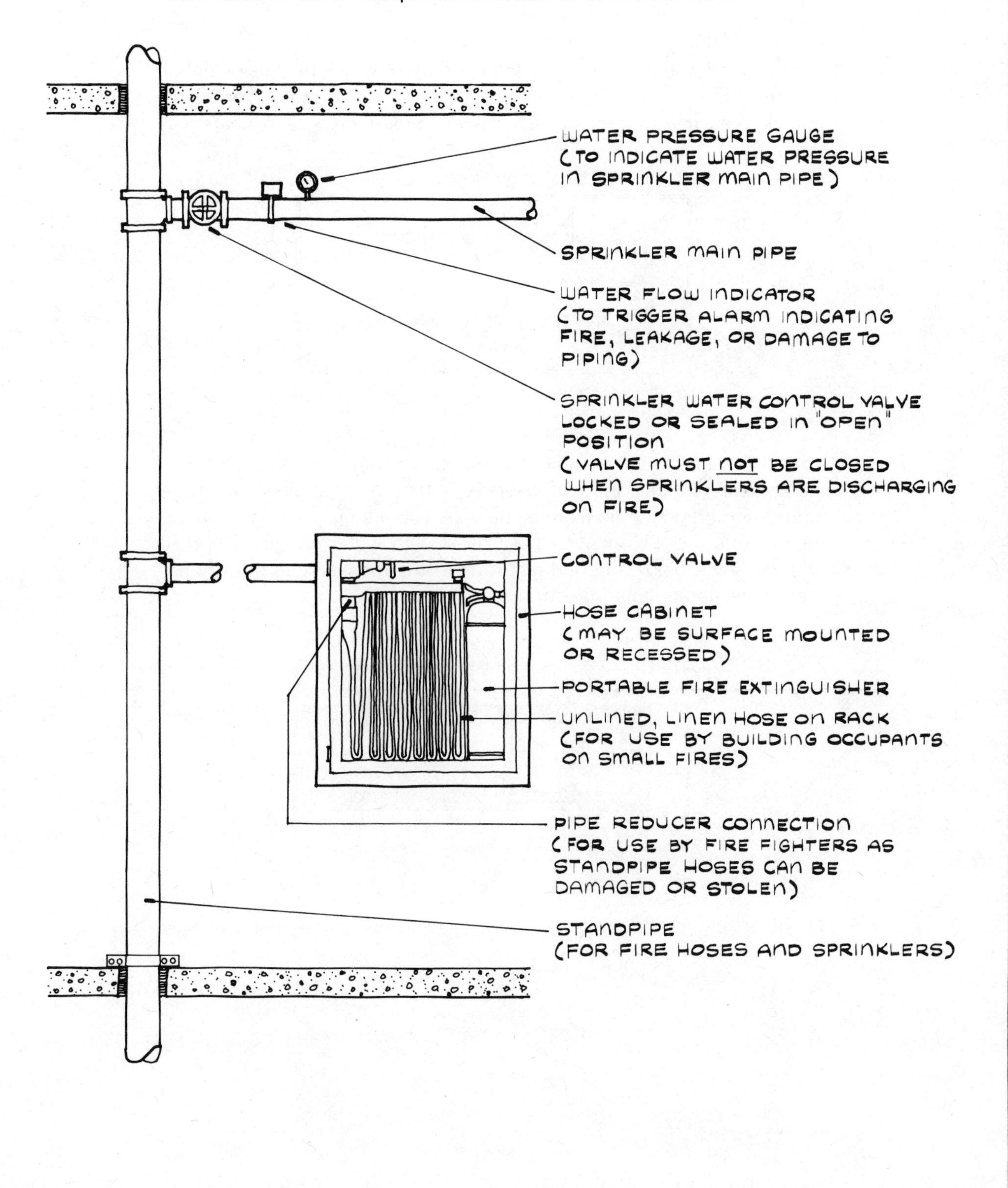

FIRE DETECTION AND SUPPRESSION—STANDPIPE HOSE COVERAGE

Shown below are the theoretical reach and the actual reach of standpipe hoses during a fire. In the theoretical situation, where the direct route could be used, only 50 ft of hose is required. However, nearly 200 ft will be required if the indirect path is used.

Theoretical

To use standpipe hoses at their full coverage (or reach), it is essential that fire fighters know the direct path to the fire and that corridors be tenable (i.e., free of dangerous levels of smoke, gases, and heat).

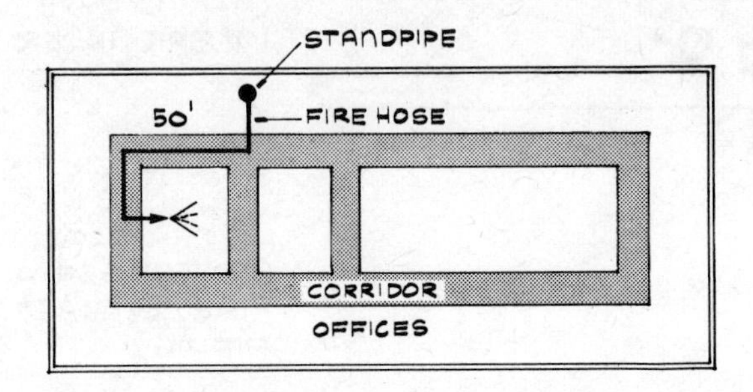

Actual

In real fire situations, fire fighters often must advance to the fire with an equipment working load of over 100 lb, move along smoke-filled corridors, and withstand extreme heat. (A 100 ft length of $1\frac{1}{2}$ in. rubber-lined, cotton, double-woven hose filled with water weighs over 150 lb.) As shown below, the indirect route actually taken can quickly use up the available length of standpipe hose.

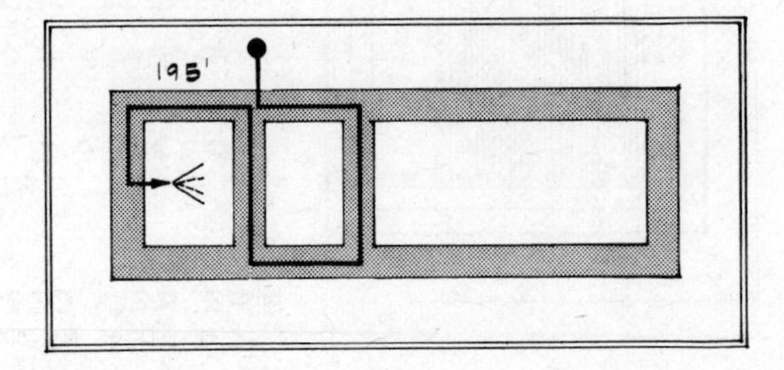

REFERENCE

Mendes, R. F., *Fighting High-Rise Building Fires*, Boston, Mass.: National Fire Protection Association (1975).

CLASSIFICATIONS OF FIRES

The National Fire Protection Association classifies portable fire extinguishers according to the types of fires for which they are designed. Portable fire extinguishers are designed only to suppress small fires. For example, the maximum discharge time span generally is about 8 to 90 seconds for typical portable units. Reference standard is "Portable Fire Extinguishers," NFPA No. 10, 1975.

Classification Symbol (and Label Color in Parentheses)	*Type of Fire*	*Description (and Method of Suppression in Parentheses)*	*Typical Extinguishers*
A (GREEN)	Ordinary combustibles	Materials such as wood, paper, cloth, fiber, rubber, and many plastics (can be quenched and cooled by water)	Water base, foam, soda acid
B (RED)	Flammable liquids	Liquids such as paint, paint thinner, gasoline, oil, tar, solvents, fat, and so on; flammable gases, greases, and similar materials (requires blanketing and smothering)	Dry chemical, foam, carbon dioxide
C (BLUE)	Electrical equipment	Energized electrical equipment such as overheated fuse boxes and other electrical sources, and wiring. Classification refers to source of ignition rather than to fuel as fires are classes A and B in terms of fuel (requires nonconductive extinguishing agent)	Dry chemical, carbon dioxide
D (YELLOW)	Combustible metals	Metals such as magnesium, titanium, zirconium, sodium-potassium alloys, and so on (requires smothering)	Dry powder with sodium chloride or graphite base

FIRE DETECTION AND SUPPRESSION—PORTABLE FIRE EXTINGUISHERS

Properly located portable fire extinguishers should be easily reached and be placed in conspicuous locations along normal paths of protected egress away from potential fire hazards. They should be in good operating condition and have the extinguishing agent capability required to suppress potential fires (e.g., using water on a grease or electrical fire may make conditions worse). Shown below is a typical layout for an "ordinary hazard" office occupancy where extinguishers should not have to be carried more than 75 ft.

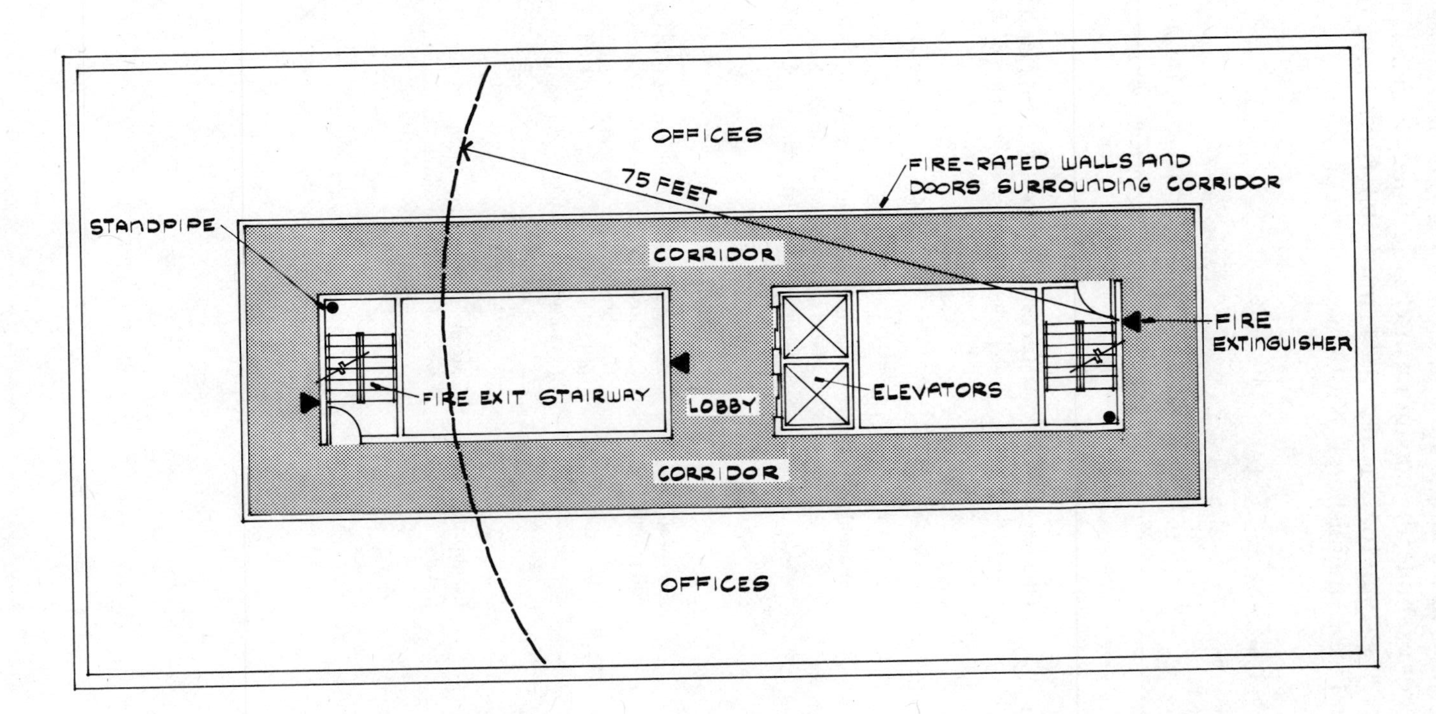

▲ FIRE EXTINGUISHER (LOCATED SO THAT NO OCCUPANT IS MORE THAN 75 FT. AWAY FROM AN EXTINGUISHER)

● STANDPIPE (SERVES AS AN EXTENSION OF FIRE-FIGHTERS' HOSES TO GROUND-LEVEL HYDRANTS)

FIRE DETECTION AND SUPPRESSION—FIRE EXTINGUISHER MOUNTING HEIGHT

Portable fire extinguishers should be mounted at positions where they can be easily reached and removed. According to Occupational Safety and Health Act (OSHA) regulations, portable extinguishers having a gross weight of 40 lb or less should be installed so the top of the unit is not more than 5 ft above the floor and extinguishers having a gross weight of more than 40 lb should be installed not more than $3\frac{1}{2}$ ft above the floor. For ease in floor cleaning operations and to avoid footfall obstructions, do not mount extinguishers closer than 6 in. to the floor. To operate most portable extinguishers, pull out the restraining pin and, by squeezing the discharge lever, apply extinguishing agent in a sweeping motion from side to side at the base of the fire. It is important that potential users learn to operate extinguishers on practice fires.

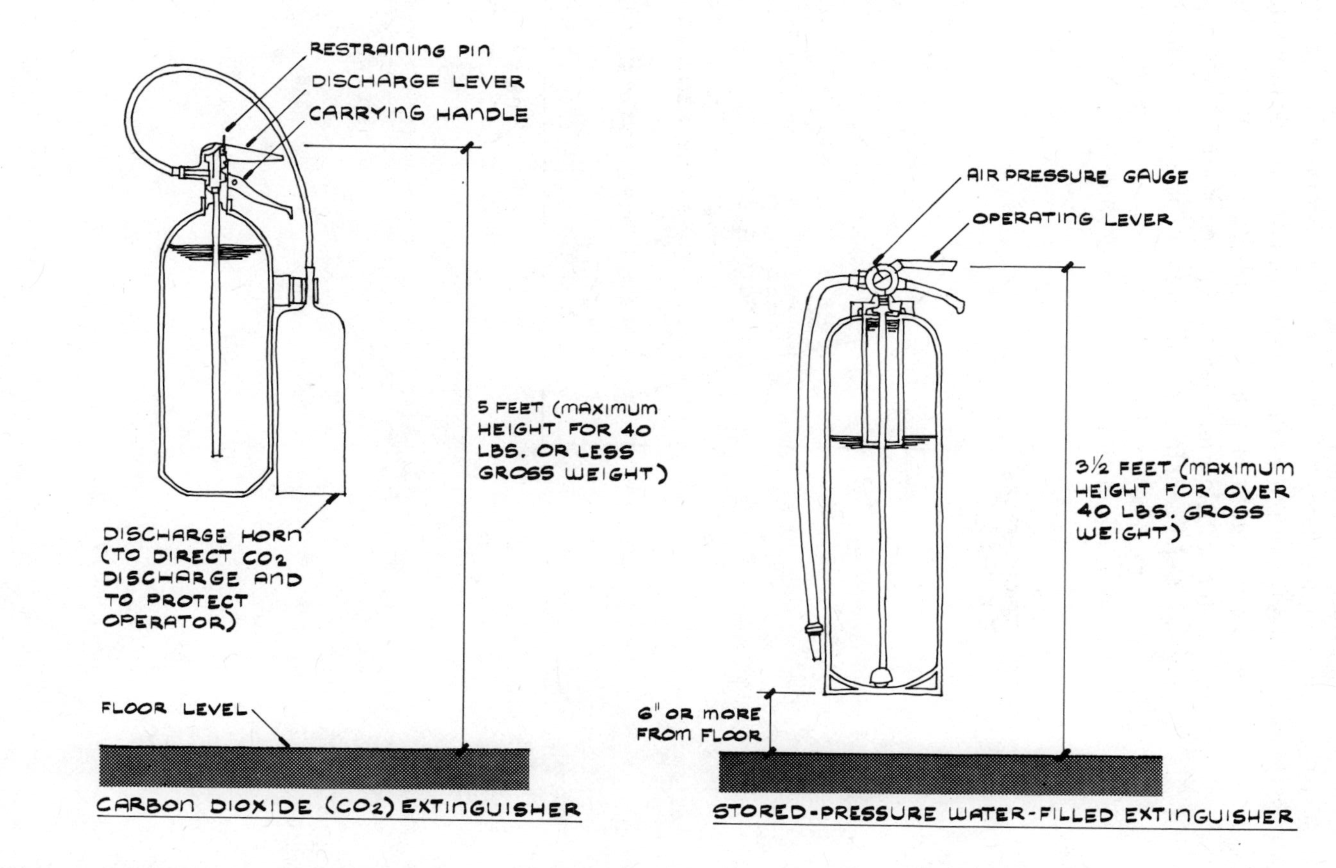

FIRE DETECTION AND SUPPRESSION—AUTOMATIC SPRINKLER SYSTEMS

National Fire Protection Association records indicate that automatic sprinkler systems are over 96% effective in suppressing or containing fires. The principal reason given for unsuccessful performance is closed water control valves before the fire occurs or before the fire is completely extinguished. In a fire, sprinkler systems discharge water to extinguish the fire or prevent its spread from the area or room of origin. The water discharged from sprinklers cools burning materials by direct contact of water particles, removes heat from the room, wets unburned combustibles to prevent fire spread, and displaces oxygen by formation of steam. The graph below shows example air temperature conditions for flaming fires in identical rooms with and without sprinklers.

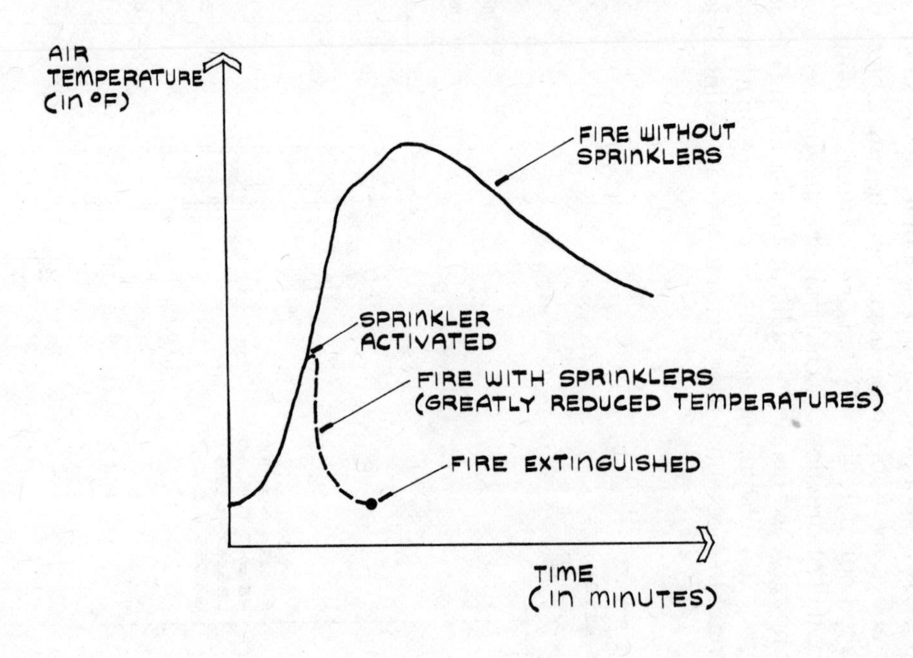

Sprinkler heads are designed to discharge finely divided water particles in the event of fire. Finely divided particles of water have increased surface area and therefore greater ability to absorb heat. The conversion of one gallon of water into steam absorbs up to 9500 Btu of heat depending on the size of the water particles. Sprinkler heads can be activated by soldered link (breaks apart when heat melts solder), glass bulb (liquid expands when heated, breaking glass), chemical (melts allowing a strut to collapse), or electrical (quick-response device discharges molten metal on heat-sensitive sprinkler element).

Sprinkler heads are constructed so they will open at predetermined temperatures within the range of 135 to 575°F, depending on application requirements. Heads are also rated for use at maximum ceiling temperatures within the corresponding range of 100 to 475°F to help prevent premature operation from extended exposure where elevated temperature is normal. The sprinkler head shown below is an upright type that uses a deflector to direct water over its area of coverage. Upright heads are less susceptible to water flow obstruction from accumulation of sediment in the water piping than are pendent heads.

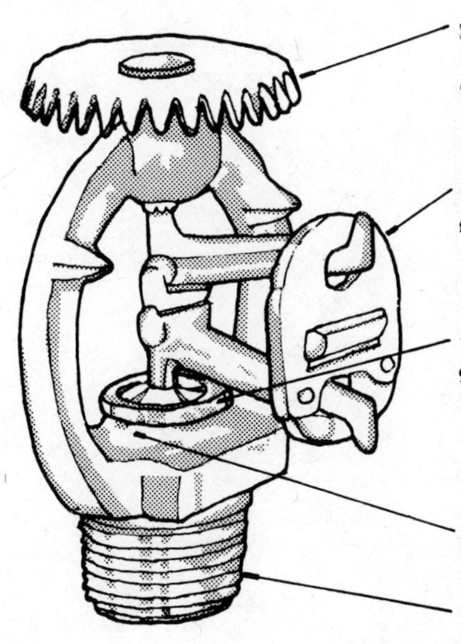

UPRIGHT SPRINKLER HEAD

FIRE DETECTION AND SUPPRESSION—SPRINKLER COVERAGE AREA

The area of sprinkler water coverage depends on the physical characteristics of the sprinkler head and the water flow and pressure available to the head. When placing sprinklers, be sure to consider the effects of obstructions such as beams, lighting fixtures, and partitions. The maximum spacing (*S*) between sprinklers is normally 15 ft. Sprinklers must not be spaced too close together, however, as the discharging water from the first operating head could wet adjacent sprinklers delaying activation in fire situations where they may be needed. Where sprinklers must be placed close together (e.g., spacings less than 6 ft around escalator openings), cross baffles or recessed baffle pockets should be provided midway between the heads. Coverage areas for light hazard occupancies vary from 225 to 130 sq ft, for ordinary hazard from 130 to 100 sq ft, and for extra hazard 90 sq ft (cf., "Standard for the Installation of Sprinkler Systems," NFPA No. 13, 1975).

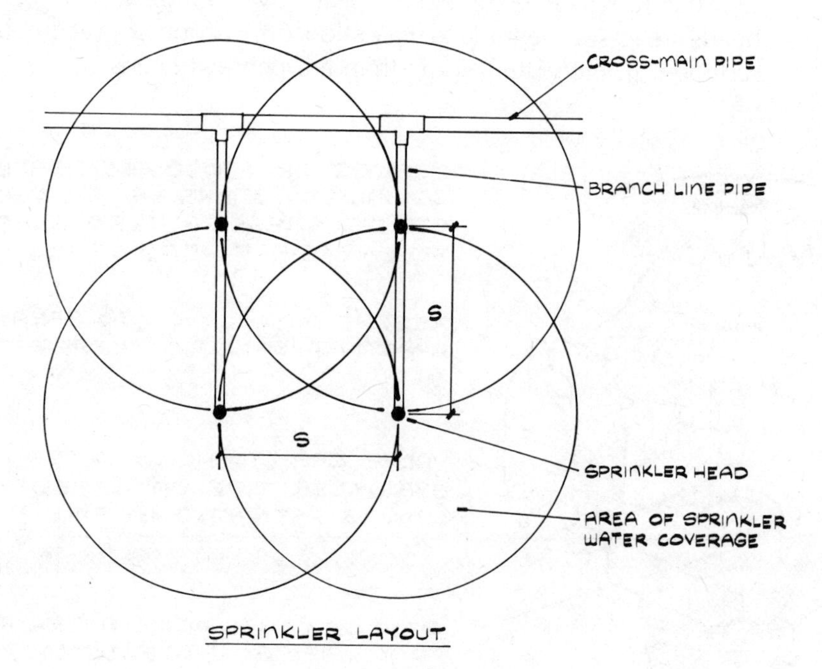

SPRINKLER LAYOUT

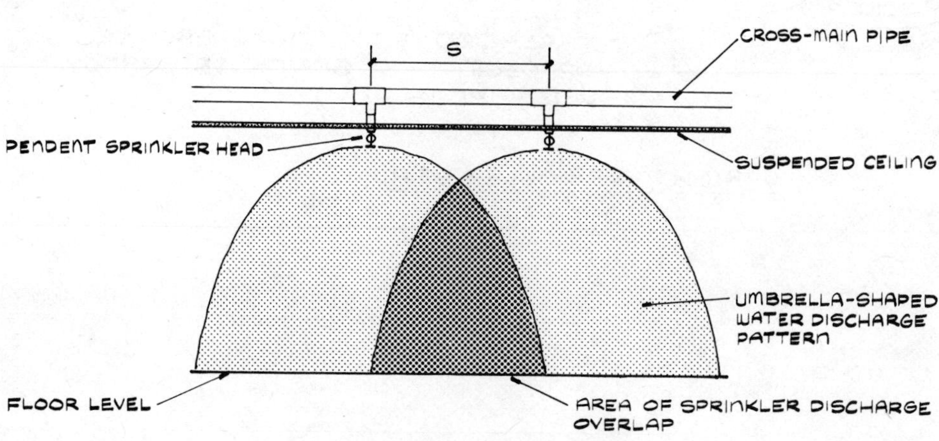

SPRINKLER DISCHARGE PATTERN

FIRE DETECTION AND SUPPRESSION—CEILING CONFIGURATIONS

Shown below are exposed standard pendent, sidewall, and hidden pendent sprinkler heads. Sidewall heads can be unobtrusively installed along the sides of exposed ceiling beams. In small rooms (e.g., hotel, apartment), sidewall heads can be used to provide uniform coverage without the overhead piping required for pendent heads. Standard pendent heads can be recessed in modeled ceilings to reduce head visibility as shown. Cover plates, matching the color of the ceiling surface, also can be used to hide pendent heads above flat ceiling surfaces. In addition, flush heads are available that do not extend below ceilings as far as standard pendent heads.

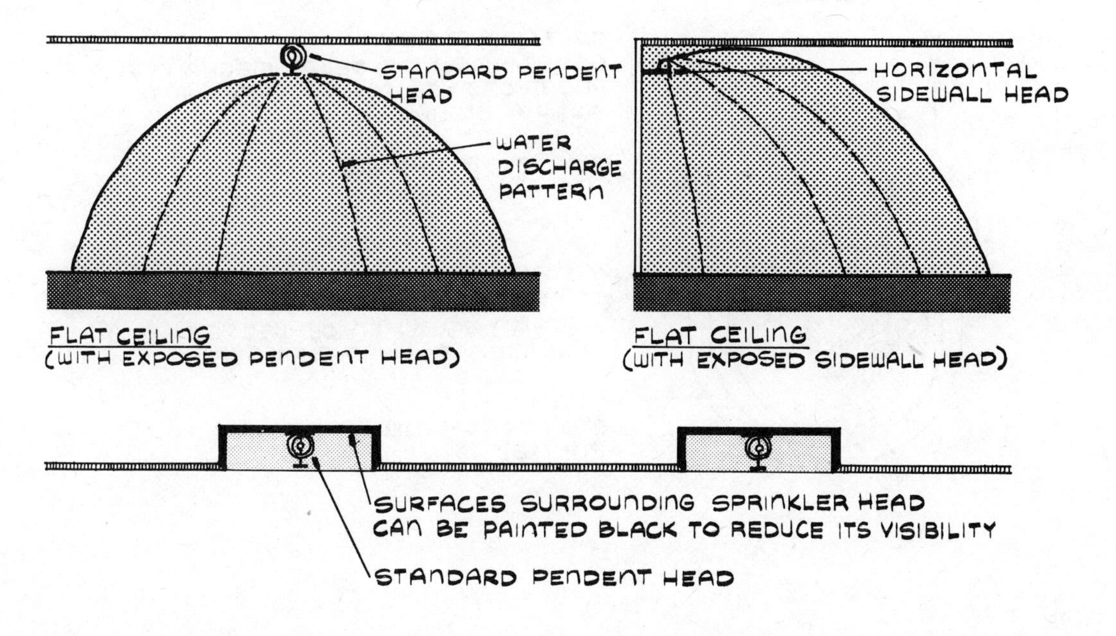

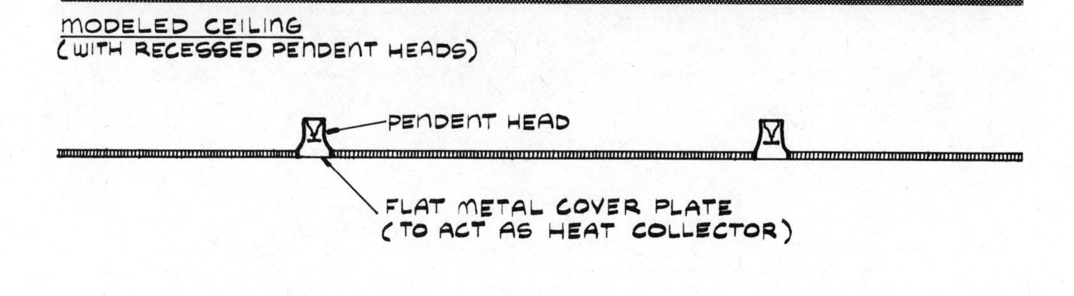

Note: Lighting systems (e.g., perimeter or central cove) can be arranged so the visibility of sprinkler heads will be low.

FIRE DETECTION AND SUPPRESSION—WATER FLOW INDICATORS

Water flow-sensitive devices in sprinkler system piping can be used to trigger alarms indicating fire, leakage, or damage. This alarm feature is especially important where water damage would be disastrous (e.g., computer equipment rooms, storage rooms for valuable documents). The devices normally detect water flow of about 10 gpm or greater. A time-delay mechanism helps prevent false alarms by absorbing fluctuations from routine water surges. When installed on each floor of a building, water flow indicators help fire fighters to locate the area of sprinkler operation. In addition, water control shut-off valves should be provided on each floor.

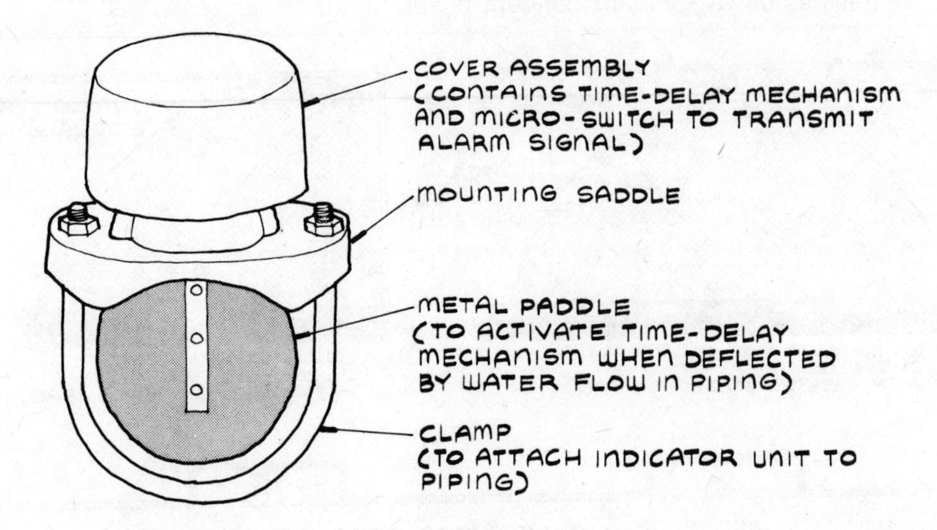

SPRINKLER SYSTEM TYPES

Sprinklers are usually required in large basement areas and windowless buildings (where fires are often smoky and difficult to reach), in hospital and related health care facilities, in areas used to store or handle hazardous materials, and in large public occupancy areas. Water sources may be city mains, gravity tanks, pressure tanks (e.g., two-thirds water-filled with one-third air under pressure), and fire pumps. The sprinkler system must be carefully integrated with the building's structure, mechanical system, and other services. Some of the various types of sprinkler systems are described below.

Dry Pipe

The dry pipe system contains air under pressure. Following loss of air pressure through an open sprinkler head, a dry pipe valve automatically opens allowing water to enter the piping network and to flow through the opened sprinkler head(s). This system is generally used only where areas are not heated and are subject to freezing temperature conditions. However, the sprinkler control room must be heated to prevent freezing of dry pipe valves, pumps, tanks, and related equipment. After a fire, the system must be completely drained to return it to the dry condition with air supplied by a compressor. Consequently, it will be out of service for a period of time.

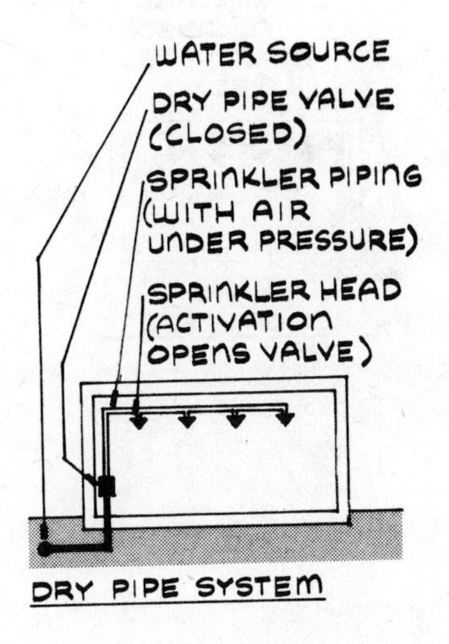

DRY PIPE SYSTEM

Wet Pipe

The wet pipe system contains water under pressure at all times for immediate release when sprinkler heads are activated. This system is used in heated areas or areas not subject to the danger of the water freezing in the piping network. With the wet pipe system, water is available from a water supply to immediately extinguish fires.

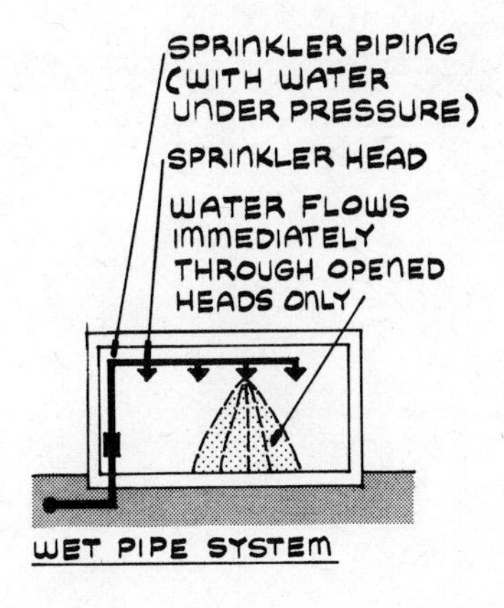

WET PIPE SYSTEM

Deluge

In the deluge system, sprinkler heads are open at all times and normally there is no water in the piping. Mechanical or hydraulic valves, operated by heat- or flame-sensitive devices, are used to control water flow to the sprinklers by opening a water control disc. The deluge system is designed to rapidly wet down an entire area in which fire may occur. This system is designed for special hazard situations involving possible flash fires where flame spread would be rapid and for buildings with especially high ceilings. Water is applied quickly with the deluge systems, but a large water supply is required.

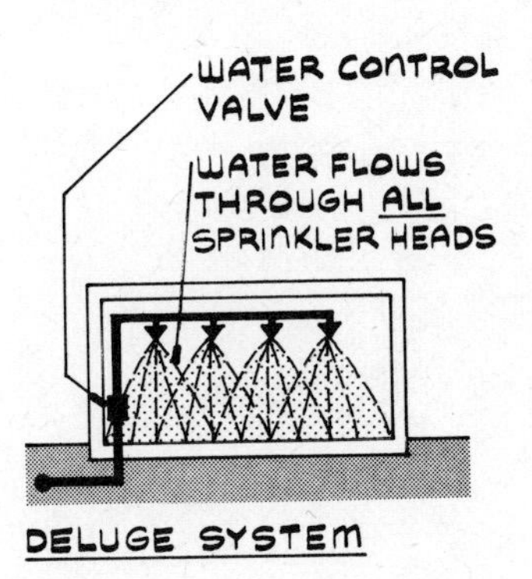

DELUGE SYSTEM

SPRINKLER SYSTEM TYPES (Continued)

Preaction

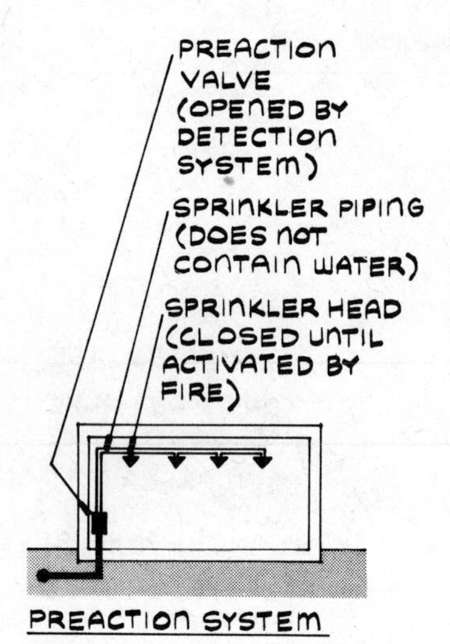

PREACTION SYSTEM

The preaction system is a closed head, dry system used where the danger of water damage is a serious problem (e.g., from frequent false alarms or equipment failures). The preaction valve is opened by an independent fire detection system more sensitive than the sprinkler heads. The detection system provides advanced warning, allowing human intervention, and opens the preaction valve. The open preaction valve permits water to fill the sprinkler system (it becomes "wet") and to discharge when sprinkler heads open in response to the heat from fire. This dry system is also used in areas where freezing may occur.

Note: Sprinkler systems use only about 10% of the water normally required to extinguish a fire by fire hoses. Where water damage must be avoided (e.g., computer equipment rooms, tape vaults), on-off bimetallic sprinkler heads, with a metal element that will turn off water flow at each sprinkler head when the temperature drops, can be used with wet pipe and preaction systems.

FIRE DETECTION AND SUPPRESSION—SPRINKLER WATER DISCHARGE DENSITY

The graph below shows "hydraulically designed" sprinkler system discharge quantity (called "water density") in gallons per minute (gpm) per square foot for area of sprinkler operation in square feet. Sprinklers should be sized to discharge sufficient water to extinguish fires which reasonably may be anticipated. In addition, the fire compartment should be enclosed by fire-resistant constructions having ratings equal to, or greater than, the duration of the sprinkler water supply. Sprinkler discharge for a 3500 sq ft compartment in an "ordinary hazard" occupancy would be 0.19 gpm/sq ft or 665 gpm (see dashed lines on graph).

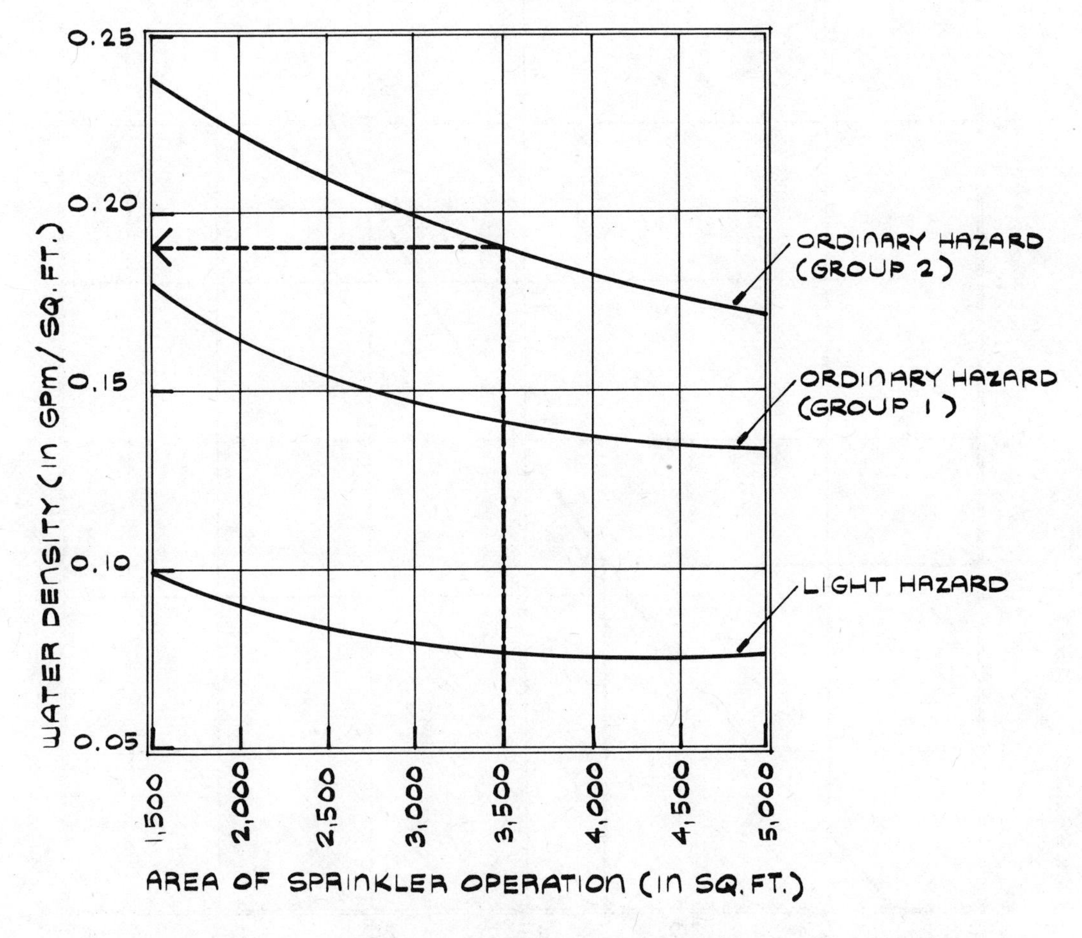

Note: Refer to current edition of NFPA No. 13, the sprinkler systems standard, for up-to-date water density recommendations.

REFERENCE

Jensen, R., "21 Ways to Better Sprinkler System Design," *Actual Specifying Engineer*, June 1973.

FIRE DETECTION AND SUPPRESSION—SPRINKLER HEAD COEFFICIENTS

The graph below shows the water pressure in pounds per square inch (psi) required to operate an example sprinkler head with a measured average coefficient (K) of 5.65 for various discharge water flow rates in gpm. For example, a design water flow rate of 22 $\frac{1}{2}$ gpm will require a sprinkler pressure of 16 psi (see dashed lines on graph). Refer to sprinkler manufacturers' catalogs for sprinkler head coefficients and other design parameters.

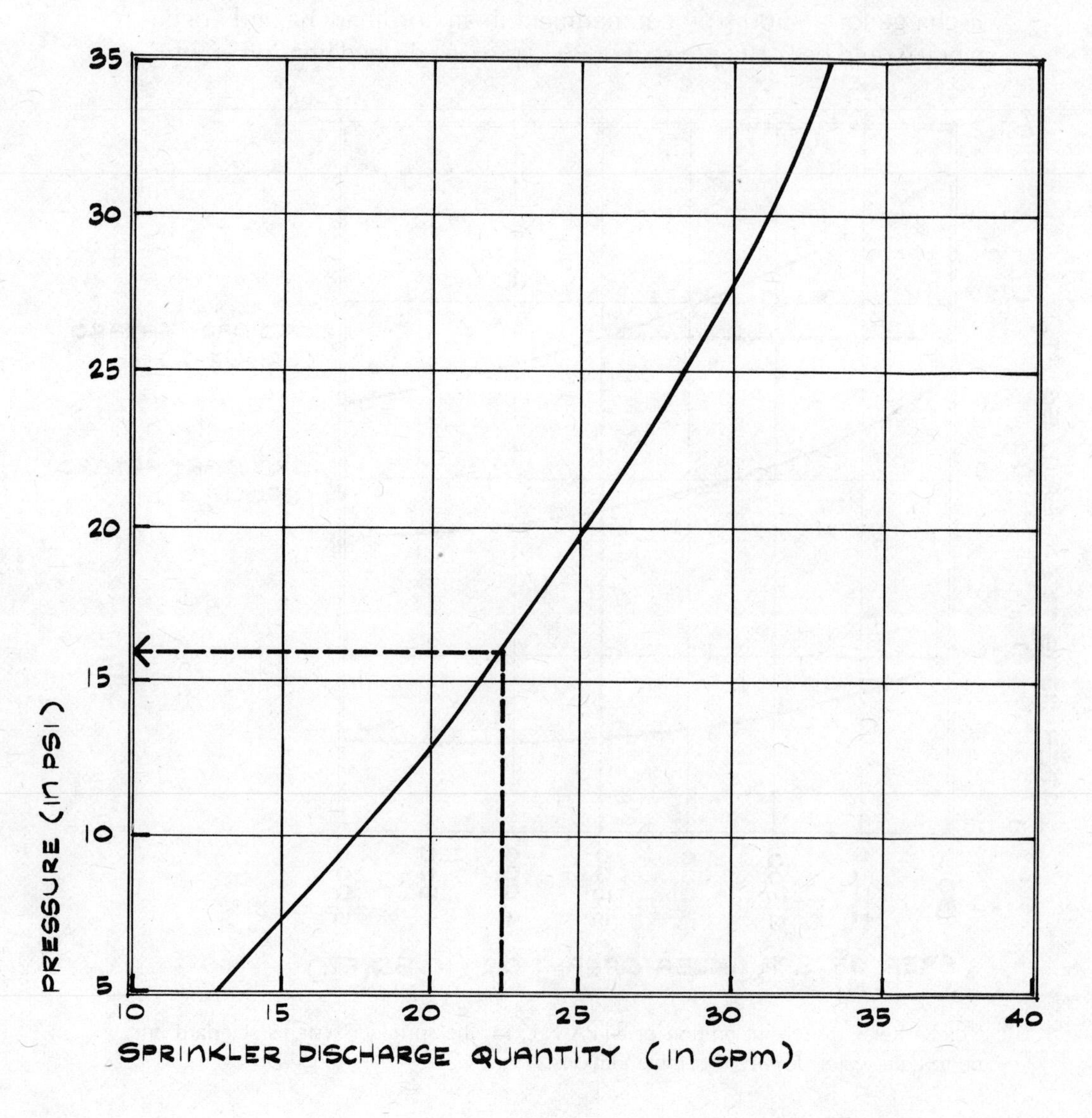

FRICTION LOSS FOR FITTINGS AND VALVES

To find the friction loss for fittings (e.g., 90 degree ell, 90 degree tee, straight run of tee) and valves, add the equivalent length in feet to the actual length of pipe section where they occur. Use the pressure loss graph for water flow to find loss in psi per 100 ft. Equivalent lengths are given in the table for streamlined fittings and recessed threaded fittings. For threaded fittings, double the table values.

	Diameter (in.)									
	1	*1¼*	*1½*	*2*	*2½*	*3*	*3½*	*4*	*5*	*6*
90 degree ell	1.5	2	2.5	3.5	4	5	6	7	8.5	10
90 degree tee	2.5	3	3.5	5	6	7.5	9	10.5	12.5	15
45 degree ell	0.9	1.2	1.5	2	2.5	3	3.5	4	5	6
Straight run of tee	0.45	0.6	0.75	1	1.25	1.5	1.8	2	2.5	3
½ reduction	1.3	1.65	2	2.5	3	3.75	4.5	5	6.5	8
Gate valve	0.6	0.8	1	1.3	1.6	2	2.4	2.7	3.3	4
Check valve	10	14	16	20	25	30	35	40	50	60

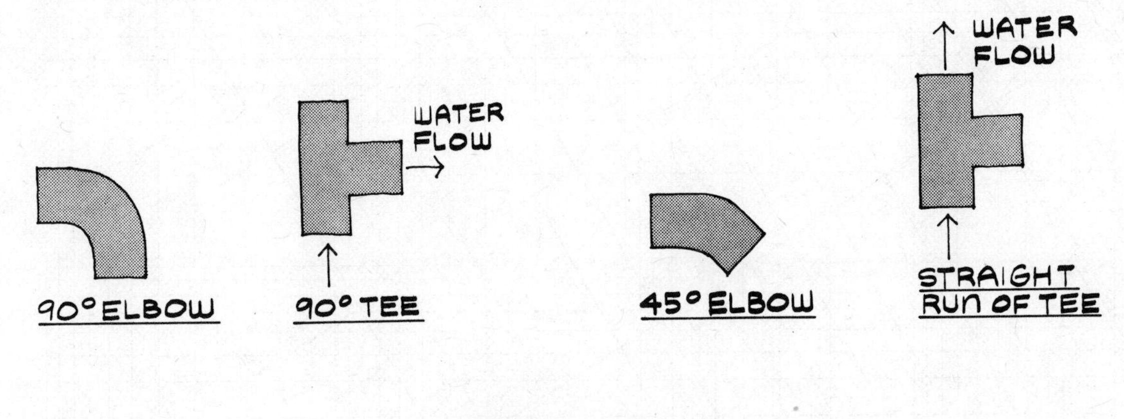

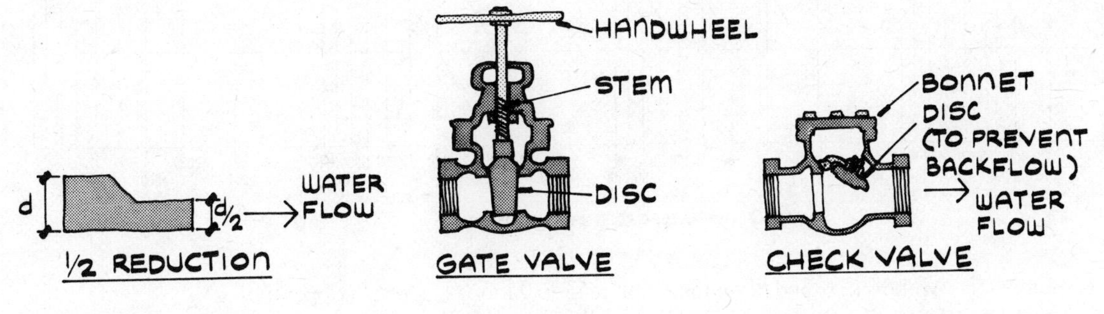

SOURCES

ASHRAE Handbook of Fundamentals, New York: American Society of Heating, Refrigerating and Air-Conditioning Engineers (1972).

Babbitt, H. E., *Plumbing*, New York: McGraw-Hill (1960).

"Piping Design," Part 3 in *System Design Manual*, Syracuse, New York: Carrier Corporation (1960).

FIRE DETECTION AND SUPPRESSION—PRESSURE LOSS GRAPH FOR WATER

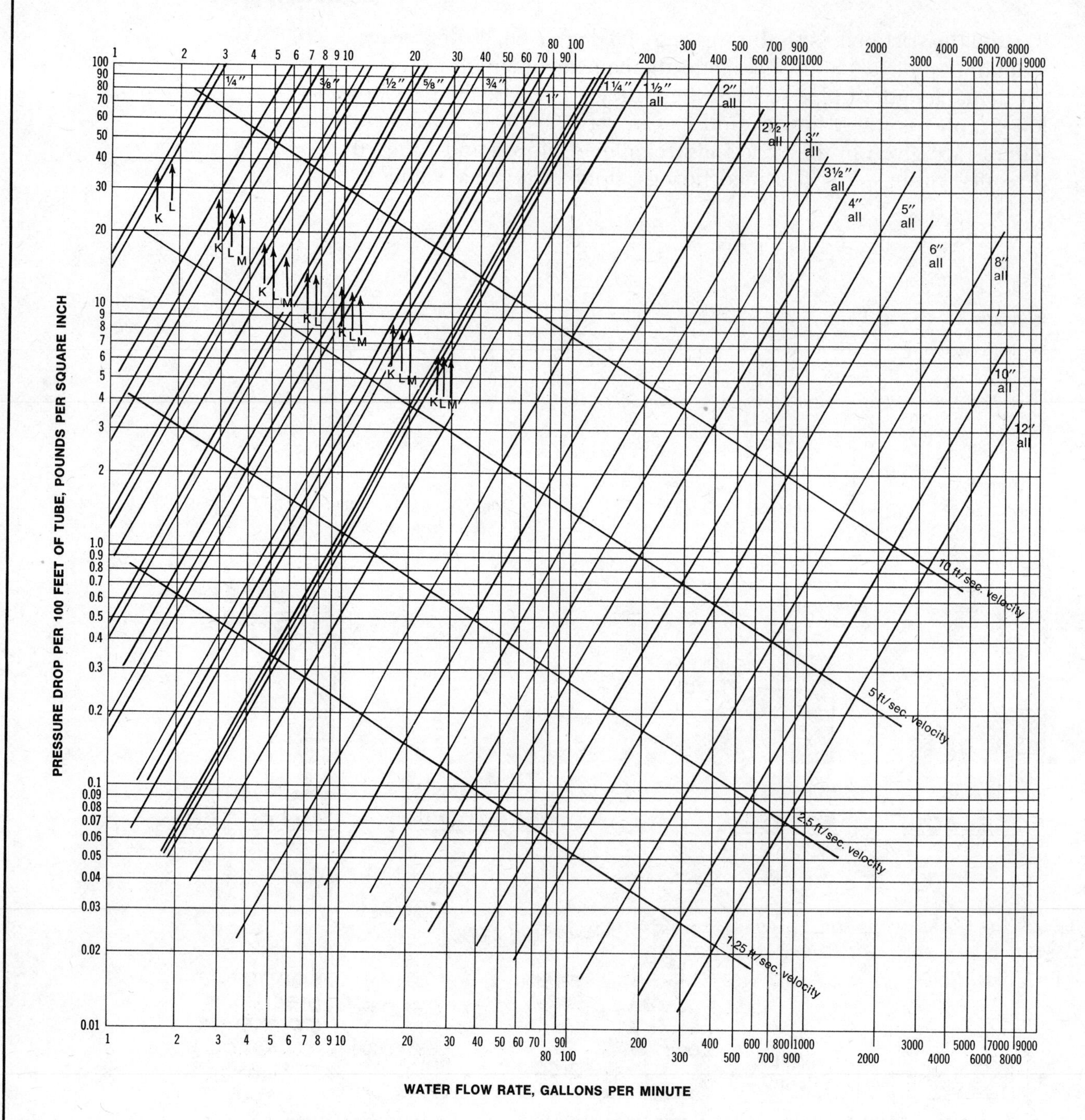

Note: The symbols *K*, *L*, and *M* on the graph refer to various grades of copper tubing according to wall thickness. "All" applies to all grades.

Courtesy of Copper Development Association.

PIPING PRESSURE LOSSES

Sprinkler Layouts

In automatic sprinkler systems, sprinkler heads are usually spaced to cover areas from 225 to 90 sq ft per head, depending on the occupancy and hazard conditions. Individual sprinkler heads are connected together along branch lines. The table on p. 141 gives a schedule of the maximum number of sprinkler heads that may be supplied by various pipe sizes for light, ordinary, and extra-hazard occupancies. In "schedule sizing" layouts, branch pipes extend from cross-main pipes in the same manner branches extend from a tree trunk. In multistory buildings, cross-main pipes in turn connect one or more branch pipes to vertical pipe risers. Use manufacturers' data to determine sprinkler water coverage patterns at recommended discharge water flow rates and pressures.

Sprinkler Operating Pressure

The required sprinkler operating water pressure in pounds per square inch can be determined by manufacturers' graphs or by the formula given below.

$$P = \left(\frac{Q}{K}\right)^2$$

where P = pressure required by sprinkler head in psi
Q = discharge water flow rate in gpm
K = sprinkler head coefficient (no units)

Pressure Losses for Pipe, Valves, and Fittings

The piping pressure loss (P_d) in pounds per square inch can be computed by summing the friction losses in the individual pipe sections making up the longest total length (i.e., actual pipe length from water source to most remote sprinkler head plus equivalent lengths). Friction losses at valves and fittings (e.g., ells, tees) can be determined from tables for equivalent length. Add this additional length in feet to the actual length of pipe in which it occurs. Use the "pressure loss graph for water" to find the pressure drops in psi per 100 ft of pipe at the required water flow rates. Pressure losses also occur due to elevation and can be found by 0.434 h (where h is height of water column in feet).

PIPING PRESSURE LOSSES (Continued)

Available Pressure

The water pressure available to the sprinkler system (P_a) must exceed the system pressure losses. If available pressure from city mains is unsatisfactory, pumps will be required. Use the formula below to evaluate the relationship between available water pressure and system pressure losses. Larger pipe sizes can be used to reduce system pressure losses.

$$P_a \geqslant P_d + 0.434h + P$$

where P_a = available water pressure in psi
P_d = pressure drop through piping (includes fittings and valves) in psi
h = height from water source to sprinkler head in ft
P = pressure required by sprinklers in psi

NUMBER OF SPRINKLERS

The table below gives the maximum number of sprinkler heads that may be supplied by various copper pipe sizes (diameter in inches) for light, ordinary, and extra-hazard occupancies. Sprinkler network pipe sizes can be reduced by using smaller pipe sizes along the cross-main run from the riser to the most remote branch pipe. The procedure of using smaller pipe where less water is to be carried is called "telescope sizing."

	Maximum Number of Sprinklers		
Pipe Diameter (in.)	*Light Hazard*	*Ordinary Hazard*	*Extra Hazard*
1	2	2	1
1¼	3	3	2
1½	5	5	5
2	12	12	8
2½	40	25	20
3	65	45	30
3½	115	75	45
4	—	115	65
5	—	180	100
6	—	300	170

SOURCE

"Installation of Sprinkler Systems," NFPA No. 13, 1975, pp. 28–31.

FIRE DETECTION AND SUPPRESSION—EXAMPLE PROBLEM—SPRINKLER PIPING

Given: A sprinklered office building has a 45 ft vertical pipe riser connected to a centrifugal water pump located 20 ft away. (See schematic piping layout sketch below.) Water pressure (P_a) available from pump will be at least 55 psi. Sprinkler heads are $\frac{1}{2}$ in. standard pendent. Design sprinkler water flow rate (Q) will be 22.5 gpm at the most remote head within the area being considered.

Find: What will be the pressure drop through the copper pipe network to the most remote sprinkler head? There will be two sprinkler heads operating as shown in the layout sketch for this office wet pipe system.

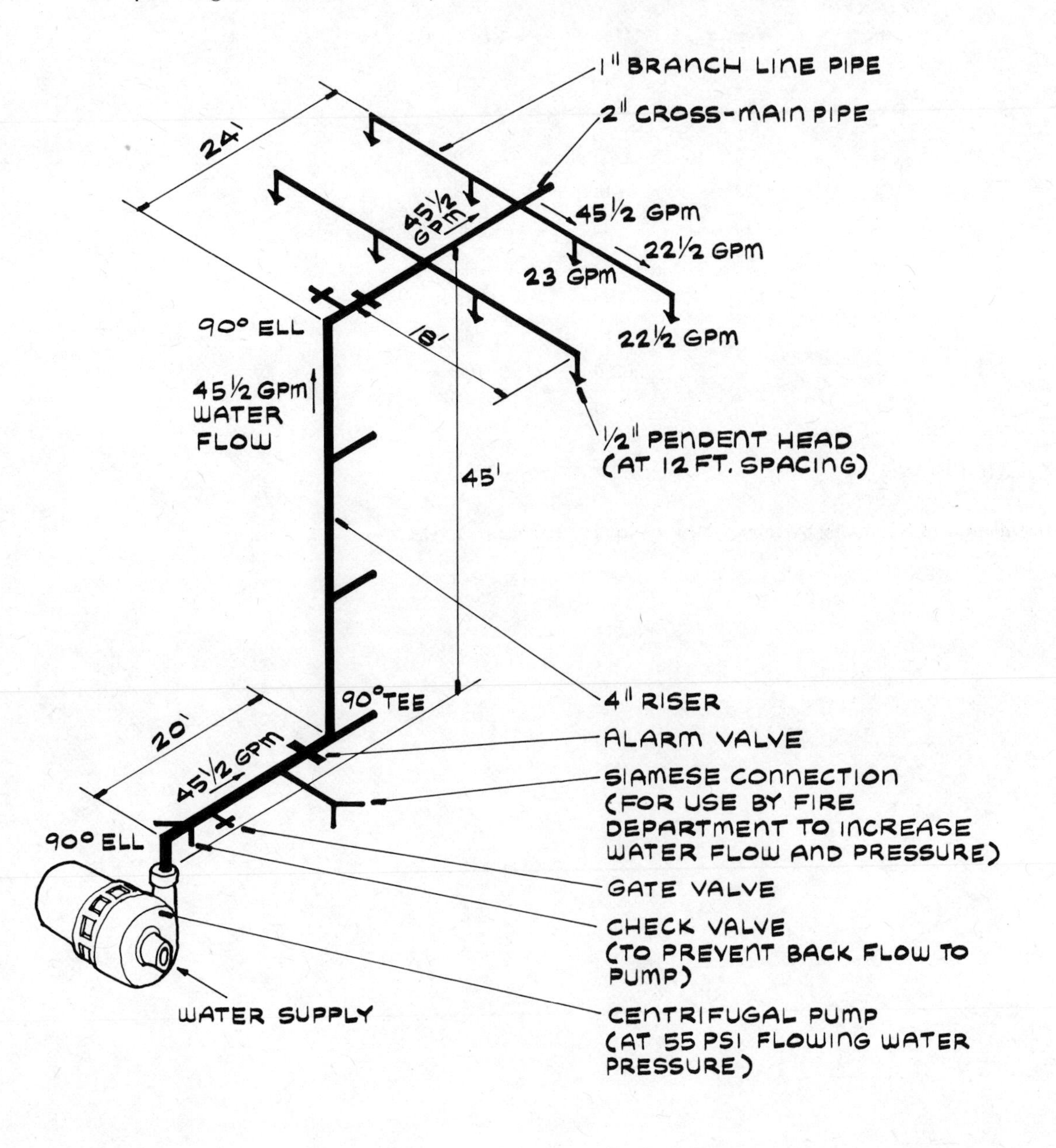

FIRE DETECTION AND SUPPRESSION—EXAMPLE PROBLEM—SPRINKLER PIPING (Continued)

First, find the required water pressure (P) to operate $\frac{1}{2}$ in. pendent sprinkler heads with a K of 5.65 at a water flow rate (Q) of 22.5 gpm.

$$P = \left(\frac{Q}{K}\right)^2 = \left(\frac{22.5}{5.65}\right)^2 \simeq \boxed{16 \text{ psi}}$$

Next, find the allowable pressure drop (P_d) through the pipe, valves, and fittings.

$$P_a \geqslant P_d + 0.434h + P$$
$$55 \geqslant P_d + 0.434(45\text{-}1) + 16$$
$$P_d \leqslant 55 - 35 = \boxed{20 \text{ psi}}$$

Use equivalent length tables to find the equivalent lengths for valves and fittings in the 4 in., 2 in., and 1 in. copper pipe. Find the pressure drops from the pressure loss graph on p. 138. For example, the pressure drop of a 4 in. pipe for a water flow rate of $45\frac{1}{2}$ gpm will be 0.07 psi/100 ft, including equivalent length of valves and fittings (see dashed lines on example problem pressure loss graph).

Finally, since the 7 psi (see "pressure drop table for example problem") total pressure drop through the pipes, valves, and fittings is less than the allowable drop of 20 psi, sufficient pressure will be available to operate the two sprinkler heads.

Note: As additional sprinkler heads open, the water flow will increase causing a corresponding increased pressure drop through the piping network which must not exceed 13 psi (i.e., 20 − 7 = 13 psi). National Fire Protection Association records for sprinklered buildings indicate that over 60% of wet pipe sprinkler systems extinguish fires with one or two sprinkler heads operating and over 80% with five or less heads. New York Board of Fire Underwriters records for fires in sprinklered high-rise office buildings indicate that over 95% of the fires were controlled by one or two sprinklers. Nevertheless, piping analysis should be based on the provision of sufficient water and pressure to all heads that may reasonably be expected to operate within a compartment enclosed by fire-rated constructions.

EQUIVALENT LENGTH TABLE FOR EXAMPLE PROBLEM

Pipe Size	*Valve or Fitting*	*Quantity*	*Equivalent Length Factor (ft)*		*Equivalent Length (ft)*
			Streamlined	*Threaded*	
1 in.	90 degree ell	1	1.5	3	3
	Straight run of tee	1	0.45	0.9	0.9
	90 degree tee	1	2.5	5	5
				Total equivalent length =	8.9 ft
2 in.	Reduction to 1 in. pipe	1	2.5	5	5
	Straight run of tee	1	1	2	2
	Gate valve	1	—	1.3	1.3
	Alarm valve	1	—	20	20
	90 degree ell	1	3.5	7	7
				Total equivalent length =	35.3 ft
4 in.	Reduction to 2 in. pipe	1	5	10	10
	Straight run of tee (includes siamese connection)	3	2	4	12
	Alarm valve	1	—	40	40
	90 degree tee	1	10.5	21	21
	Gate valve	1	—	2.7	2.7
	Check valve	1	—	40	40
	90 degree ell	1	7	14	14
				Total equivalent length =	139.7 ft

PRESSURE DROP TABLE FOR EXAMPLE PROBLEM

Pipe Size	*Length (ft)*	*Equivalent Length (ft)*	*Total Length (ft)*	*Pressure Drop per 100 ft (psi)*	*Pressure Drop (psi)*
1 in.	12 + 1 = 13	3.9	16.9	10	2
	6	5	11	33	3.6
2 in.	24	35.3	59.3	2	1.2
4 in.	45 + 20 = 65	139.7	204.7	0.07	0.1

Total pressure drop = 6.9 psi

FIRE DETECTION AND SUPPRESSION—EXAMPLE PROBLEM—SPRINKLER PIPING (Continued)

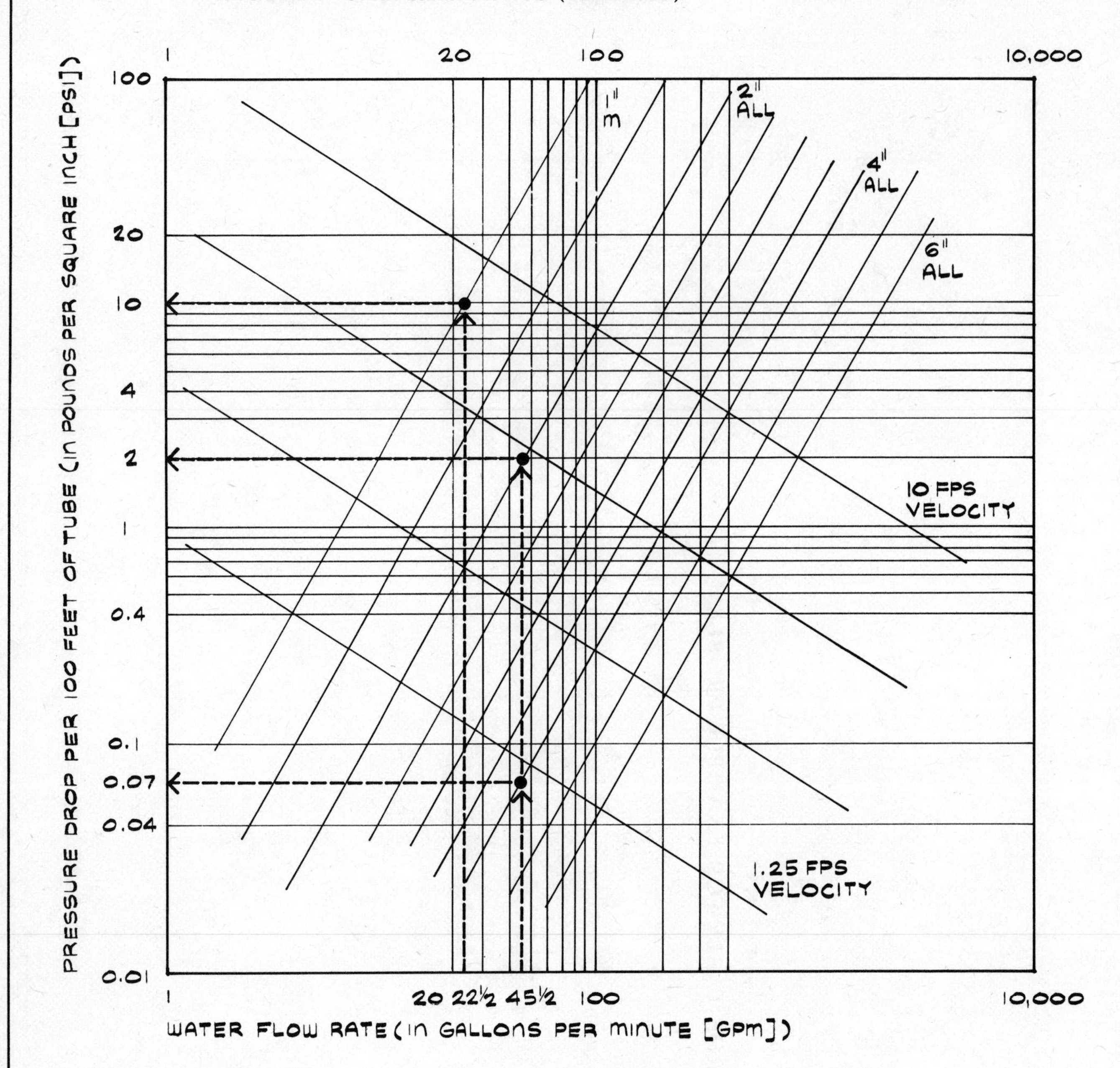

FIRE DETECTION AND SUPPRESSION—SPRINKLER RISER LOCATIONS

Risers should be of sufficient size to supply the sprinklers in areas between fire-rated walls on any one floor. Cross-mains supply water to the branch lines. The "center central" and "side central" riser locations are preferred as they eliminate long dead-end pipe runs.

Center Central

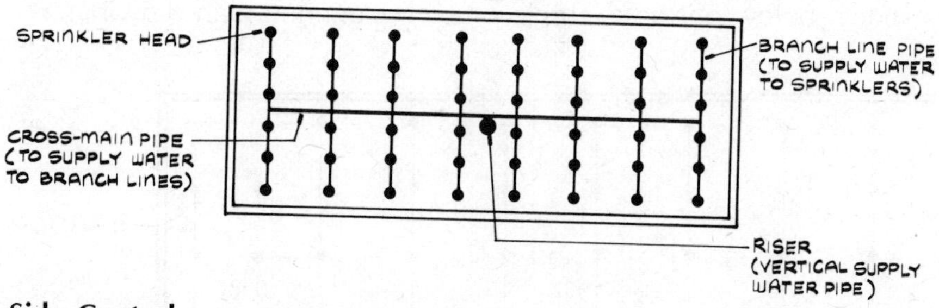

Side Central

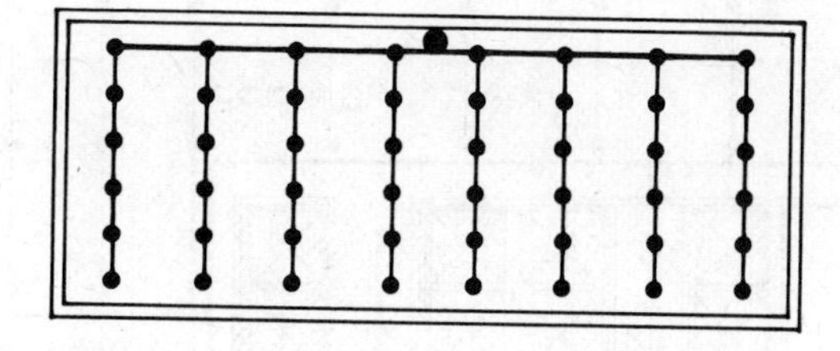

Central End

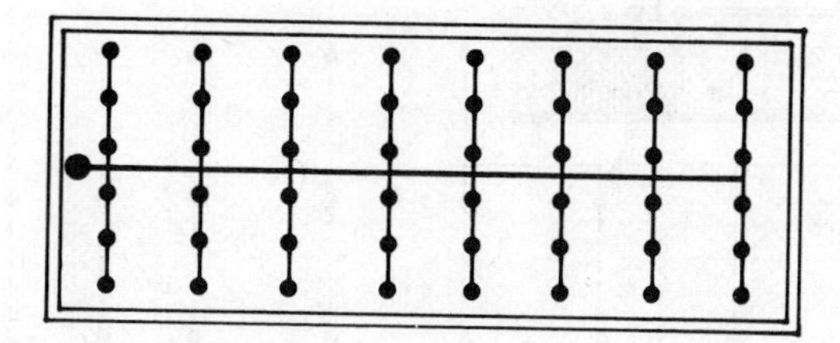

Side End

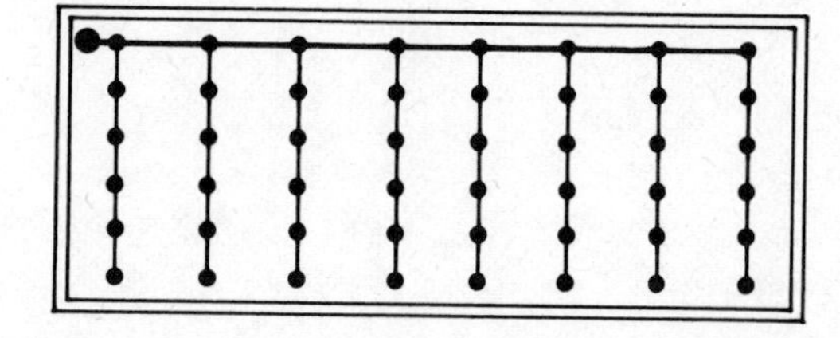

Note: The sprinkler head layout should be integrated within the ceiling system so that coverage requirements will be achieved with the minimum number of sprinkler heads.

FIRE DETECTION AND SUPPRESSION—EXAMPLE LAYOUT OF HYDRAULICALLY DESIGNED SPRINKLER SYSTEM

In hydraulically designed sprinkler systems, the precise amount of water for the area of sprinkler coverage forms the basis for sizing the system's pipes. Hydraulic calculations are used to determine how water is distributed throughout the piping network. When available water pressure is low, however, larger pipe sizes are often used to achieve lower friction losses. Where available water pressure is high, smaller pipe sizes are often used with an interconnected grid or loop of cross-mains as shown below on the example office occupancy plan drawing.

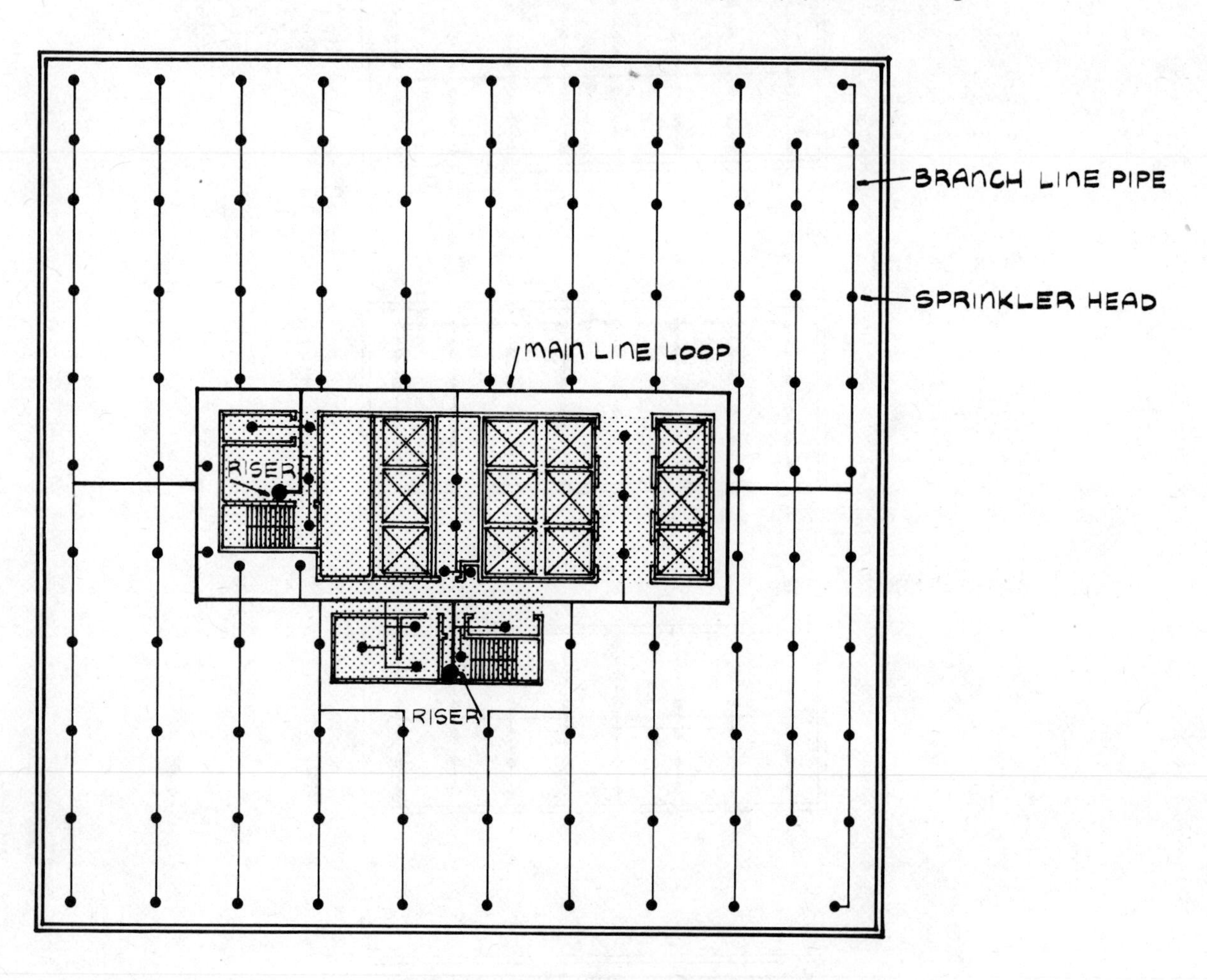

FIRE DETECTION AND SUPPRESSION—SPRINKLER SYSTEMS FOR THEATERS

Fire in theaters, where conditions generally are crowded, could cause the audience to panic. It is vital, therefore, that theater fires be quickly detected and extinguished. Noncombustible proscenium curtains (e.g., flexible wire-woven asbestos, asbestos cloth, or sheet steel on a rigid frame) and sprinklers can be used to separate the stage and audience areas. Sprinkler heads should be protected from accidental impact of moving scenery, battens, and so on by structural members or sprinkler guard wire cage assemblies.

Proscenium Stage

Proscenium curtains and smoke vents located in the stage house can be used to prevent the spread of smoke and hot gases into the auditorium seating areas.

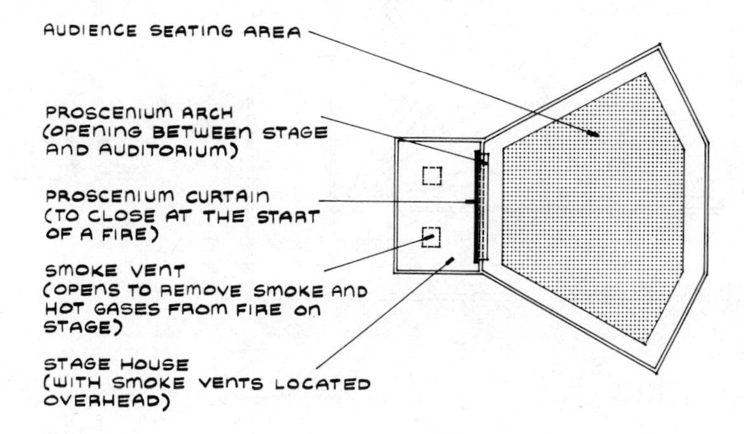

Arena Stage

In the arena stage theater, there is no proscenium arch. Consequently, it is impractical to close off the stage with a noncombustible curtain.

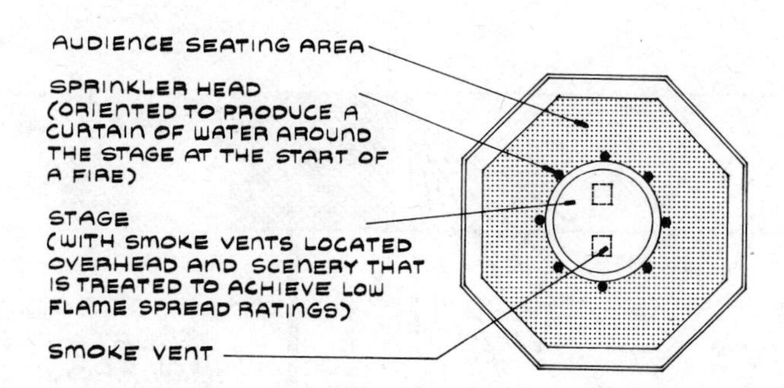

Note: Floor drains can be used to prevent water accumulation near the stage from activated deluge sprinkler systems.

REFERENCE

Thompson, R. D., "A Proposed System for Fire Protection of the Proscenium Arch in the Flexible Theater," *Journal of the American Institute of Architects*, January 1961.

FIRE DETECTION AND SUPPRESSION—WATER SPRAY PROTECTION OF FLOOR OPENINGS

Floor openings (e.g., stairwells, escalators) can be protected by a sprinkler system of high-velocity water spray nozzles activated by fire detectors. Open type nozzles are used to immediately blanket the entire opening with water at the start of a fire. The jet of water impinges on a nozzle deflector to produce the desired spray pattern. An adequate water drainage system also must be provided where these systems are used.

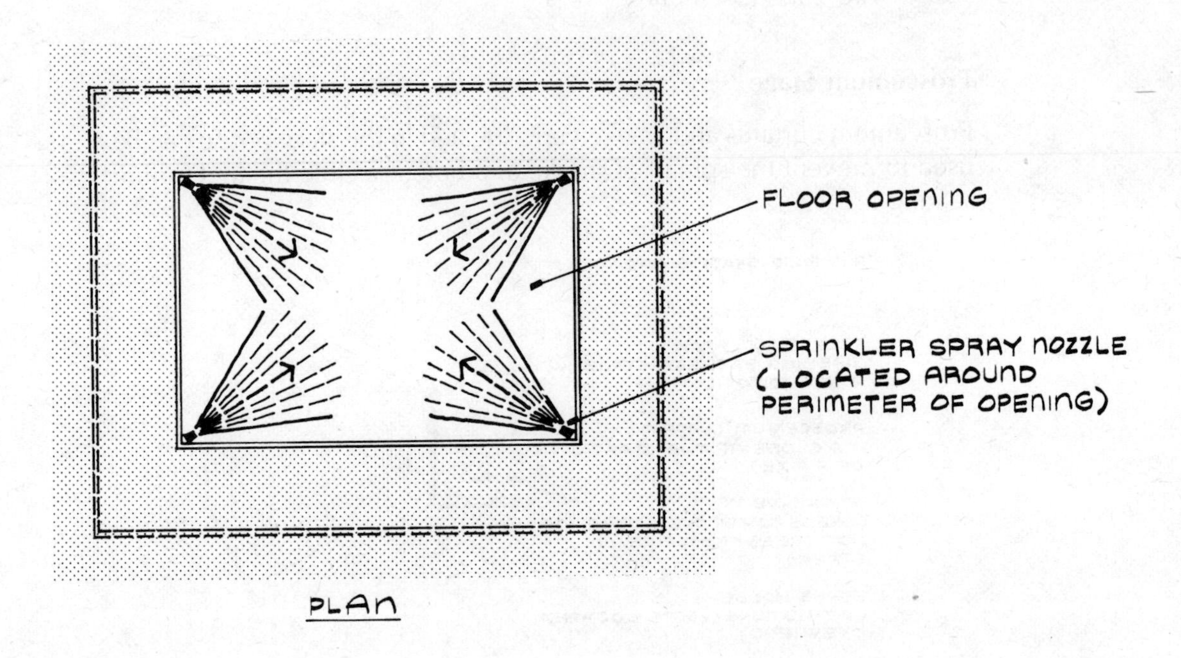

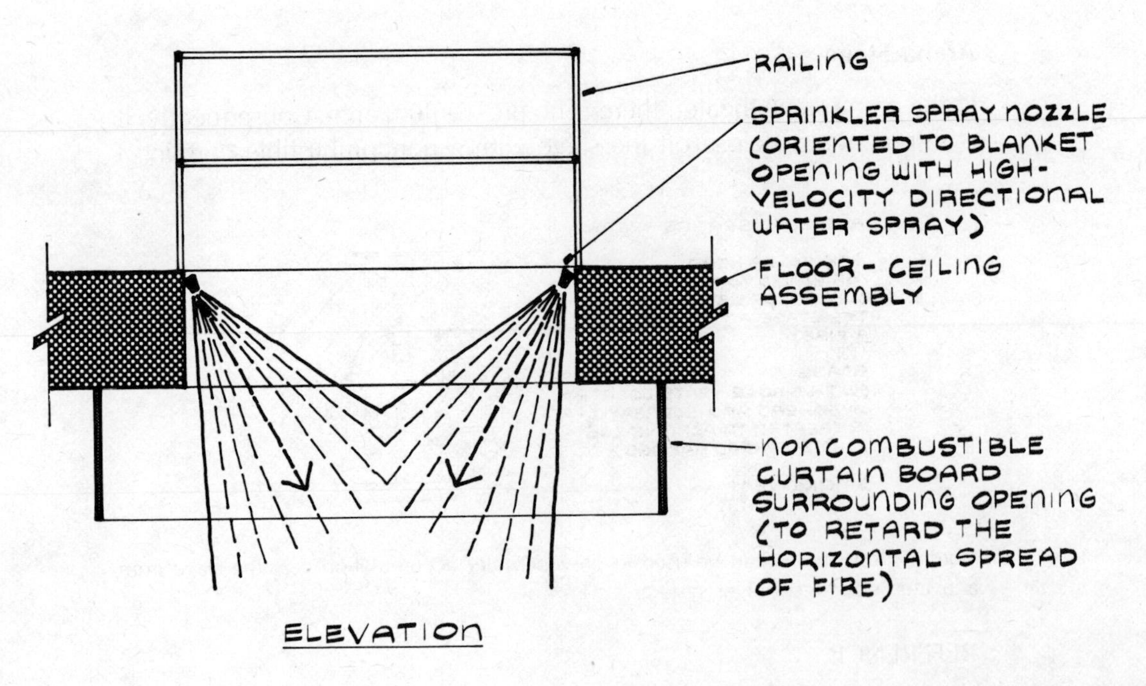

WATER PIPING SYSTEM NOISE CONTROL

Pipe Isolation

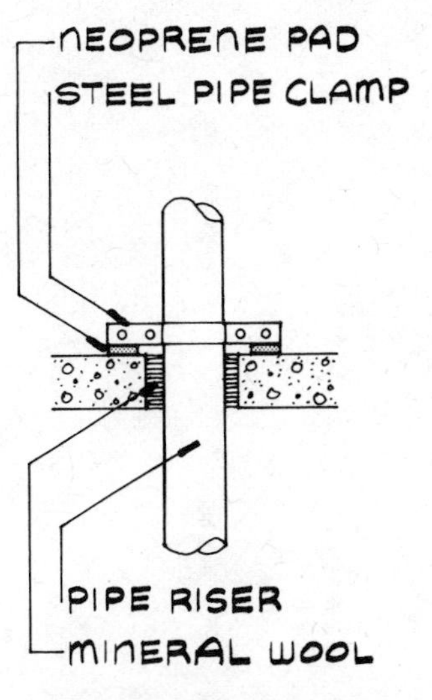

PIPE PENETRATION

Where water pipes are located adjacent to acoustically sensitive areas, the piping runs should be resiliently isolated from the structure to prevent transmission of noise and vibrations. Example sensitive applications would be auditoriums, theaters, conference rooms, private offices, hospitals, apartments, hotels, and libraries. Piping should be supported by resilient ceiling hangers or floor mounted resilient supports (e.g., rubber, neoprene, steel springs, or glass-fiber lined metal clamps). Butyl rubber expansion joints can be used to reduce the noise and vibrations transmitted through the pipe walls. All penetration openings through the building structure should be oversized and the pipe perimeter packed with mineral wool and caulked airtight. Avoid solid anchorage of long vertical pipe risers to floor slabs. Use steel clamps which in turn are supported by springs or neoprene pads as shown in the margin.

Water Flow Velocities

To prevent the generation of water flow noise by turbulence, velocities should be controlled by using large pipe sizes, pressure reducing devices, and similar measures. The table on the following page presents recommended maximum water flow velocities for domestic water systems.

Water Hammer

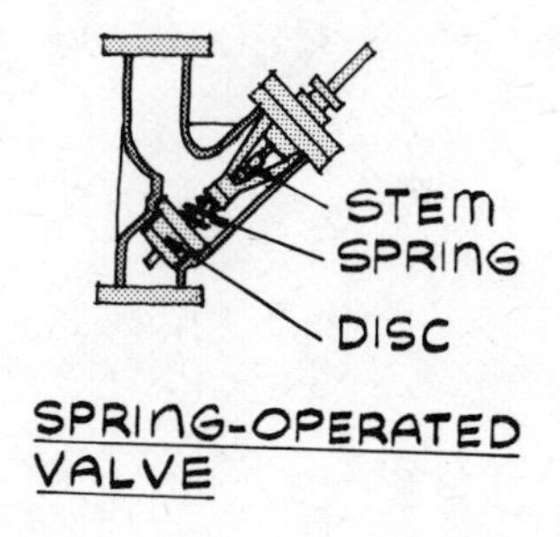

SPRING-OPERATED VALVE

"Water hammer" in piping systems is caused by the rapid closing of a valve which abruptly stops a moving column of water (e.g., by quickly closing a sink faucet). The resulting forward and backward water surge within the piping produces pounding noises called water hammer. Water hammer noise can be prevented by using spring-operated valves (see sketch in margin) to slowly close valve stems. In addition, gas-filled stainless steel bellows units can be installed to absorb the surge of water at abrupt pipe turns.

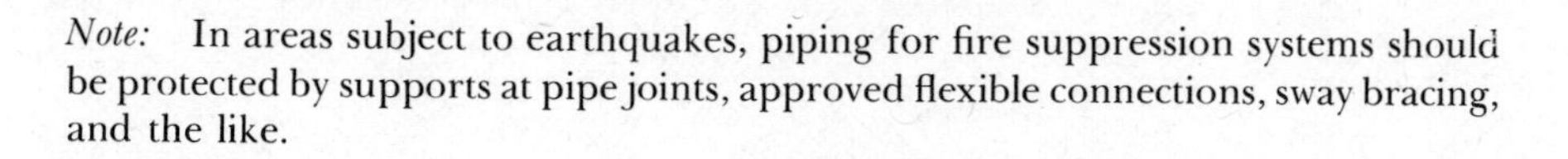
Note: In areas subject to earthquakes, piping for fire suppression systems should be protected by supports at pipe joints, approved flexible connections, sway bracing, and the like.

REFERENCES

Bell, L. H., *Fundamentals of Industrial Noise Control,* Trumbull, Conn.: Harmony Publications (1973).

Rettinger, M., *Acoustic Design and Noise Control,* New York: Chemical Publishing Co. (1973), pp. 339–345.

SUGGESTED WATER FLOW VELOCITIES

The table below shows maximum water flow velocities in feet per second (fps) that should not be exceeded by water piping systems adjacent to acoustically sensitive spaces.

Pipe Diameter (in.)	*Maximum Flow Velocity (fps)*
1	3
1½	4
2	4.5
2½	5
3	6
3½	6.5
4	7
5	8
6	8.5
8	10
10	12

To meet stringent acoustical criteria, pipes carrying water at high flow velocities can be wrapped with a layer of glass-fiber or mineral wool inside a cover of sheet lead, dense vinyl, or sheet metal.

Example

A 4 in. riser in a shaft near an acoustically sensitive area has a water flow rate (Q) of 250 gpm. What will be the flow velocity (v) in fps? Is this flow velocity satisfactory for routine acoustical considerations?

$$v = 0.4\frac{Q}{d^2} = 0.4\frac{250}{(4)^2} = \boxed{6.25 \text{ fps}}$$

The water flow velocity is less than 7 fps (see table above), so water flow requirements are satisfied.

FIRE DETECTION AND SUPPRESSION—VIBRATION ISOLATION OF PUMPS

Water pumps for fire protection normally operate for brief test periods on a weekly basis and for fire suppression operations. Pumps should have standby power for use in blackouts or local electrical power failures. An example vibration-isolated water pump is shown below.

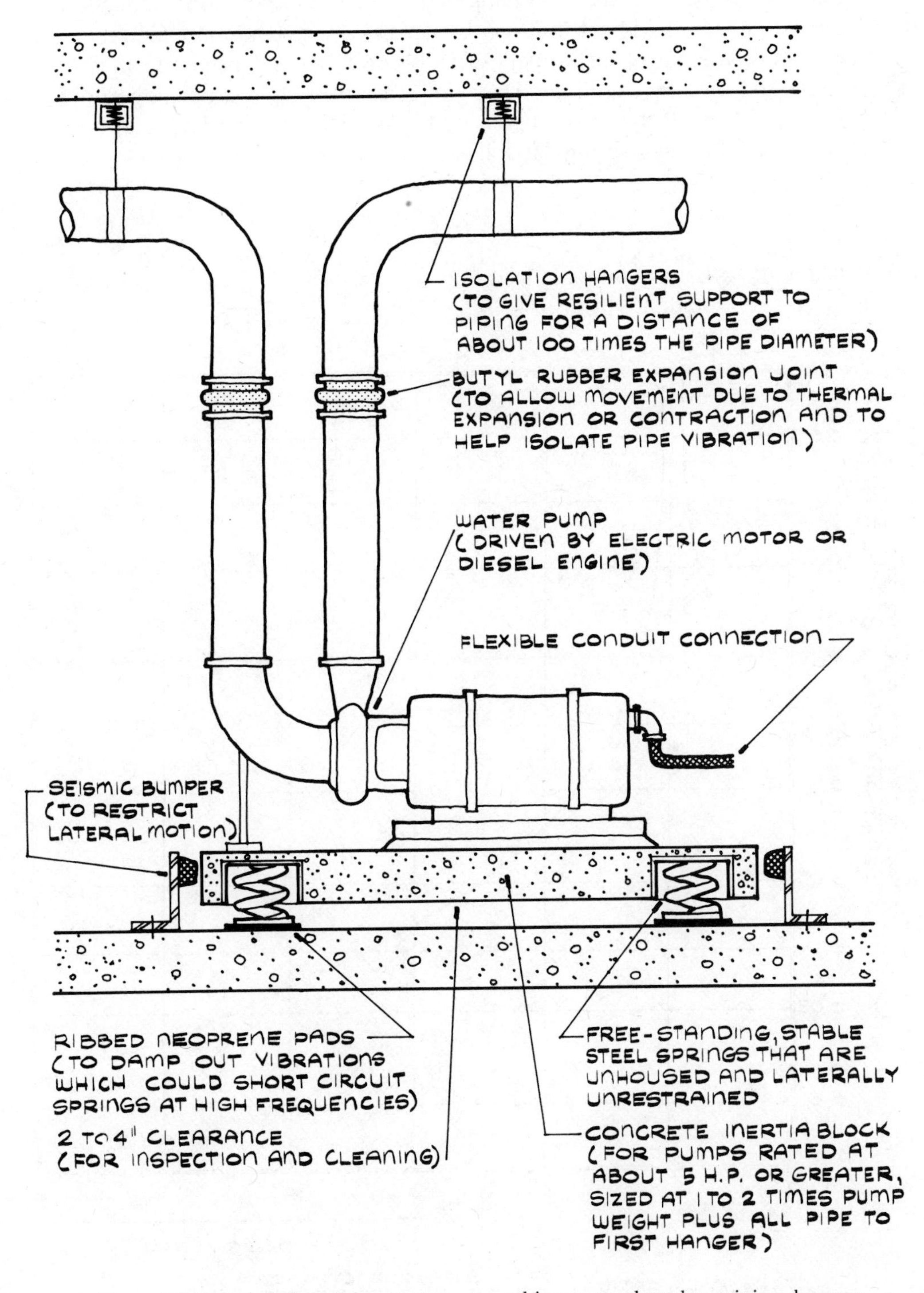

Note: Where sprinkler piping systems are subject to earthquakes, piping damage due to building movement can be lessened by using approved flexible connections.

The principle of vibration isolation involves supporting the vibrating equipment by resilient materials such as ribbed neoprene pads, precompressed glass-fiber pads, and steel springs. The goal is to choose a proper resilient material that, when loaded, will provide a "natural frequency" that is three or more times lower than the "driving frequency" of the equipment. Driving frequency is an operational characteristic of the equipment that can be obtained from the manufacturer. Natural frequency is the frequency of vibration that occurs when a mass supported by a resilient material is deflected from its rest position and released. Natural frequency (f_n) in hertz (or cycles per second) can be calculated for steel springs as follows:

$$f_n = 3.13\sqrt{1/y}$$

where f_n = natural frequency in Hz
y = static deflection in in.

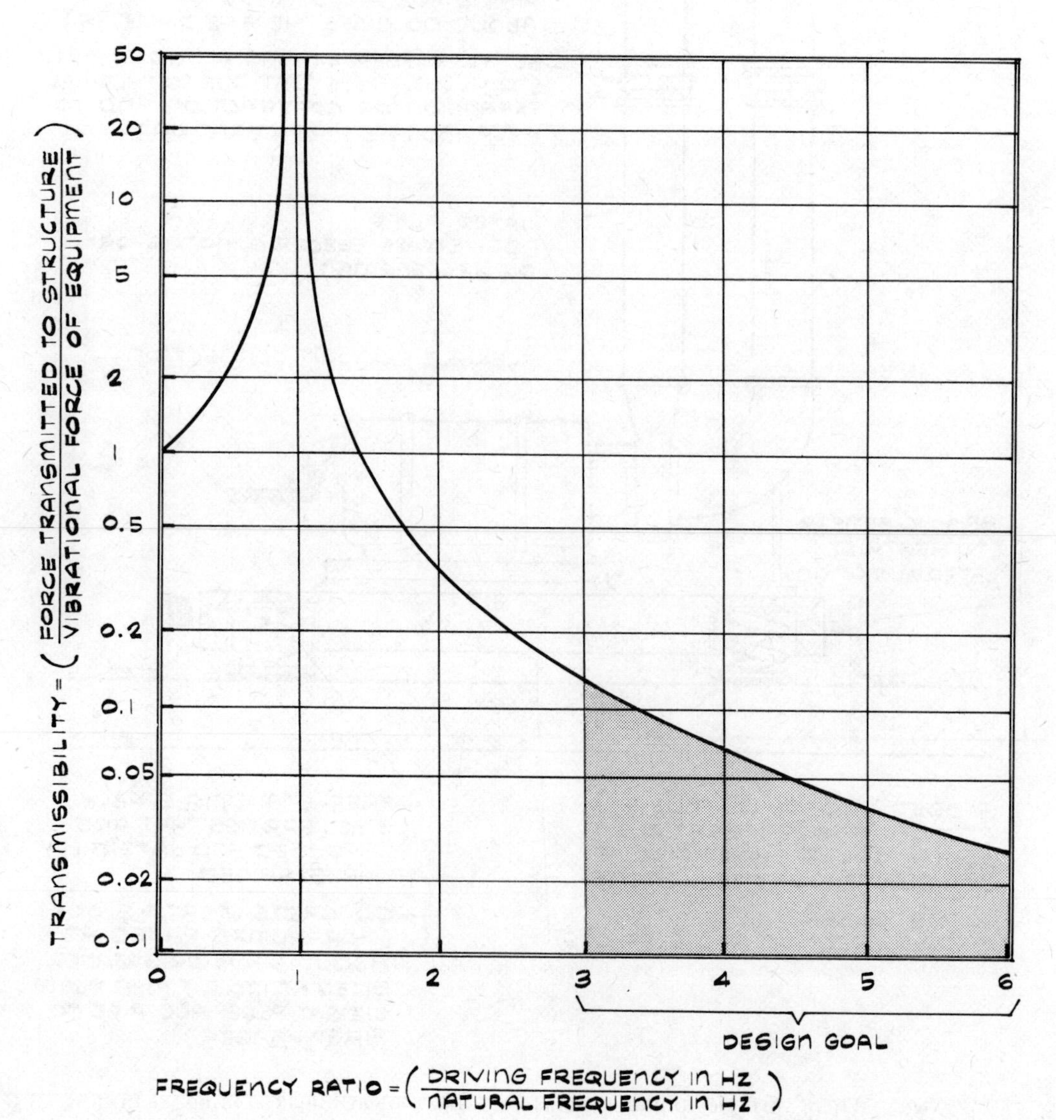

FIRE DETECTION AND SUPPRESSION—NATURAL FREQUENCY AND DEFLECTION

Curves below show natural frequency (f_n) for free-standing steel springs, precompressed glass-fiber, rubber, and cork. The most resilient isolators are springs because they have the largest deflections. Static deflection is the distance an isolator will compress (or deflect) when weight is applied to it.

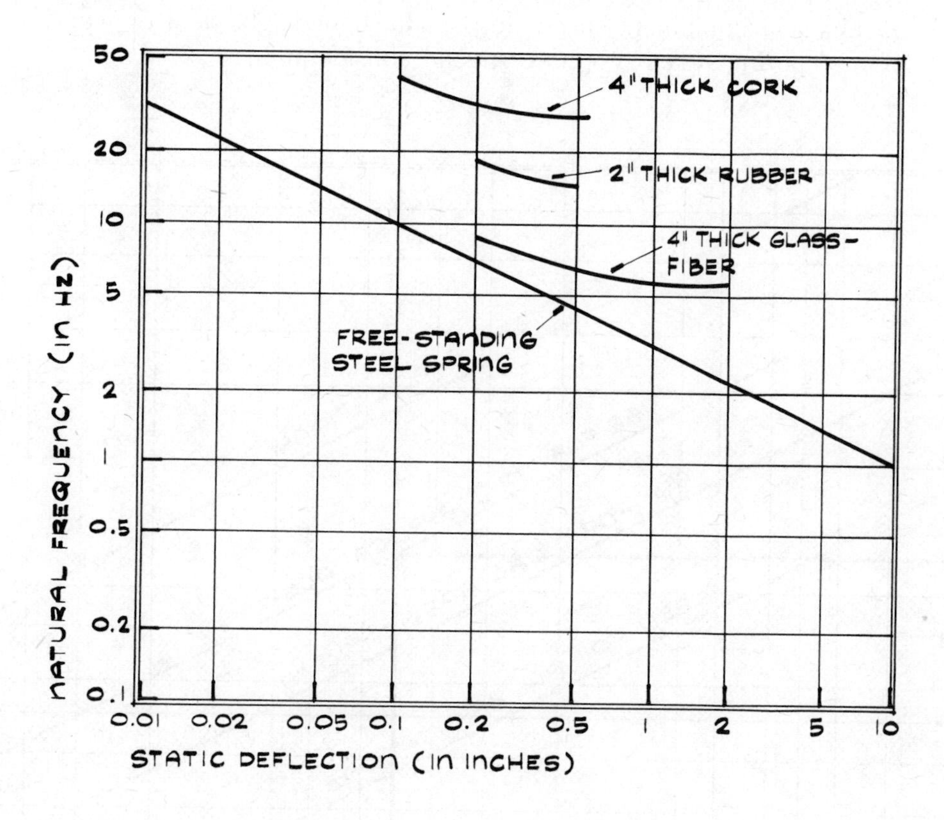

FIRE DETECTION AND SUPPRESSION—VIBRATION ISOLATION DESIGN GRAPH

The graph below can be used to find the static deflection required for vibrating equipment isolators. For example, a motor that drives a pump at 1200 rpm will require resilient isolators with a static deflection of at least $\frac{1}{2}$ in. (See dashed lines on graph.) The graph is based on the assumption that the resilient materials in turn will be supported by a rigid base. Nonrigid, lightweight base supports such as above-grade mechanical equipment spaces (especially lightweight steel or wood-frame flooring systems) require special consideration.

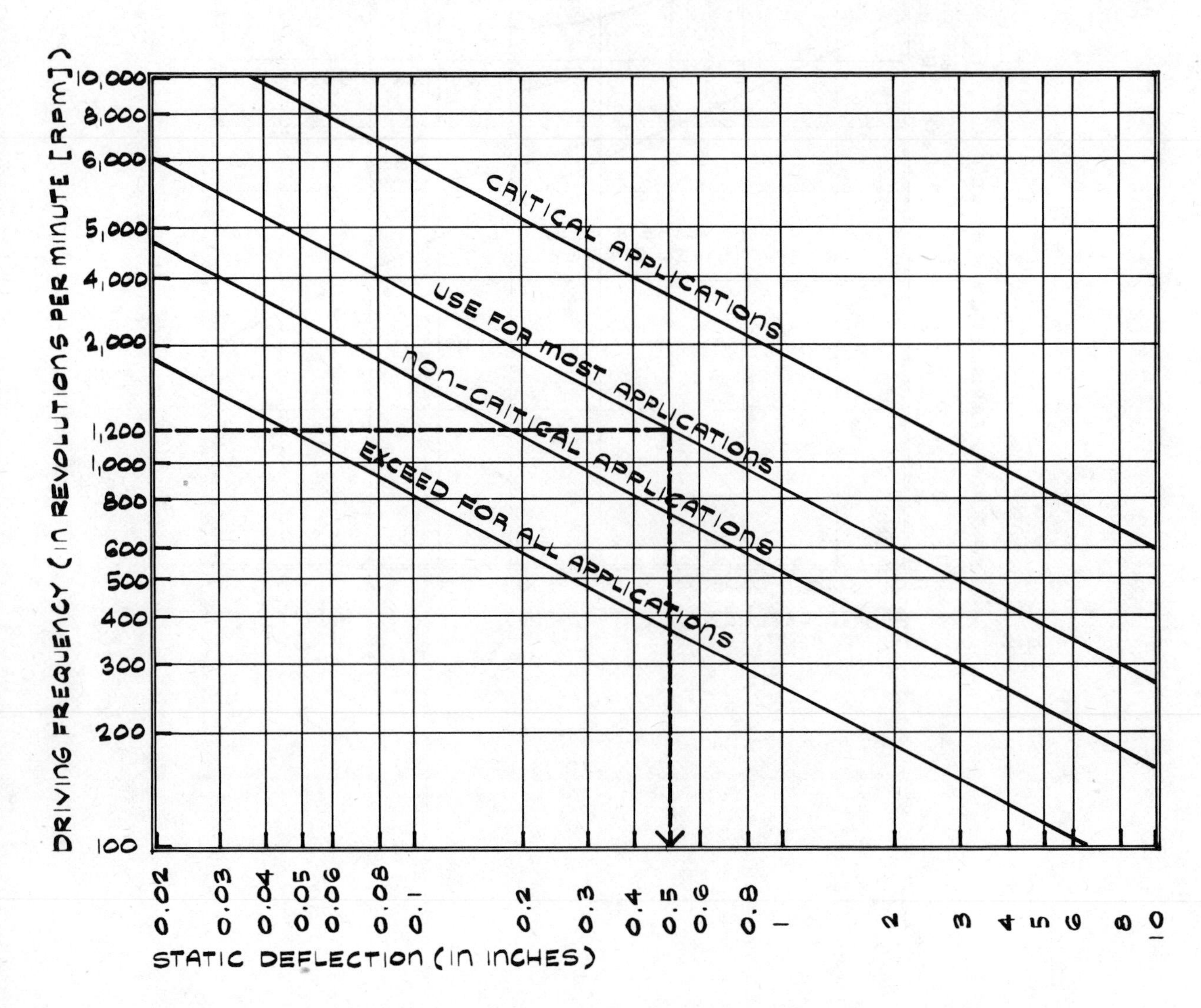

Note: For a discussion of noise and vibrations, see M. D. Egan, *Concepts in Architectural Acoustics*, New York: McGraw-Hill (1972), pp. 121–127.

FIRE DETECTION AND SUPPRESSION—RUBBER MOUNTS

The graph below shows the static deflection under load for typical rubber mounts of various hardness measured by durometer. Durometer is expressed by a scale of 0 (soft) to 100 (hard) from surface indentation measurements. Rubber mounts are often used to vibration isolate relatively small mechanical equipment operating at high speeds. For example, a small engine weighing 500 lb, with a driving frequency of 2000 rpm, will require a static deflection of 0.20 in. for "most applications." (See graph on preceding page.) If the engine is to be supported by four corner mounts, a mount whose static deflection is 0.20 in. for a load of 500/4 = 125 lb will be required. Use a rubber mount with a durometer of 30. (See dashed lines on graph below.)

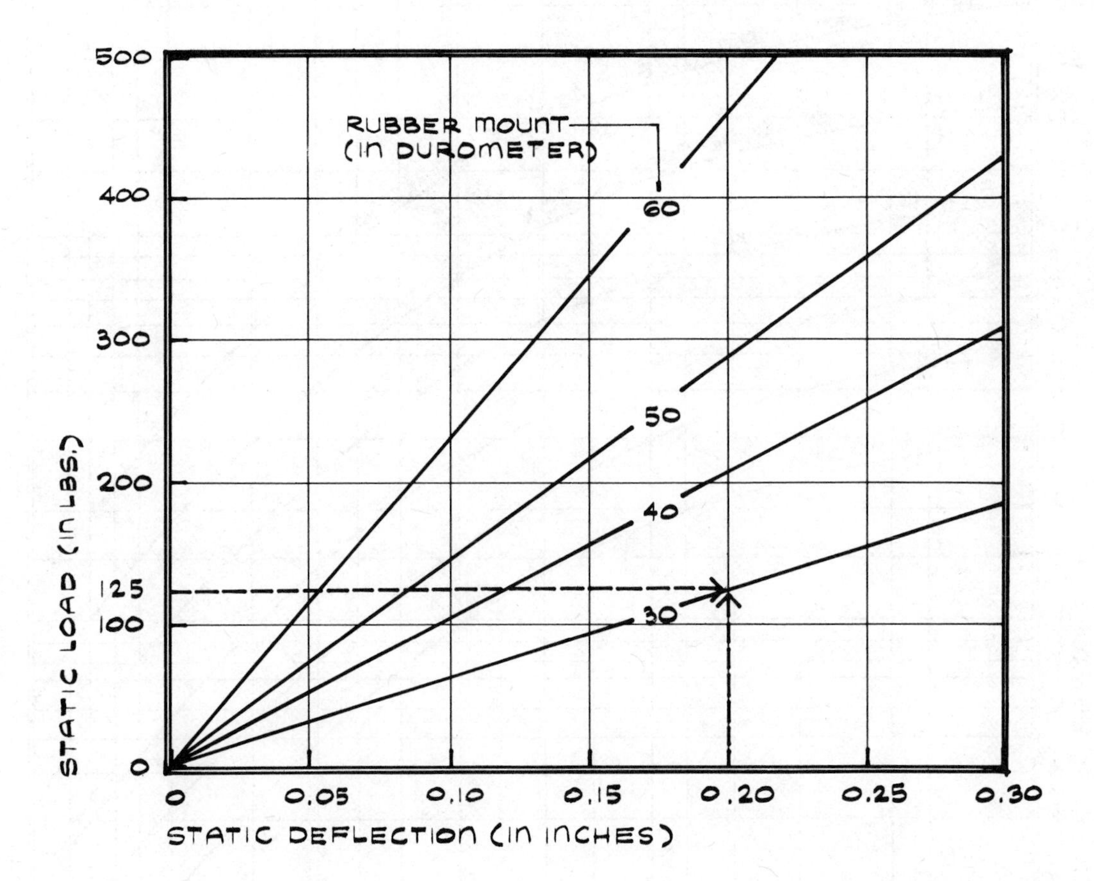

FIRE DETECTION AND SUPPRESSION—VIBRATION ANALYSIS GRAPH

Typical vibration displacements in mils under normal operating conditions are shown by the graph. One mil is equal to 1/1000 of an inch. During start-up and shut-down operations, however, much greater displacements will normally occur.

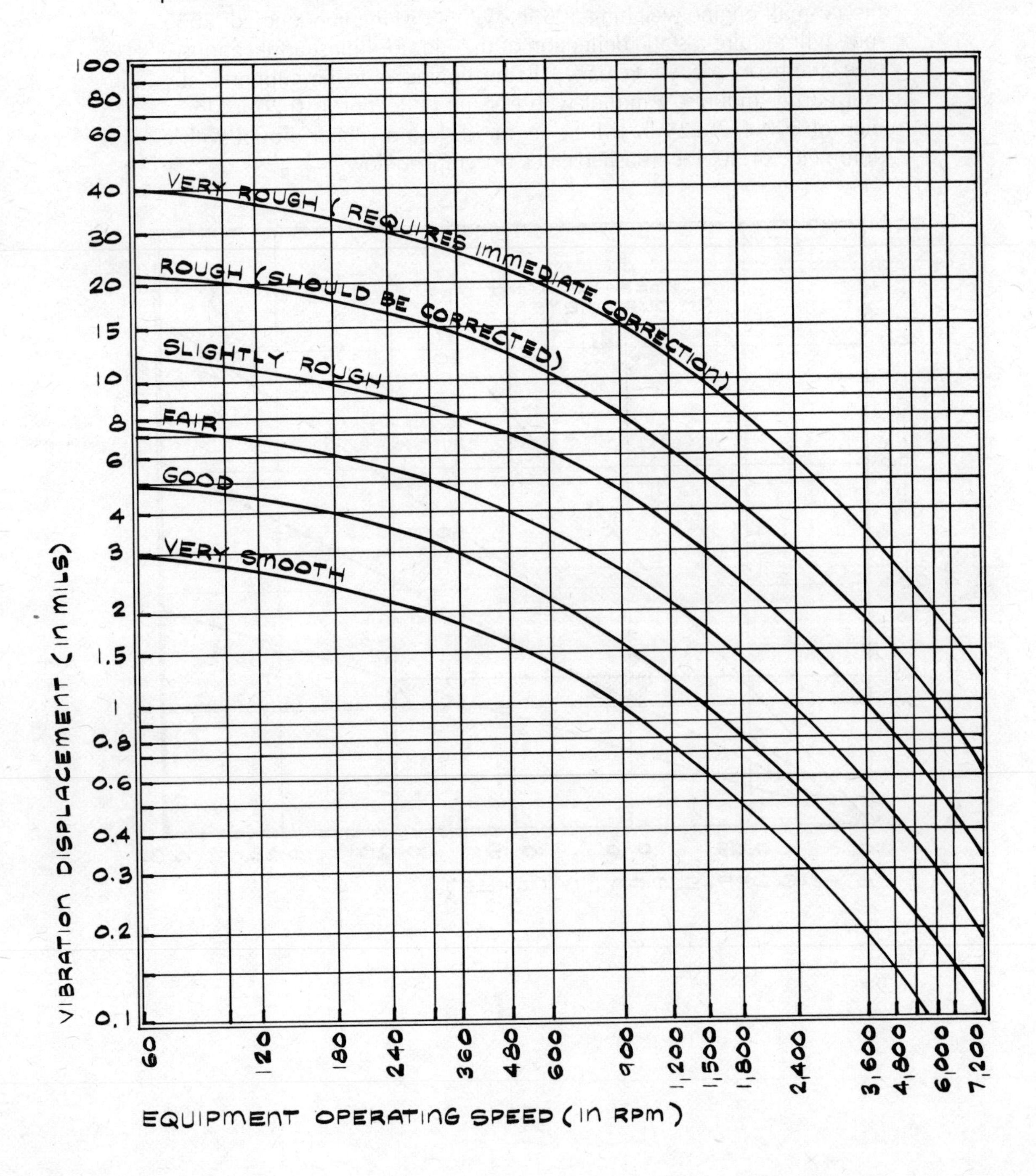

FIRE DETECTION AND SUPPRESSION—CARBON DIOXIDE SYSTEMS

Carbon dioxide (CO_2) does not leave a residue after its use, is nontoxic, and usually causes no damage to electrical equipment. Under normal conditions, carbon dioxide is about 50% heavier than air. Example CO_2 system applications include computer equipment rooms, electronic installations, food preparation areas, fur vaults, film vaults, and record documents storage rooms. Shown below is an example "local application" system in which carbon dioxide is discharged directly on the burning materials (e.g., deep fat fryer in food preparation area). This extinguishing method is similar to blowing out a match flame by human breath (about 4% carbon dioxide).

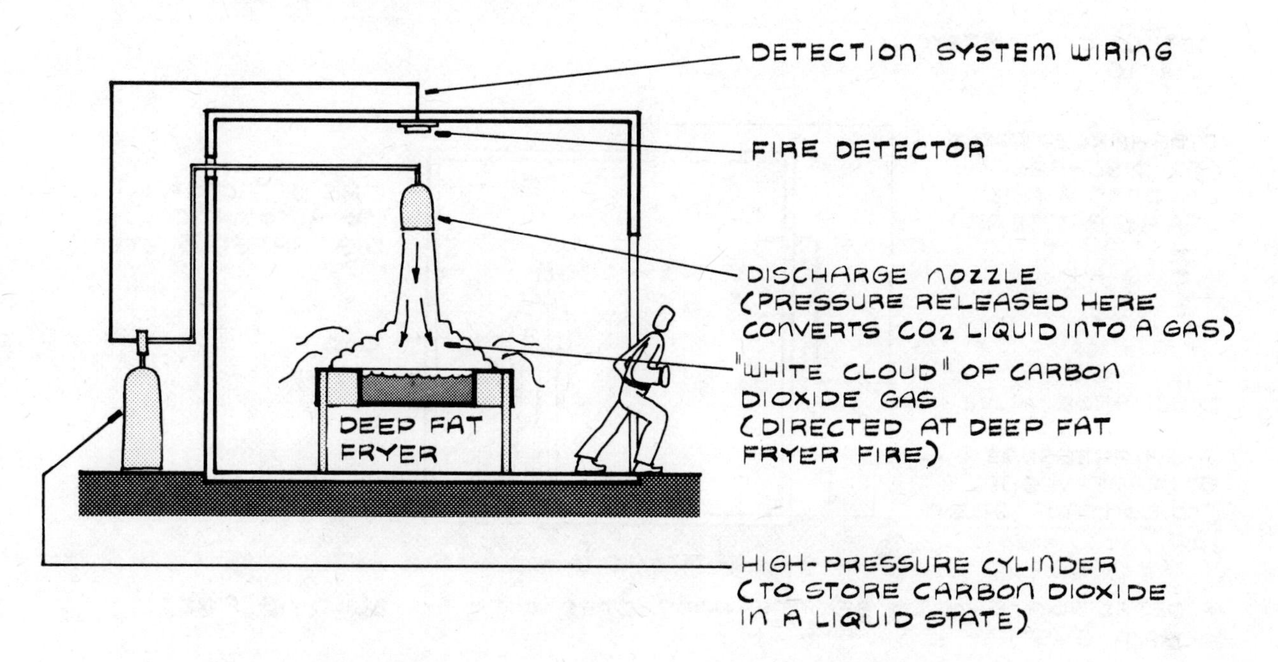

Note: "Total flooding" systems (where CO_2 is discharged uniformly throughout a room) must not be used where building occupants will be unable to evacuate prior to system discharge. Typical concentrations of 30 to 60% carbon dioxide for total flooding systems are a threat to human life. Consequently, the fire detection system should activate a predischarge alarm warning and discharge the system after a delay.

REFERENCE

Bryan, J. L., *Fire Suppression and Detection Systems,* Beverly Hills, Calif.: Glencoe Press (1974).

FIRE DETECTION AND SUPPRESSION—HALON SYSTEMS

Halon (i.e., halogenated hydrocarbon) does not wet or leave a residue after its use. Under normal conditions, halon is about five times heavier than air. Example halon system applications, where rapid fire suppression is required, include bank vaults, computer equipment rooms, electronic installations, fur vaults, libraries, and museums. Shown below is a "total flooding" system in which halon is discharged to create a uniform, low concentration throughout the room. For effective total flooding, doors to the fire area should automatically close and the air distribution system should shut down. Where high halon concentrations are used (greater than 10%), self-contained breathing units must be used.

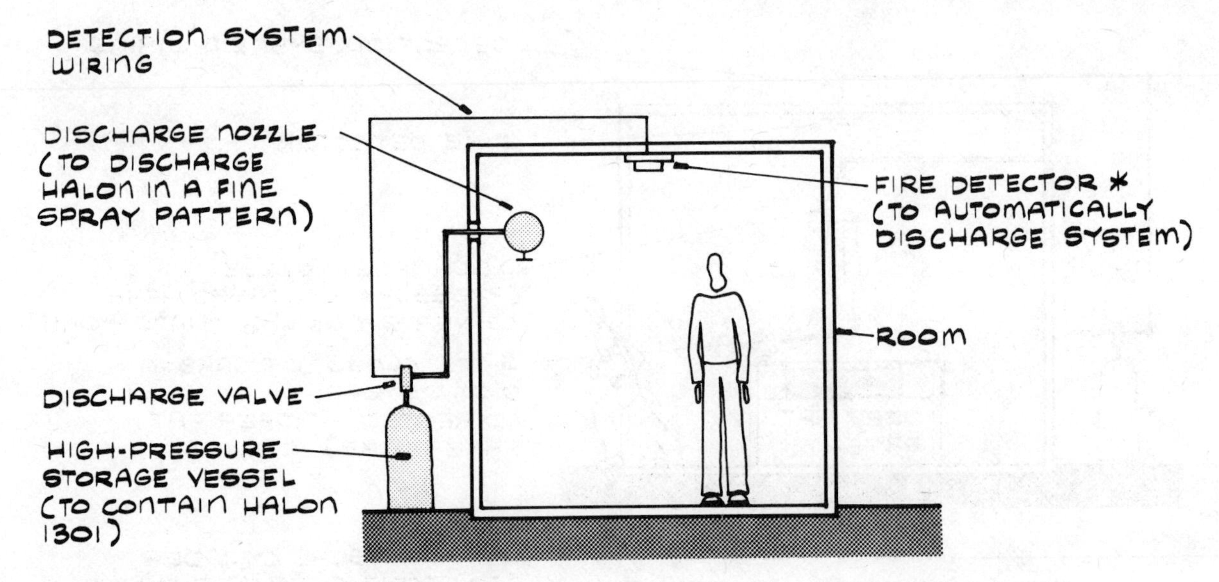

* DETECTION SHOULD BE INTERCONNECTED WITH THE BUILDING FIRE ALARM SYSTEM.

REFERENCES

An Appraisal of Halogenated Fire Extinguishing Agents, National Academy of Sciences, Washington, D.C., 1972.

Ford, C. L., "Halon 1301 Systems," in R. Jensen (ed.), *Fire Protection for the Design Professional*, Boston, Mass.: Cahners Books (1975).

STANDPIPE AND SPRINKLER SYSTEMS CHECKLIST

Base standpipe and sprinkler system water flow computations on the water that actually will be available to suppress fires. For example, public water supplies frequently decrease due to aging water mains or greater demand from increased numbers of system users, and the maximum use also varies throughout the day. Dependable water supplies can be provided in buildings by gravity tanks, pressure tanks, or fire pumps with suction tanks or reservoirs.

Standpipes used to supply fire hoses in buildings and to supply sprinkler systems must be sized for the water flow and pressure demand of both systems. Separate water control valves at each floor level also should be provided.

Sprinkler systems can operate as a heat detector, provide water flow alarm to alert building occupants and the fire department, and extinguish or retard fire growth by discharging water on the fire. There are four basic sprinkler system types: dry pipe, wet pipe, deluge, and preaction. The type used should be the system best suited for the fire hazard and occupancy.

Sprinkler head layouts should be integrated with the ceiling system so that coverage requirements will be achieved with the minimum number of heads. Avoid designs that do not consider structural and mechanical systems. This kind of approach often results in the need to relocate piping runs during sprinkler installation to avoid beams, joists, ducts, and other obstructions. Consideration also should be given during the design stages of a project to the possible future alterations in building space use.

Openings through floors, conveyor openings in walls, and shafts can be protected by sprinklers. Spray nozzles should be oriented to blanket openings with water during a fire. Sprinklers also can be installed at fire door openings to help achieve greater fire resistance.

Use dry-extinguishing systems such as carbon dioxide or halon for supplementary protection of building areas where equipment or materials could be damaged by water or for special fire hazards such as flammable liquids. Nevertheless, sprinklers used to protect computer equipment operations can limit a fire to individual units and can protect the computer room from fires in adjacent areas. For extinguishing flammable liquid fires, foam, carbon dioxide, or dry chemical systems can be used.

SMOKE VENTING

SMOKE VENTING—INTRODUCTION TO VENTING

Venting is the removal of smoke, toxic gases, and heat from buildings to provide tolerable conditions close to the fire.

Unvented Building

As shown by the example fire in the unvented single-story building below, smoke reduces visibility preventing effective rescue and fire-fighting operations. In addition, heat build-up can cause serious damage to the building structure.

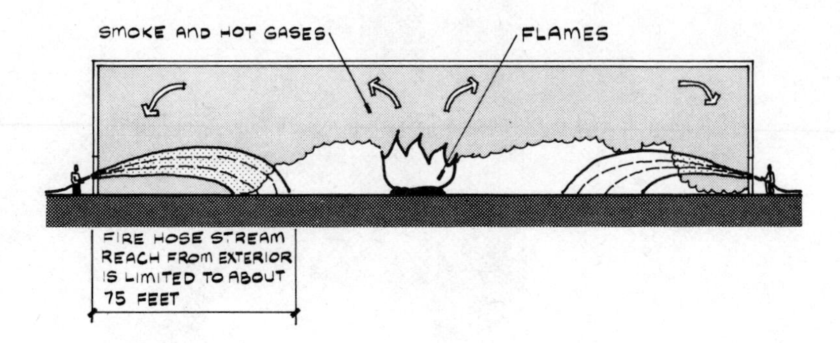

Vented Building

In the fire in the vented building shown below, smoke and heat are removed allowing fire fighters to enter the building and to suppress the fire without excessive use of water.

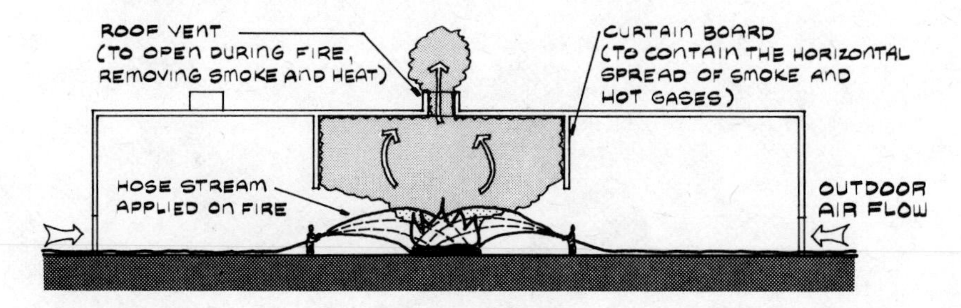

Note: Vents can also provide protection against tornadoes, hurricane force winds, and other extreme wind conditions by diminishing the air pressure difference across the building's exterior.

SMOKE VENTING—ROOF VENTING

Openings in roofs can be used to remove smoke and hot gases so that exit routes will not become blocked and fire fighters can locate the fire. Although venting generally increases fire severity, it provides better conditions for effective fire fighting. For example, smoldering fires produce little or no flames to indicate their location in the building.

The required area of vent (A_v) in square feet for the roof shapes shown below can be found by:

$$A_v \simeq 0.14\,P\,h\,\sqrt{h/d}$$

where P = perimeter of fire (estimated from location and amount of combustibles) in ft

h = distance between bottom of curtain board and floor in ft

d = distance between vent and bottom of curtain board in ft

Gable Roof

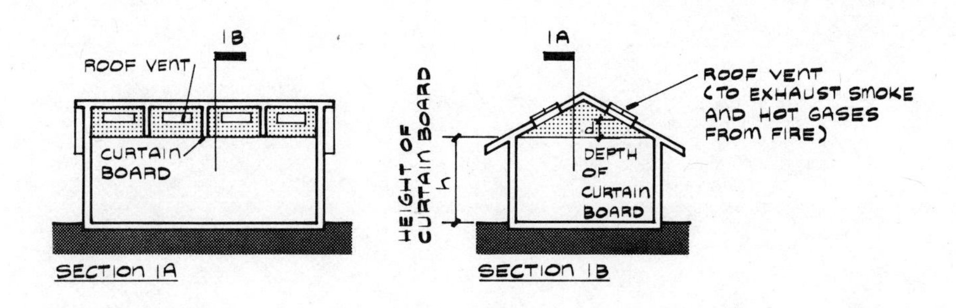

Butterfly Roof

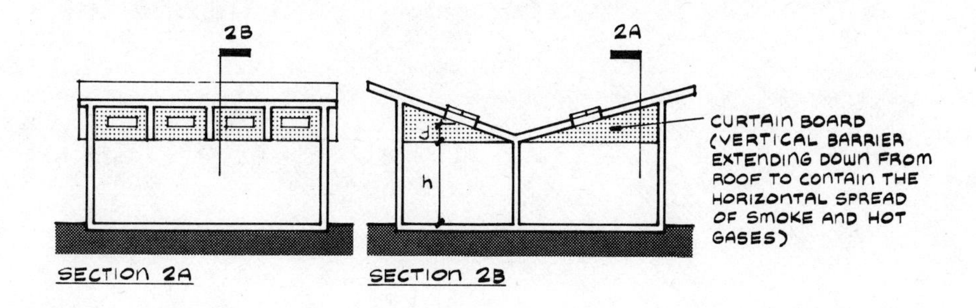

REFERENCE

Lie, T. T., *Fire and Buildings,* London, England: Applied Science Publishers, Ltd. (1972).

EXAMPLE PROBLEM—ROOF VENTING

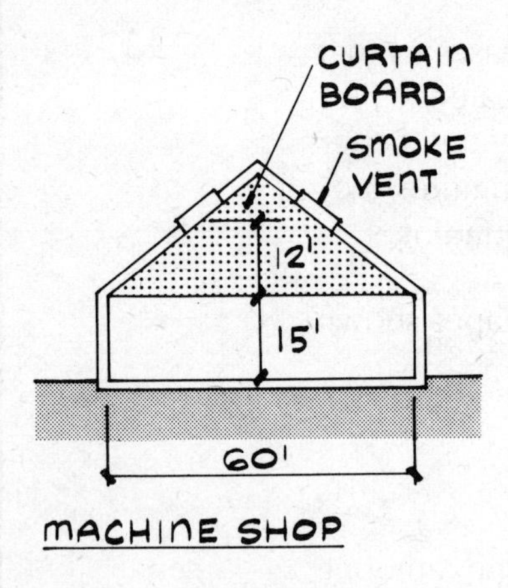

Given: A single-story machine shop is divided into compartments by curtain boards spaced 175 ft apart. The width of the shop is 60 ft. Distance between the bottom of the curtain board and the floor (h) is 15 ft. Depth of the curtain board (d) below the roof vents is 12 ft. Combustible materials in the shop are distributed so that a 35 ft × 40 ft fire area should be the largest extent of fire spread between the curtain boards.

Find: What will be the required area of vents (A_v) in square feet between curtain boards to prevent the spread of smoke and hot gases beyond this compartment?

First, find the anticipated perimeter of fire (P) in feet.

$$P = (2 \times 35) + (2 \times 40) = \boxed{150 \text{ ft.}}$$

Next, find the required area of vents.

$$A_v \simeq 0.14\, P h \sqrt{h/d}$$
$$A_v \simeq 0.14 \times 150 \times 15 \sqrt{15/12} \simeq \boxed{352 \text{ sq ft}}$$

SMOKE VENTING—SMOKE VENTING FOR CORRIDORS

Access to exits must be free of smoke and gases. During the smoldering phase of fires, corridors should be pressurized or kept free of smoke by wall vents located along the perimeter of the building. The sketches below show example corridor venting by wall vents.

Wall Vents Closed

Fires in small rooms can fill corridors with toxic gases and smoke within a few minutes after ignition (e.g., less than 4 minutes in the Henry Grady Hotel fire tests).

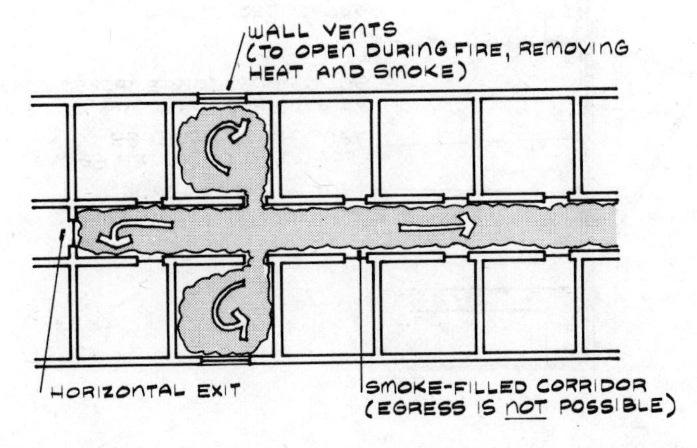

Wall Vents Open

In less than 1 minute after opening wall vents, corridors can be cleared of smoke.

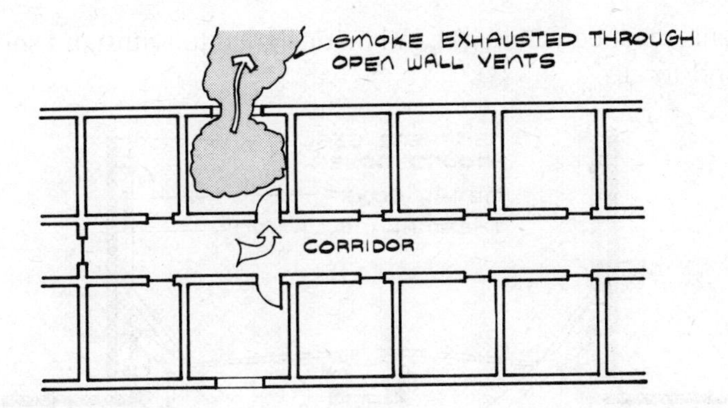

REFERENCE

Koplon, N. A., "Report of the Henry Grady Fire Tests," City of Atlanta Building Department, Atlanta, Ga., January 1973.

SMOKE VENTING—VENT DEVICE DETAILS

Wall Vent

Panel wall vents, distributed along the perimeter of a building, can be opened to remove heat and smoke. The open vent area should generally be 20 sq ft / 50 linear ft of exterior wall in each story. Vents generally should be spaced not more than 50 ft apart and the end vents not more than 25 ft from the corners of the building. Vents can be equipped with safety bars where needed and can be locked with devices that require an opening tool.

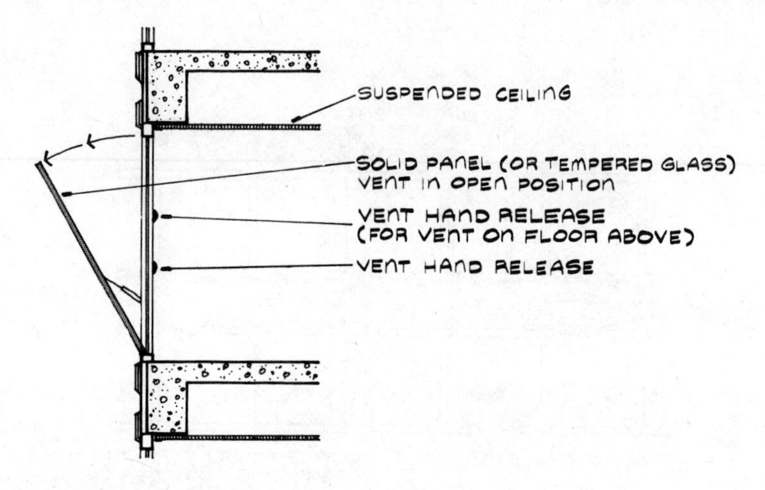

Note: Fire fighters should be able to open vents from the floor below the fire floor or at another remote location.

Roof Vent

Roof vents must be watertight and be designed to withstand snow loads and wind forces.

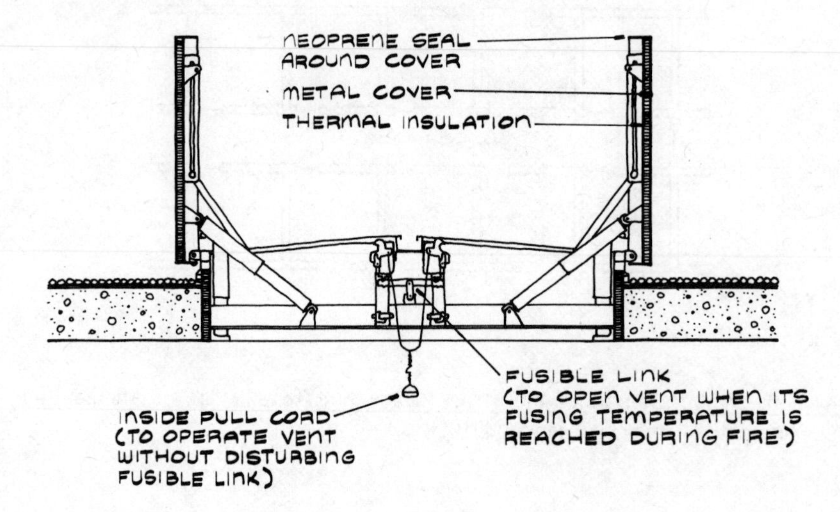

SMOKE VENTING—ESCALATOR FIRE PROTECTION (SPRINKLER/VENTING)

When a fire is detected by a smoke detector located near the stair opening in the example building shown below, the automatic damper opens to exhaust smoke and gases to the outdoors. Replacement air is drawn through the outdoor air intake located above the escalator floor openings. To effectively control smoke movement, a typical exhaust rate of 60 air changes per hour is required (i.e., about 10 times the normal return air rate). To provide a thermal barrier, sprinklers (or water spray nozzles) are also provided.

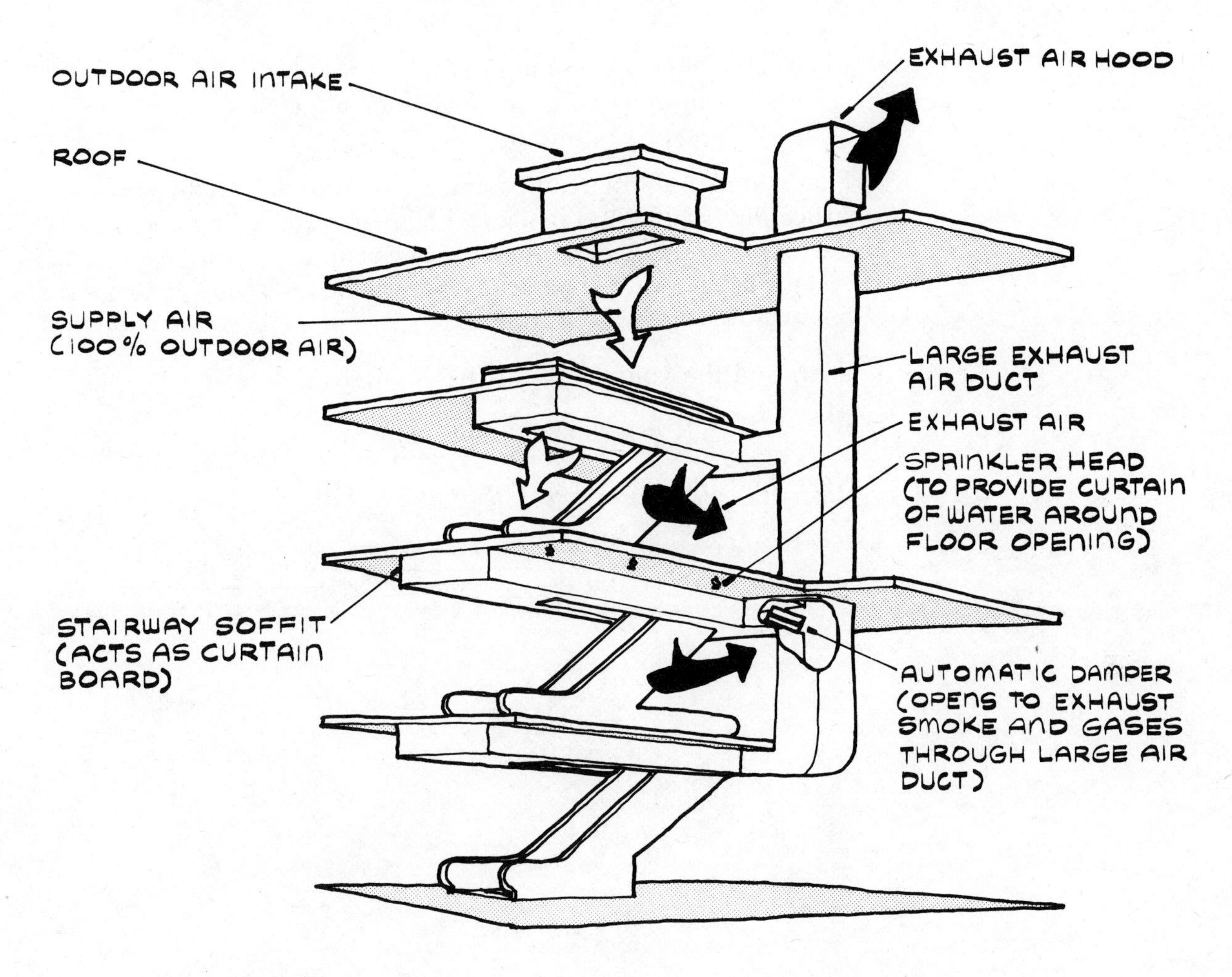

REFERENCE

Handbook of Industrial Loss Prevention, New York: McGraw-Hill (1967).

AIR CHANGE

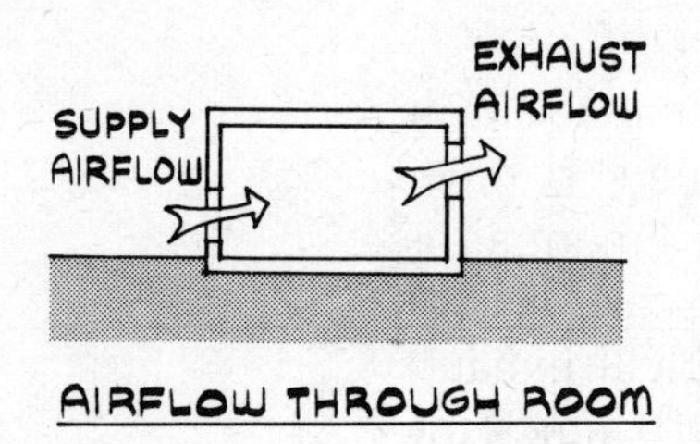

Room air change is the complete displacement of room air. As shown by the sketch in the margin, removed air (called "exhaust airflow") by natural convection or mechanical exhaust fans is normally replaced by "supply airflow." Airflow volume in cubic feet per minute (cfm) for a specific air change per hour (ach) can be determined by the formula:

$$Q = \frac{NV}{60}$$

where Q = airflow volume in cfm
N = number of air changes per hour (ach)
V = room volume in cu ft

Example: A building compartment, surrounded by fire-rated constructions, is 5000 sq ft in area with an average ceiling height of 9 ft. For smoke control, exhaust air will be removed through a mechanical air duct at a rate of 60 ach. What will be the required exhaust air fan airflow volume (Q_e) in cfm?

First, find the volume (V) of the compartment.

$$V = 5000 \times 9 = \boxed{45{,}000 \text{ cu ft}}$$

Next, find the exhaust airflow volume (Q_e).

$$Q_e = \frac{NV}{60} = \frac{60 \times 45{,}000}{60} = \boxed{45{,}000 \text{ cfm}}$$

SMOKE VENTING—AUDITORIUM STAGE HOUSE VENTING

Smoke vents can prevent the spread of smoke and hot gases to auditorium seating areas as shown by the sketches below.

Unvented Stage House

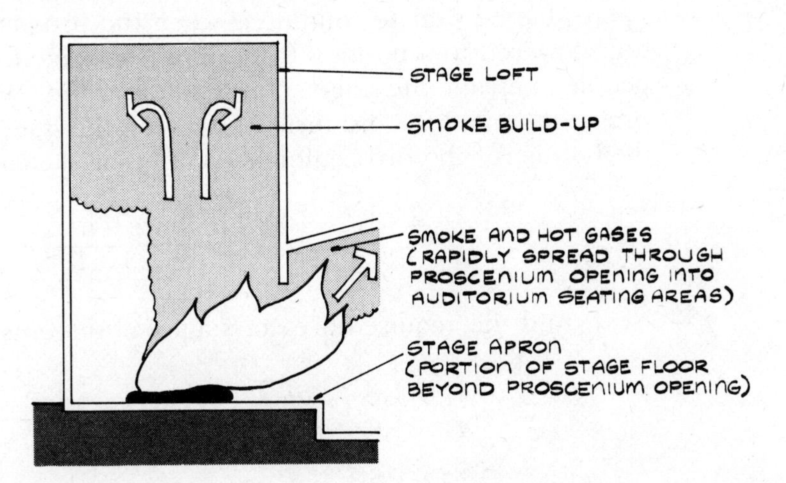

Vented Stage House

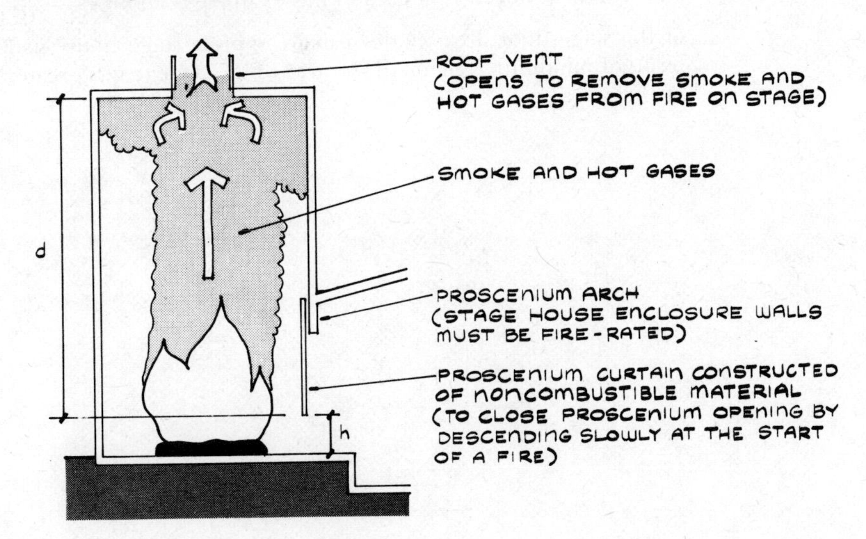

The required area of stage house vents (A_v) in square feet can be found by:

$$A_v \simeq 1.6\, A_o \sqrt{h/d}$$

where A_0 = open area of proscenium in sq ft
h = distance between bottom of proscenium curtain and stage floor in ft
d = distance between roof vent and bottom of proscenium curtain in ft

EXAMPLE PROBLEM—STAGE HOUSE VENTING

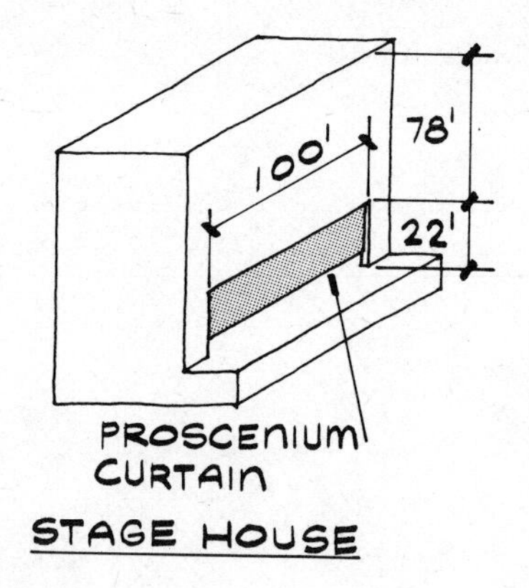

Given: An auditorium stage house has a floor area of 50 ft × 150 ft and a loft height of 78 ft. The proscenium opening is 100 ft wide × 22 ft high.

Find: What will be the required area of stage house vents (A_v) in square feet? Evaluate conditions when the proscenium curtain has been lowered to a position $6\frac{1}{2}$ ft above the stage floor (i.e., to allow people to escape the stage area). First, find the open area (A_o at the proscenium and the distance (d) between the stage house vents and the bottom of the proscenium curtain. (See sketch in the margin.)

$$A_o = 6.5 \times 100 = \boxed{650 \text{ sq ft}}$$
$$d = 78 + (22 - 6.5) = \boxed{93.5 \text{ ft}}$$

Next, find the required area of stage house vents (A_v).

$$A_v \simeq 1.6 A_o \sqrt{h/d}$$
$$A_v \simeq 1.6 \times 650 \sqrt{6.5/93.5} \simeq \boxed{274 \text{ sq ft}}$$

Note: The required vent area in this example is about $\frac{274}{50 \times 150} \times 100 \simeq 3\frac{1}{2}\%$ of the stage floor area. Codes usually express stage house vent requirements in terms of minimum fraction of the area of the stage (e.g., "greater than $\frac{1}{8}$ of the area of the stage").

SMOKE VENTING—CURTAIN BOARDS

Curtains of sheet metal, asbestos cement board, or gypsum plaster, which extend down from the ceiling level, can be used in large single-story buildings to contain the spread of smoke and hot gases. The table below gives suggested maximum curtain board spacings and maximum area within the boundary of the curtains (cf., "Guide for Smoke and Heat Venting," NFPA No. 204, 1968).

Heat Release Occupancy	*Curtain Board Spacing (ft)*	*Maximum Area Within Curtain Boards (sq ft)*
Low	250	50,000
Moderate	250	50,000
High	100	10,000

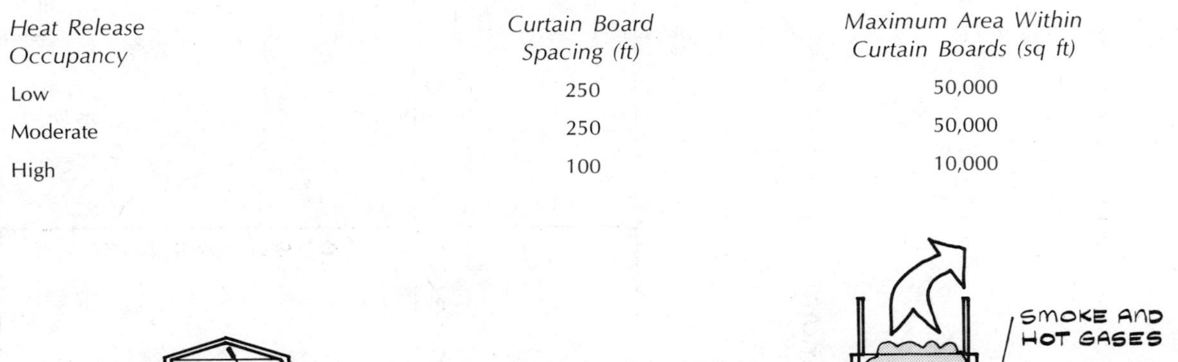
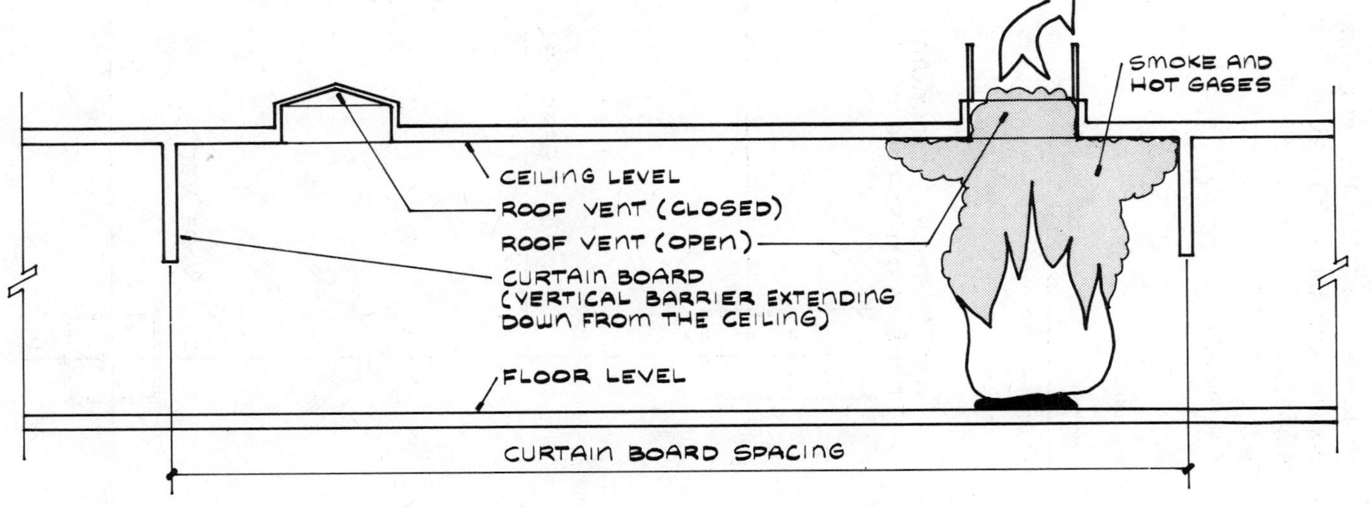

Note: Consult manufacturers' catalogs for vent open or "free" area. In general, several small vents are more effective than a large vent of equal area. Well-distributed smaller vents also can lessen the risk to adjacent buildings from burning materials carried by convection and wind forces.

SMOKE VENTING—ROOF VENT LAYOUTS

Based on anticipated heat release from combustible materials in single-story buildings, suggested roof vent spacing distance (*D*) is given in the table below (cf., "Guide for Smoke and Heat Venting," NFPA No. 204, 1968). Properly vented compartments remove smoke and gases and can prevent "smoke explosions" which occur when doors are opened to smoke-logged spaces. The fresh supply of oxygen forms an explosive mixture with the unburned gases that are already at their ignition temperatures.

Heat Release Occupancy	*Vent Spacing (ft)*
Low	150
Moderate	120
High	100 to 75*

*Depends on anticipated fire severity.

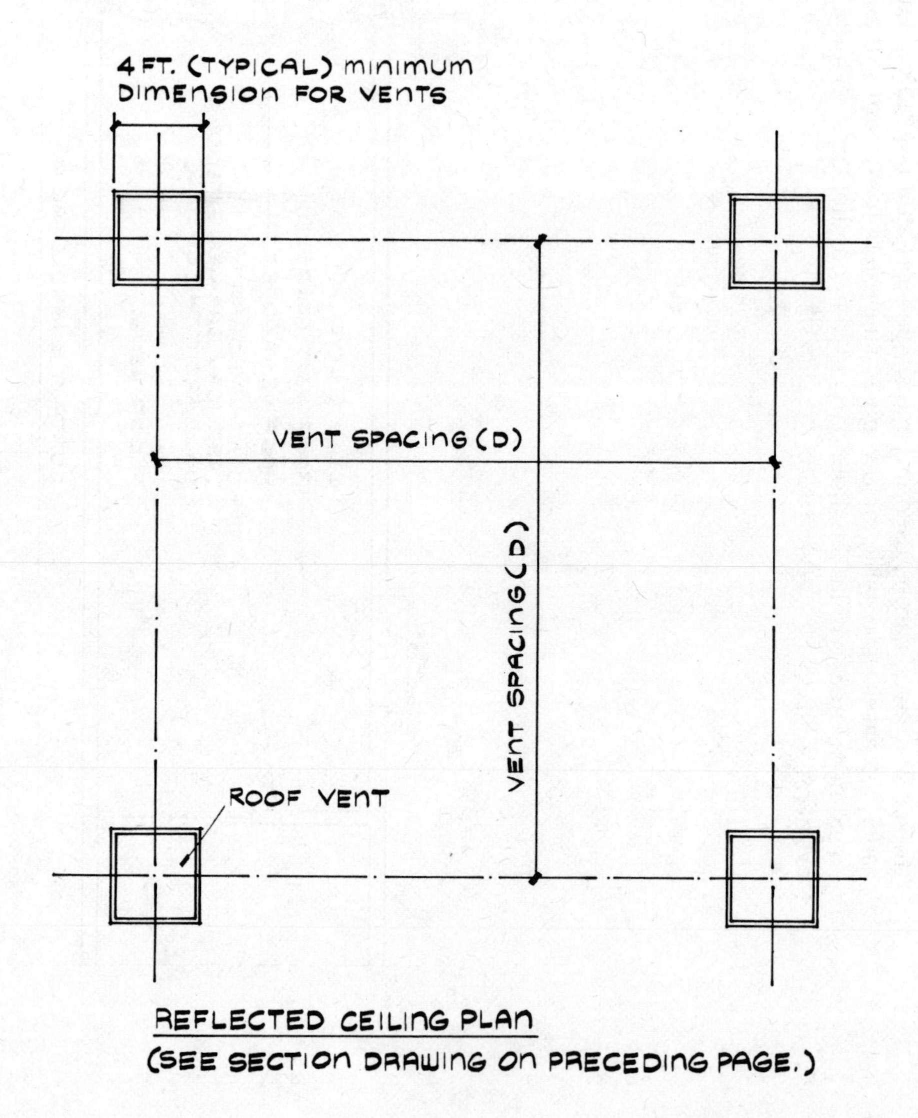

Note: Exhaust fans can be used to reduce the need for large vent areas.

HEAT RELEASE CLASSIFICATIONS

The "heat release" categories given below are based on anticipated rate of heat release from combustible materials or flammable liquids stored in buildings.

Low Heat Release

Containing scattered small quantities of combustibles.

Bakeries
Breweries
Dairy product processing plants
Foundries
Machine shops such as for dry machining and similar operations
Meat packing plants
Metal stamping plants

Moderate Heat Release

Containing moderate quantities of combustibles, fairly uniformly distributed.

Automobile assembly plants
Leather goods manufacturing
Machine shops using combustible oil coolants, hydraulic fluids, and other combustible materials
Printing and publishing plants

High Heat Release

Containing either hazardous operations or concentrated quantities of combustibles or both.

Chemical plants
General warehouses
Painting departments
Paper mills
Rubber products manufacturing plants

REFERENCE

"Guide for Smoke and Heat Venting," NFPA No. 204, 1968.

VENT AREAS

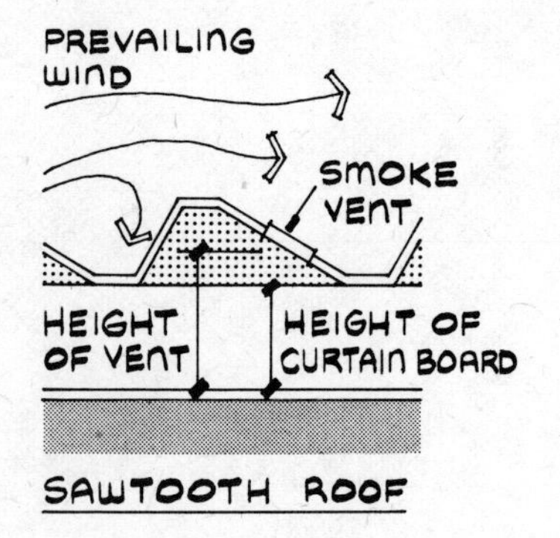

To find the required vent area (A_v) in square feet, multiply the estimated perimeter of fire (P) in feet by the factor (k) from the table below. For example, a storage building with sawtooth roof construction is subdivided by curtain boards 20 ft above the floor. Vents are located 30 ft above the floor. The perimeter of fire is estimated as 100 ft. The required vent area is $A_v = kP = 4 \times 100 = 400$ sq ft. (Enter table at 30 ft row and read $k = 4$ at intersection of 20 ft column.)

Height of	*Height of Curtain Board (ft)*							
Vent (ft)	*10*	*15*	*20*	*25*	*30*	*35*	*40*	*45*
15	2.0							
20	1.4	3.6						
25	1.2	2.6	5.6					
30	1.0	2.1	4.0	7.9				
35	0.89	1.8	3.2	5.6	10			
40	0.81	1.6	2.8	4.5	7.3	13		
45	0.75	1.5	2.5	3.9	6.0	9.2	16	
50	0.70	1.4	2.3	3.5	5.1	7.5	11	19

REFERENCE

McKinnon, G. P. (ed.), *Fire Protection Handbook,* Boston, Mass.: National Fire Protection Association (1976).

SMOKE VENTING—ATRIUM VENTING

An atrium (i.e., covered court extending to the roof) can allow smoke and gases to spread vertically to adjoining building spaces. Effective smoke removal is therefore essential to limit smoke damage, allow occupants safe access to exits, and enable fire fighters to locate and suppress the fire. In the atrium example below, smoke is mechanically exhausted through open smoke vents at the top of the atrium. Replacement air enters through ground-level air intake fans. The vents are operated by smoke detectors located in the atrium. In addition, combustibles should be controlled within the atrium so that it will be virtually noncombustible.

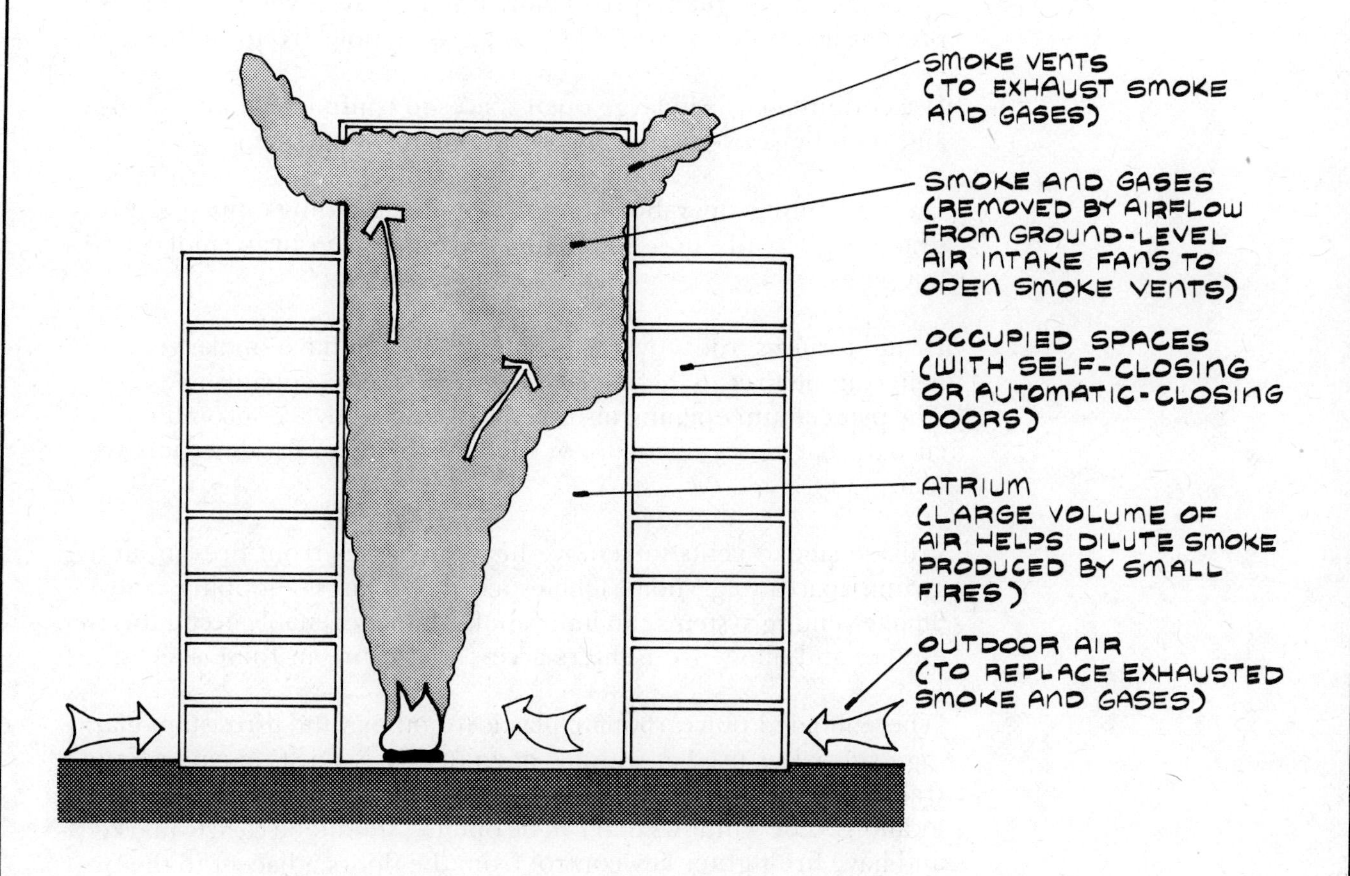

Note: To assure safe evacuation routes, pressurized enclosed stairways can be provided.

REFERENCES

Degenkolb, J. G., "Atriums and Fire Safety. . . Are They Compatible?," *The Building Official and Code Administrator,* March 1975.

Stevens, R. E., "The Problems of Atriums," *Fire Journal,* September 1968.

SMOKE VENTING CHECKLIST

Smoke vents can be used to contain the spread of smoke and gases so that egress routes will not become blocked. Smoke reduces visibility and causes irritation of the eyes and nasal passages. Escape through building areas having hydrogen chloride (HCl) concentrations greater than 100 ppm, or carbon monoxide (CO) greater than 4000 ppm, normally would not be possible.

Smoke venting provides a means of heat release from buildings, removes smoke and gases, and helps fire fighters locate the fire. With open vents, heat and smoke from fire will rise toward the openings rather than spread horizontally. Heat venting can also prevent activation of sprinklers in areas remote from the fire.

Use curtain boards in large open spaces to confine heat and smoke, and thereby prevent lateral spread along ceiling surfaces.

During venting operations, adjacent taller buildings must be protected (e.g., by fire hose water spray) from vented heat, smoke, and hot gases.

In auditoriums, roofs over stage houses should have smoke vents to help confine fires to the stage area and to allow occupants to escape. The proscenium opening also can be protected by a noncombustible curtain that closes manually or operates by a smoke detection system.

Provide smoke vents to remove heat and gases from fires in large atrium spaces (e.g., hotel lobbies, civic buildings, shopping malls). Smoke venting systems can limit smoke damage, enable occupants to escape, and allow fire fighters access to the fire in these spaces.

Where molded polycarbonate plastic (or other vandal-resistant glazing material) is used at window openings for security purposes, the framing system should allow opening for smoke venting at sufficient locations. The windows that can be opened should be clearly marked and have fire-fighter key control from the floors adjacent to the fire floor.

ESCAPE AND REFUGE

ESCAPE AND REFUGE

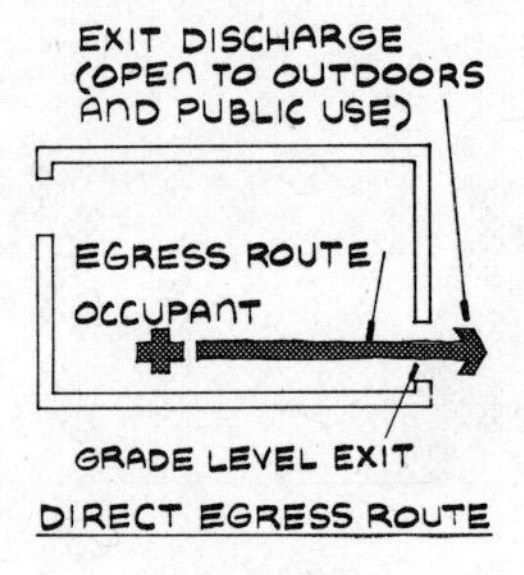

DIRECT EGRESS ROUTE

"Escape" is total separation of the occupant from a burning building. Egress to a public way through grade-level exits (called exit discharge) is shown by the sketch in the margin. "Taking refuge" means retreat to a safe area somewhere in the building. The provision of exits (i.e., protected ways of travel to an exit discharge) is a vital aspect of building design as nearly one out of every four persons who dies in a fire dies as a result of the inability to reach an exit.

COMPLEX EGRESS ROUTE

The egress routes (called exit access) for escape or to a refuge area should not be tortuous or involve several rooms, long corridors, and many flights of stairs. A complex route, which must be avoided in building designs, is also shown in the margin.

EGRESS BLOCKED BY SMOKE

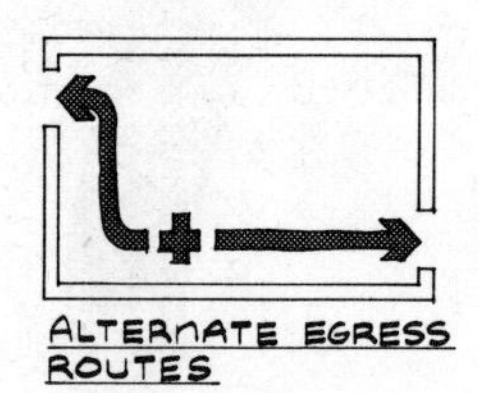

ALTERNATE EGRESS ROUTES

Even direct and well-marked egress routes could become blocked by smoke and gases during a fire. Consequently, life safety design must provide alternate routes so occupants can move to a safe area during a fire emergency (see sketches). For a comprehensive discussion of escape and refuge principles, see R. D. Caravaty, and D. S. Haviland, "Life Safety from Fire: A Guide for Housing The Elderly," Federal Housing Administration, 1968, pp. 58–63.

WALKING (4FPS)

JAMMING AT EXITS (2½FPS)

Persons with normal physical ability move forward horizontally at a speed of about 4 feet per second (fps) through unobstructed passageways. A "shuffle," caused by jamming at exits or overcrowded corridors, can be about $2\frac{1}{2}$ fps or less. For most nonhazardous situations, $1\frac{1}{2}$ to 2 minutes can be sufficient time to move to a safe area after receipt of an alarm warning.* For this situation, the travel distance to exits can be estimated by: distance = velocity × time = $2\frac{1}{2}$ fps × 90 seconds = 225 ft. Most of the maximum travel distances permitted by codes are less than 200 ft to account for the variations in occupant densities, age, and mental and physical conditions (e.g., people may use walking aids).

Exits also must be sized so that there will be sufficient width to prevent jamming. A personal space of about 5 to 10 sq ft represents the interpersonal separation most occupants consider comfortable during egress. It is influenced by clothing worn, cultural background, sex, and other factors.

WALKING UP STAIRS

For vertical movement to refuge areas, fatigue can become a significant factor if the time required to walk upstairs exceeds 1 minute, or time to walk downstairs exceeds 5 minutes. In addition, when occupants wear or carry bulky outdoor clothing, evacuation can take 50% longer than during warm weather conditions.

WALKING DOWN STAIRS

*For a discussion of alarm warning signal criteria, see "A Proposed Standard Fire-Alarm Signal," *Journal of the Acoustical Society of America,* Vol. 57, No. 3, March 1975, pp. 756–757.

ESCAPE AND REFUGE—LIFE SAFETY CONCEPTS

During a fire, the building occupant's age, health, and mental capabilities affect the ability to react and proceed to a safe location. Building height also affects the ability to reach safety as it is difficult to evacuate tall buildings. Shown below are four approaches to life safety (escape, slow escape, refuge, and minimal disruption) for various conditions of occupant mobility and building height.

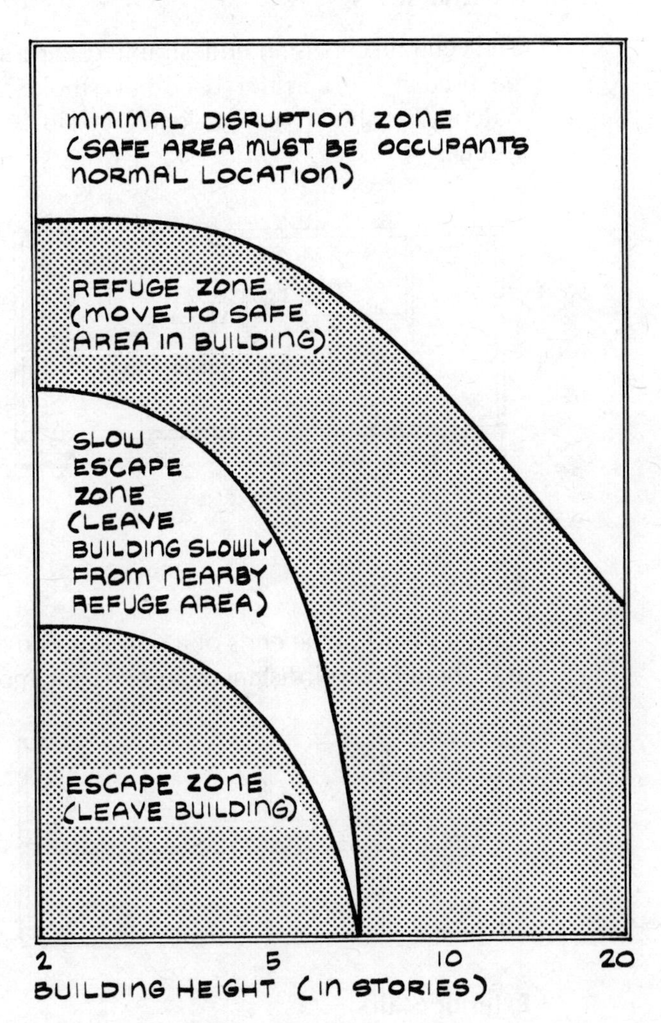

REFERENCE

Caravaty, R. D. and W. F. Winslow, "A New Approach to Fire Codes," *Architectural & Engineering News,* March 1970.

ESCAPE AND REFUGE—BUILDING ESCAPE STRATEGIES

Escape strategies for linear corridors in low-rise buildings are described below. Escape plans should be evaluated to see if alternate routes are available for use in the event that one route becomes untenable due to fire or smoke.

Central Stairs

The central stairs is an undesirable escape strategy because exits cannot be used if the corridor becomes smoke-filled. However, an exterior balcony, as shown on the sketch, could be used to increase chances of rescue.

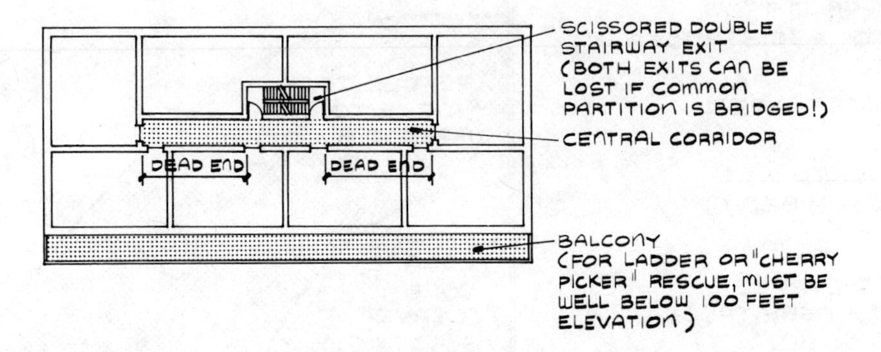

Remote Stairs

Stairway exits at the ends of a corridor can offer alternate escape routes and shorter travel distances to exits for most occupants.

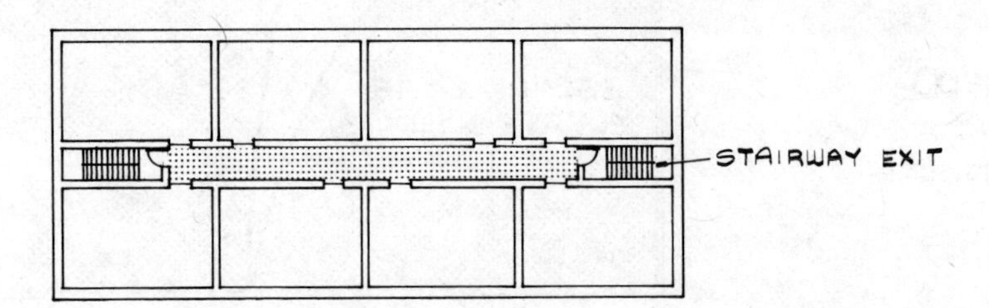

Exterior Stairs

Protected outdoor stairway exits can provide smoke-free escape routes. There are alternate escape routes to remotely located stairs as shown by the sketch.

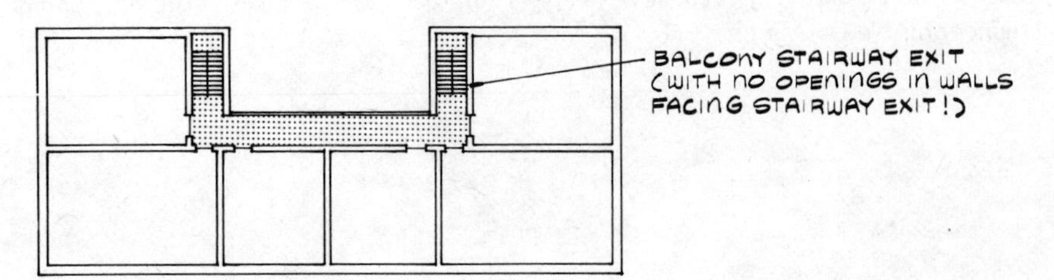

NUMBER OF EXITS

When a fire occurs, building occupants must be evacuated from the building or move to a refuge area into which fire and smoke cannot penetrate. Every exit should lead directly to a protected passageway, protected stairway, or the outdoors. Example minimum requirements for number of exits are given in the table below. For proposed minimum numbers of exits, see "Fire Protection Through Modern Building Codes," American Iron and Steel Institute, 1971, pp. 298–299.

Room or Enclosed Space	*Example Minimum Exit Requirements*
Basements	2
50 occupants or more	2
500 occupants or more	3
1000 occupants or more	4
Every building floor	2

Note: Consult with local authority having jurisdiction for minimum exit requirements.

ESCAPE AND REFUGE—EXIT DISTRIBUTION

Shown below are exit locations for an example office occupancy.

Poor

The number of exits are adequate, but they are located too close together. In addition, the access routes to the exits for occupants at the remote point shown pass through each other which could allow smoke and fire to block both routes.

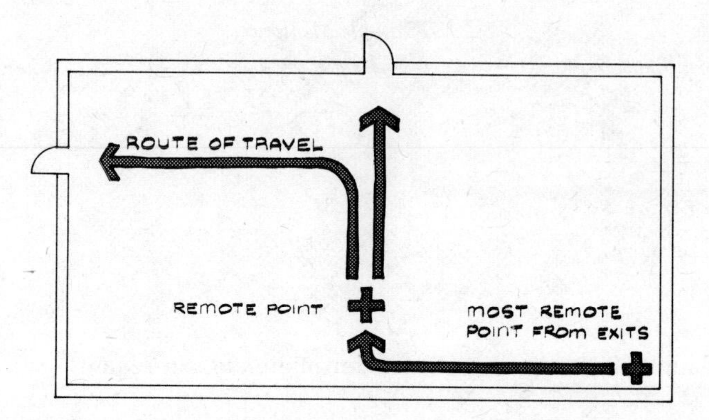

Better

The sketch below shows remotely located exits with access routes in opposite directions.

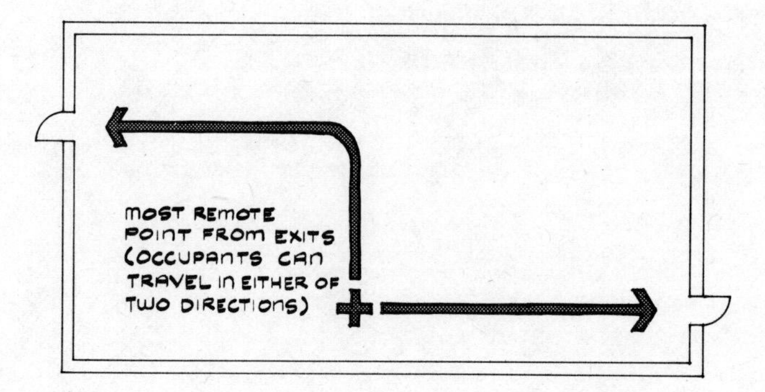

ESCAPE AND REFUGE—EXITS AND ACCESS TO EXITS

Access to exits must be free of fire and smoke and be available to building occupants at all times. Exits should be remotely located so they cannot be blocked by a single fire. Locate exits at the ends of linear circulation paths as shown by the floor plan below or at points of confluence such as courts or vestibules.

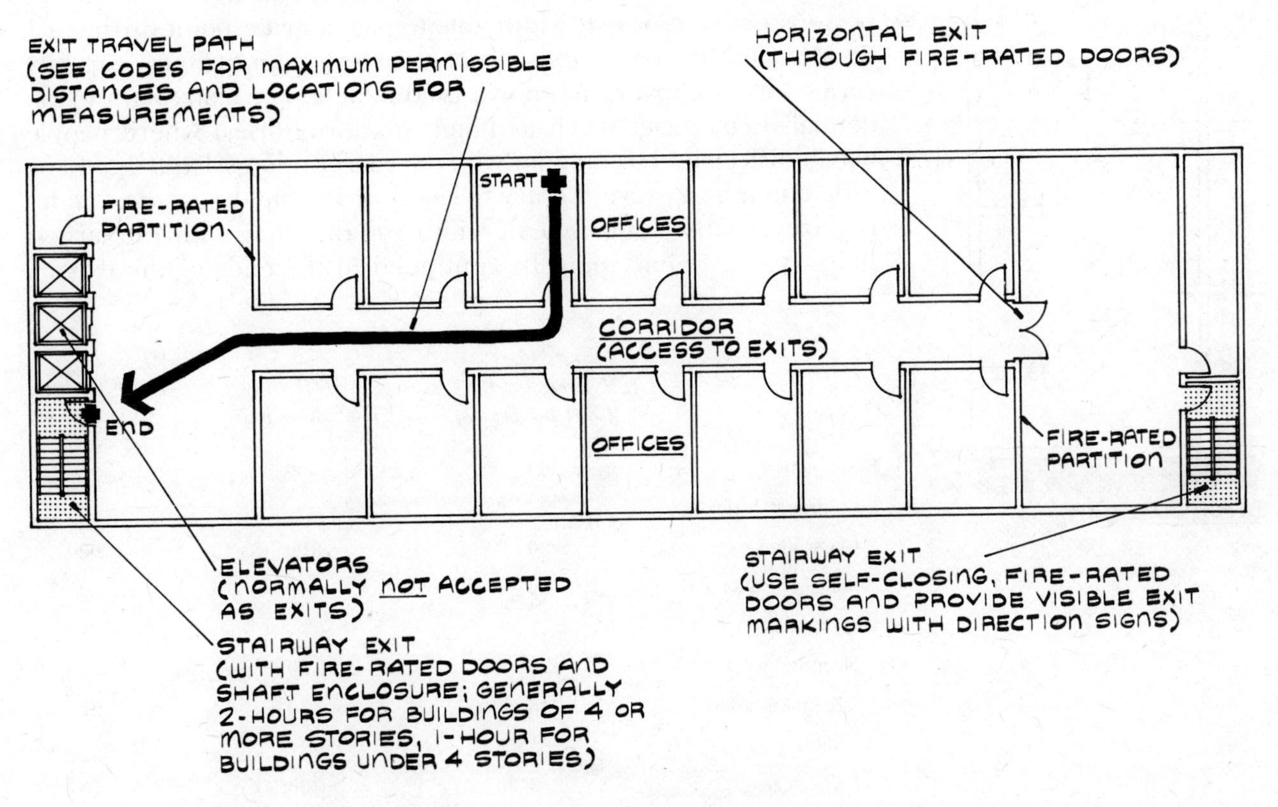

Find required exit unit width (W) by the formula:

$$W = \frac{A}{dc}$$

where A = floor area in sq ft

d = occupant density in sq ft per person (from codes or actual conditions, whichever yields greatest width requirement)

c = capacity per unit of exit width (i.e., number of persons able to pass a given point during a 1 minute period)

EXIT REQUIREMENTS

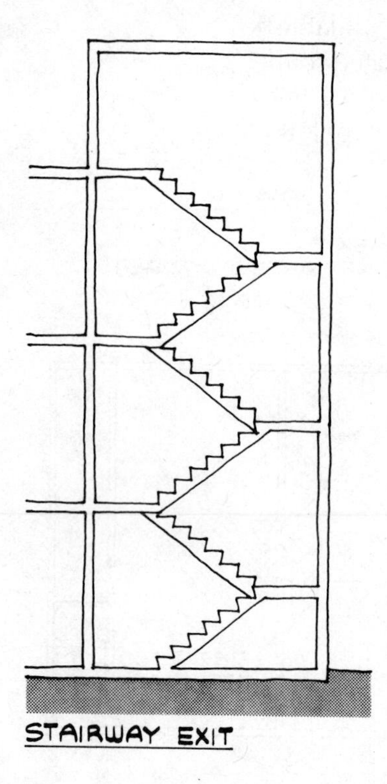

Occupant density is the number of people in a building for whom exits must be provided. The table gives density (*d*) in square feet per person for various occupancies. When the actual number of occupants is available during the design stages of a project, this figure can be used to estimate exit requirements. Capacity (*c*) is the number of persons per unit of exit width able to pass a given point during a 1 minute period. To establish consistent evacuation times, capacity varies with occupancy as shown by the table. For example, in residential occupancies (such as hotels or dormitories) where people may be asleep and unfamiliar with the surroundings, exits should be wide enough to prevent entry delays. For tall buildings, most codes require that the stairway exit width for the floor with the largest stairway exit requirement be continued to the ends of the exit.

Occupancy	*Density (d) (sq ft per person)*	*Capacity (c) (persons per unit of exit width)*	
		Horizontal Exits	*Stairway Exits*
Residential	200	60	45
Educational			
Classrooms	20	100	60
Shops, vocational	50	100	60
Institutional			
Sleeping areas	120	30	22
Treatment areas	240	30	22
Assembly			
Fixed seats	15	100	75
Without fixed seats	7	100	75
Standing areas	3	100	75
Office	100	100	60
Business	100	100	60
Mercantile			
First floor	30	100	60
Other floors	60	100	60
Industrial	100	100	60
Storage	300	60	45
Hazardous	100	60	45

SOURCES

Fire Protection Through Modern Building Codes, New York: American Iron and Steel Institute (1971).

McKinnon, G. P. (ed.), *Fire Protection Handbook,* Boston, Mass.: National Fire Protection Association (1976).

ESCAPE AND REFUGE—EXIT WIDTH

The standard unit for exit width computations is 22 in. which represents the clearance width of adults as shown below.* Note that walking adults need about 28 in. clearance to account for body sway. The sketches also show handicapped adults in a wheelchair and using crutches. The preferred minimum clearance width of openings for wheelchairs is 32 in. (i.e., 27 in. average wheelchair width plus 5 in. for hand maneuvers) and 33 in. minimum passageway width for crutch use to allow the crutch spread needed for a comfortable gait. To accommodate adults using various walking aids, corridors should be at least 36 in. wide.

---

* One-half unit of exit width is 12 in. or more as the only allowable fraction is ½.

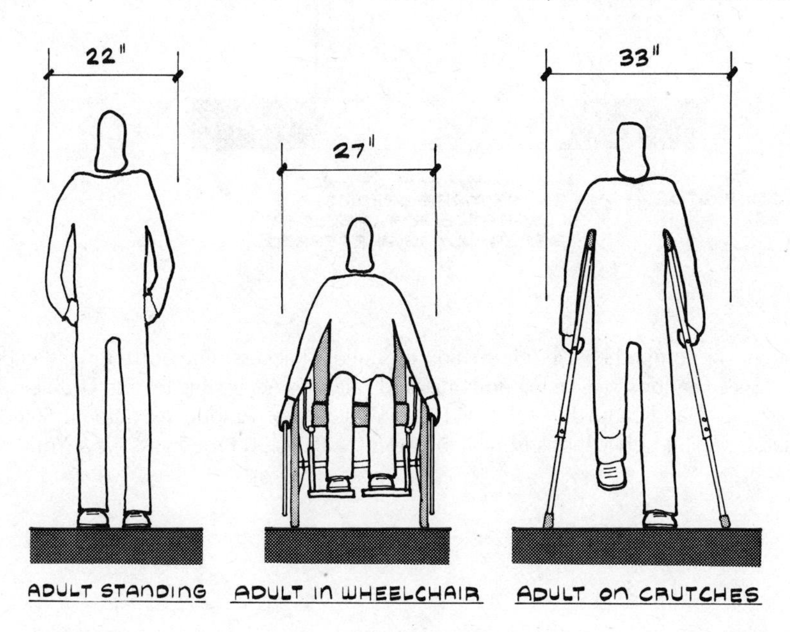

ESCAPE AND REFUGE—EXIT WIDTH (Continued)

The plan view of the door below shows the preferred 32 in. minimum clearance width of opening for wheelchairs. Clear opening requirements for an adult in a wheelchair should be measured from the door in a 90° open position to the face of the doorstop (or to the edge of the other door for a pair of doors).

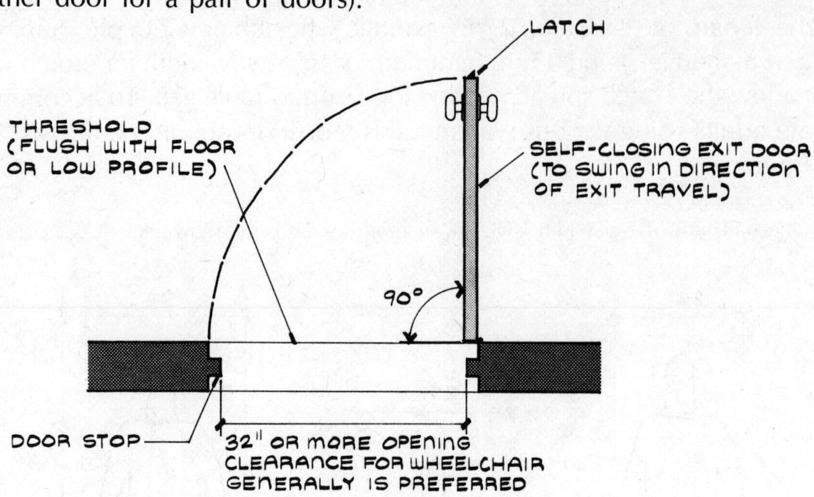

Note: For comprehensive information on building accessibility for the handicapped, see "Specifications for Making Buildings and Facilities Accessible to, and Usable by, the Physically Handicapped," ANSI A117.1. A catalog of American National Standards Institute (ANSI) publications is available from ANSI, 1430 Broadway, New York, N. Y. 10018.

EXAMPLE PROBLEM—EXIT REQUIREMENTS

Given: An office building has a floor area of 9000 sq ft per floor. The occupant density (d) is 100 sq ft per person and the stairway exit capacity (c) is 60 persons per unit of exit width. (The exit capacity factor is lower for residential and institutional occupancies than for offices where occupants are usually awake, alert, and assumed to be physically able to leave without assistance.)

Find: What will be the required stairway exit unit width?

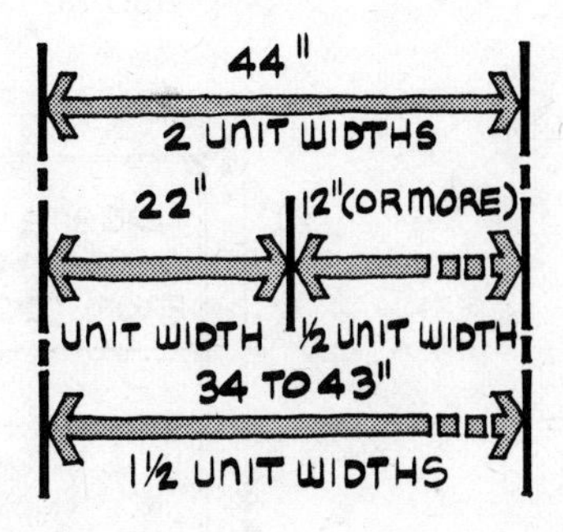

$W = \frac{A}{dc} = \frac{9000}{100 \times 60} = \boxed{1\frac{1}{2}}$ stairway exit unit widths (or 34 to 43 in. as shown in the margin)

ESCAPE AND REFUGE—FIRE DOORS

Fire doors must be in the fully closed position and latched in order to provide an effective barrier to the spread of fire. Doors should be "self-closing" (they return to closed position after being opened) or "automatic closing" (they are normally held open by an electromagnetic device that, when activated by a smoke detector, releases the door). Wired-glass panels are limited in size by codes as they radiate heat in proportion to their area (cf., "Fire Doors and Windows," NFPA No. 80).

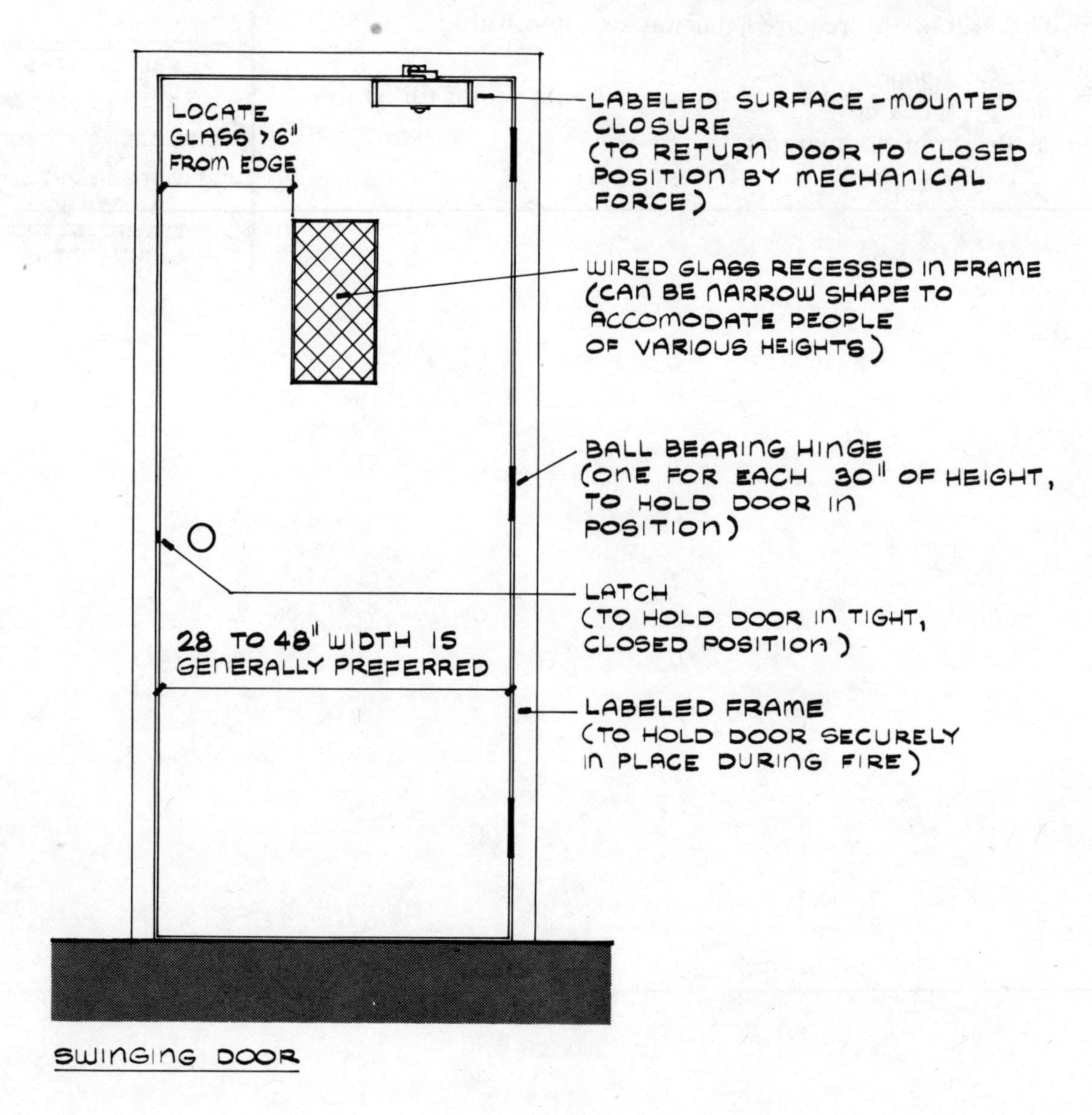

SWINGING DOOR

ESCAPE AND REFUGE—EXIT DOORS (WITH FIRE EXIT HARDWARE)

To provide rapid opening of exit doors (even in the dark), "panic hardware" can be used on outward swinging exterior doors. When a minimum force is applied to the horizontal bar, shown in the sketch below, it disengages the latch from the strike allowing rapid egress from inside the building. When the door closes, by means of various types of door closures, the latch bolt reengages the strike in the door frame to secure against unwanted ingress. Exit alarms that indicate the door has been opened can be located at the door or be integrated within the building security alarm system at a central location.

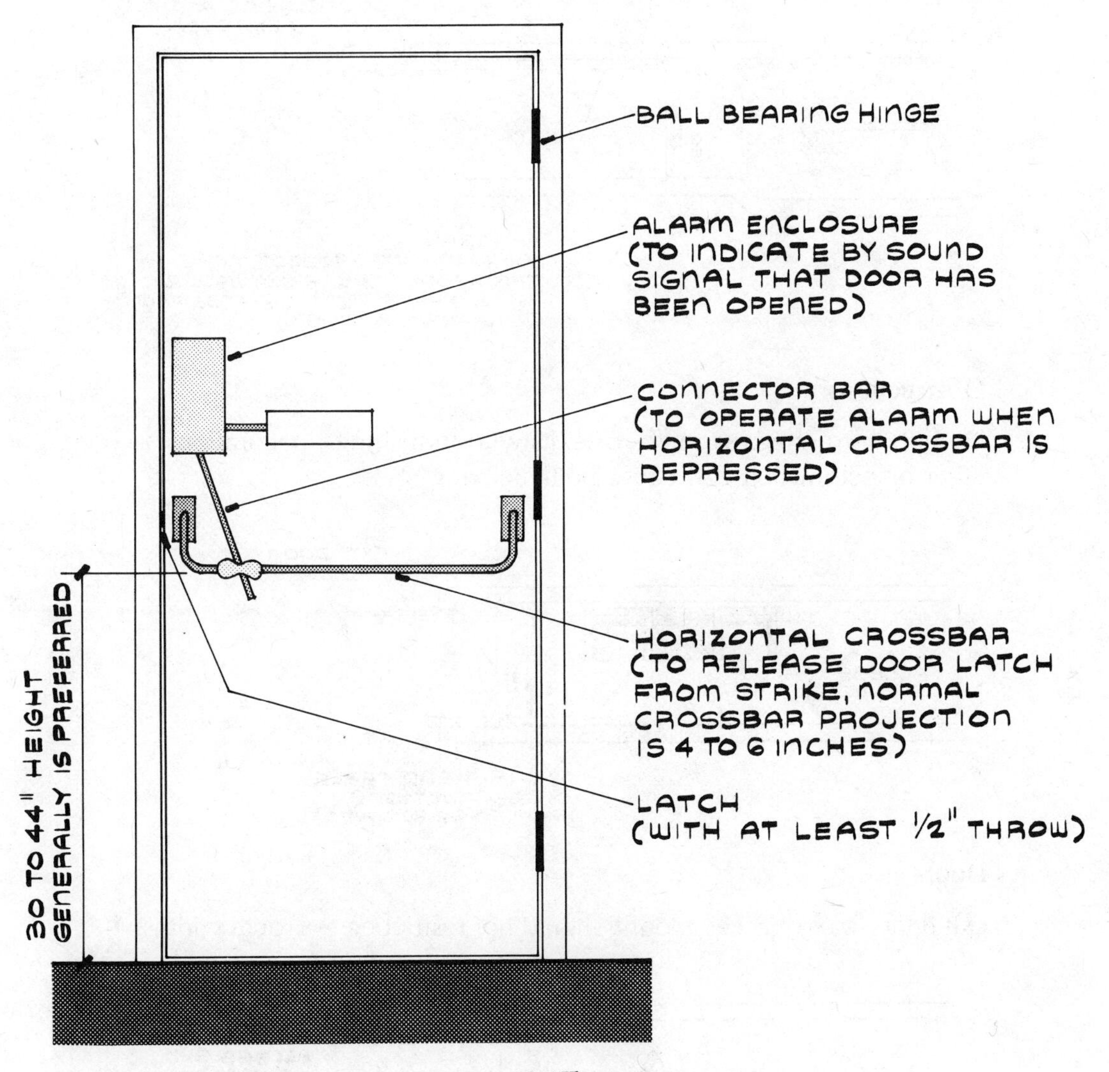

SWINGING DOOR WITH FIRE EXIT HARDWARE

Note: The gap between meeting edges of a pair of exit doors can allow burglars or vandals to enter buildings by using a coat hanger or other device to trip the horizontal bar. Use removable mullions, astragal edge projections, or movable astragal bars that block the gap when the doors are closed.

ESCAPE AND REFUGE—EGRESS ROUTES

Egress routes must provide continuous and unobstructed travel from any location in a building to the outdoors or to a refuge area. Exit access is that portion of the egress route which leads to an entrance to an exit.

Width Transitions

Measure required exit width at the narrowest point along the egress route.

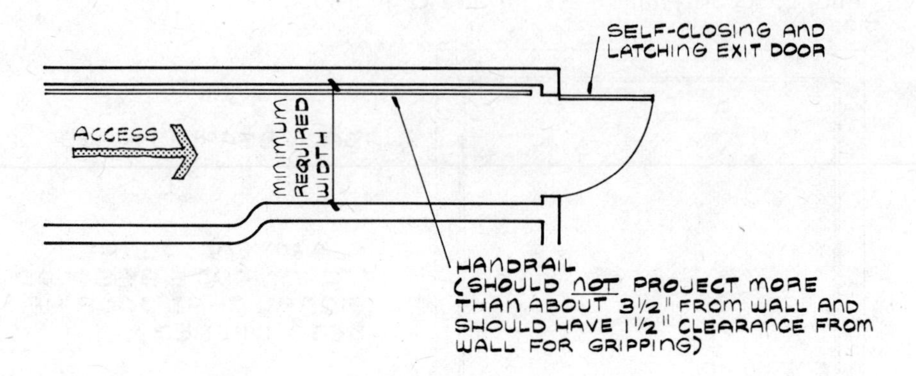

Obstructions

Do not obstruct egress routes or exits with furnishings, decorations, or other objects which can cause bottlenecking.

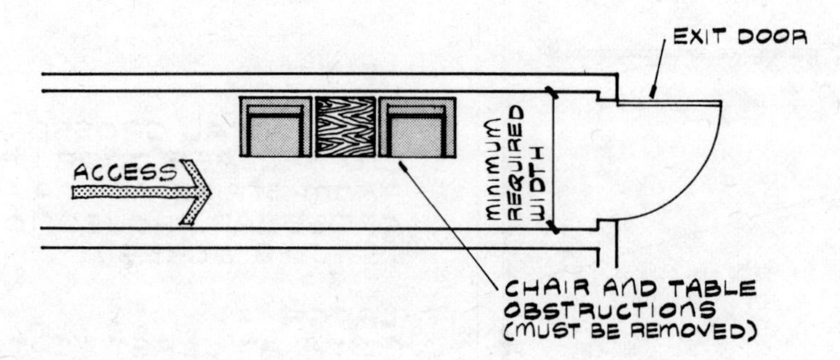

Doors

Exit doors, or exit access doors, should not restrict egress route widths.

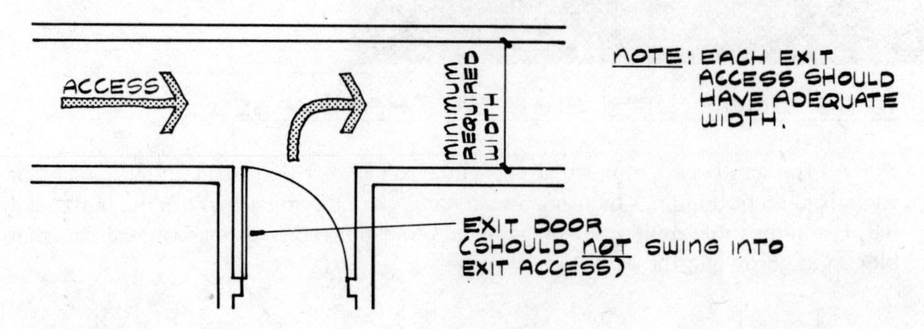

REFERENCE

Hopf, P. S., *Designer's Guide to OSHA,* New York: McGraw-Hill (1975).

ESCAPE AND REFUGE—EXIT STAIRS

Tread width and riser height should not vary along the exit route as uneven widths and heights can break the gait of users resulting in trips or falls. Stair treads should be wide enough to provide stable foot contact (e.g., tread width should be at least 11 in. for use by the handicapped). Treads should not be slippery or have cracks or crevices which could catch heels. A slight stair tread slope (or "wash") can help prevent water accumulation from cleaning operations. Risers should be vertical or slanted (as shown below) and projecting nosings should not excessively overhang risers. Where stairs are unusually wide, center handrails should be used.

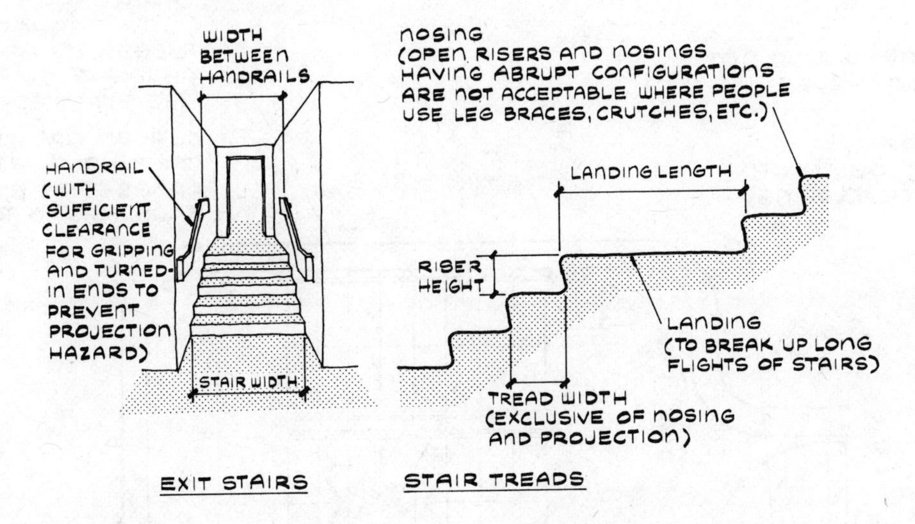

The table below describes the two classes of stairs recognized by the "Life Safety Code," NFPA No. 101, 1973.

Stair Requirements Table

DIMENSIONS	CLASS A	CLASS B
STAIR WIDTH (MINIMUM)	44"	44" OR 36"
WIDTH BETWEEN HANDRAILS (MINIMUM)	37"	37" OR 29"
TREAD WIDTH (MINIMUM)	10"	9"
RISER HEIGHT (MAXIMUM)	7½"	8"
LANDING LENGTH (MAXIMUM)	44"	44"
LANDING HEIGHT (MAXIMUM)	9'	12'

ESCAPE AND REFUGE—STAIRWAY EXIT LAYOUT

An example stairway layout is shown below in plan and section views. Exit doors to stairways should swing in the direction of egress. They should not swing into the stairway egress route obstructing the movement of occupants from other floors. Stairways that are poorly illuminated and excessively reverberant (i.e., sounds persist due to multiple reflections from sound-reflecting surfaces) will generally be avoided by building occupants. As a consequence, these exits will be unfamiliar when needed for emergency egress.

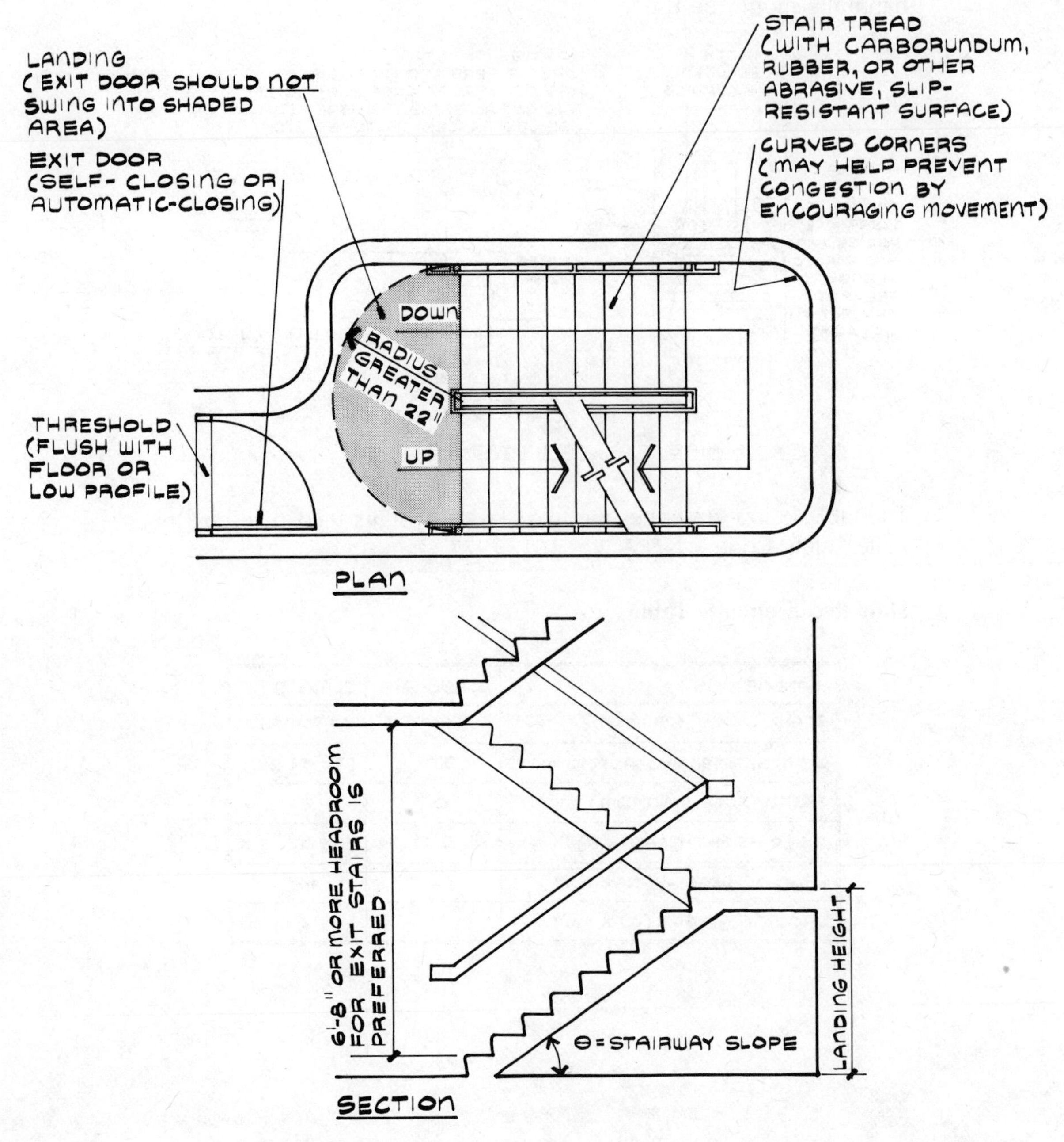

Note: For greatest comfort and safety, stairway slope (Θ) of 25 to 35 degrees is generally preferred.

ESCAPE AND REFUGE—EXIT RAMPS

Ramps are used where floor level transitions would need less than three risers. They help avoid the situations in crowded corridors where people cannot see abrupt elevation changes and could trip and fall. For use by the handicapped, ramp width should be at least 36 in. for traffic in one direction and 48 in. for two-directional traffic. The preferred slope is 5% (10/16 in. rise per 12 in.) or less. To accommodate wheelchairs, ramps should have straight level bottom platforms to provide sufficient stopping distance and should be finished in a nonslip surface (e.g., carborundum grit, rubber). Ramps can allow the physically handicapped to exit buildings at ground level without assistance. Always provide a handrail on at least one side that extends beyond the top and bottom of the ramp. Note that ramps which are too short require ambulatory users to quickly vary their stride lengths.

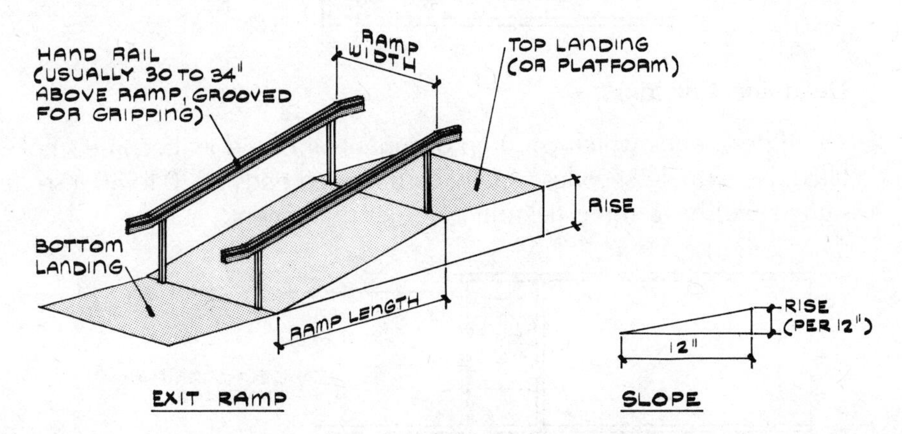

The table below describes the two classes of ramps recognized by the "Life Safety Code," NFPA No. 101, 1973.

Ramp Requirements Table

DIMENSIONS	CLASS A	CLASS B
RAMP WIDTH (MINIMUM)	44"	30 TO 44"
RISE (MAXIMUM IN 12")	1 TO 1 3/16"	1 3/16 TO 2"
HEIGHT BETWEEN LANDINGS (MAXIMUM)	NO LIMIT	12'

Note: The 1976 edition of the "Life Safety Code" designates a maximum slope of 1 in 10 for class A ramps and 1 in 8 for class B ramps.

ESCAPE AND REFUGE—RESTRICTED ACCESS AND DEAD ENDS

Restricted Access

Do not provide access to exits through rooms which could be locked or be the source of fires. Occupants must be able to go through exits without having to use a key.

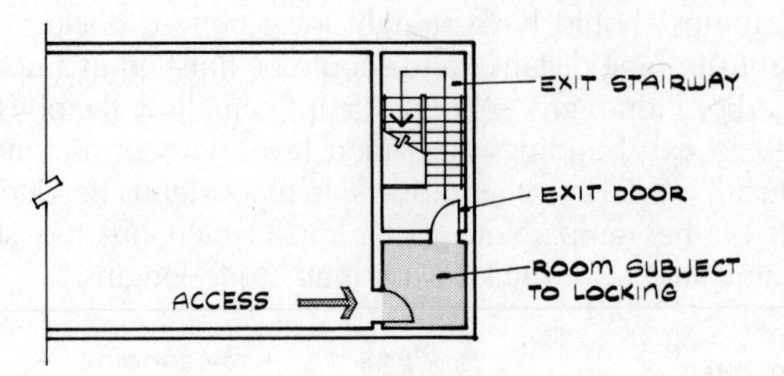

Dead-End Corridor

Avoid dead ends which can trap occupants if corridors become smoke-filled. Nevertheless, codes usually permit dead ends of 20 to 50 ft so designers will have some flexibility in exit placement.

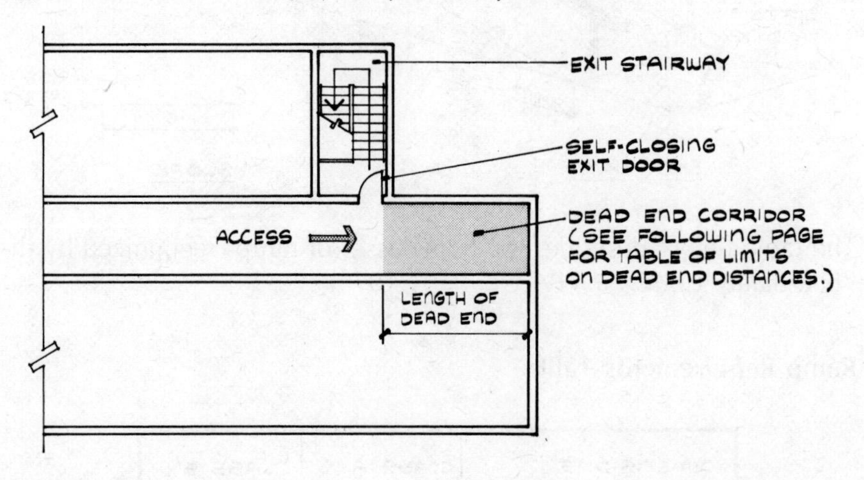

Note: For comprehensive, illustrated information on means of egress, see P. S. Hopf, *Designer's Guide to OSHA,* New York: McGraw-Hill (1975), pp. 84–100.

TRAVEL DISTANCES TO EXITS

The time required by occupants to reach an exit during a fire is critical. Consequently, codes establish maximum limits on travel distances to exits. Maximum permissible travel distances to exits depend on occupancy, fire hazard, and physical ability and anticipated alertness of occupants. For example, residential occupancies, where people are asleep for lengthy time periods, typically require shorter travel distances than office occupancies where people are assumed to be awake and familiar with egress routes. Dead ends, which could trap occupants in smoke-filled corridors, also are limited.

	Maximum Travel Distance (ft)		*Limit for*
Occupancy	*Not Sprinklered*	*Sprinklered*	*Dead Ends (ft)*
Residential			
Hotels, apartments	100	150	35
Dormitories	100	150	0
Educational			
Enclosed plan	150	200	20
Open plan	100	—	—
Institutional	100	150	30
Assembly	150	200	—
Office	200	300	50
Mercantile	100	150	50
Industrial	100	150	50
Storage	75	75	—
Hazardous	75	75	0

SOURCES

Fire Protection Through Modern Building Codes, New York: American Iron and Steel Institute (1971).

McKinnon, G. P. (ed.), *Fire Protection Handbook,* Boston, Mass.: National Fire Protection Association (1976).

ESCAPE AND REFUGE—ANALYSIS OF EXAMPLE EXIT LAYOUTS

Exit computations for the example residential, assembly, and office occupancies shown below are summarized in the table on the following page.

Residential

For apartments, hotels, and so on, travel distances to exits are usually measured from the door of individual units as there are few occupants in a relatively small area.

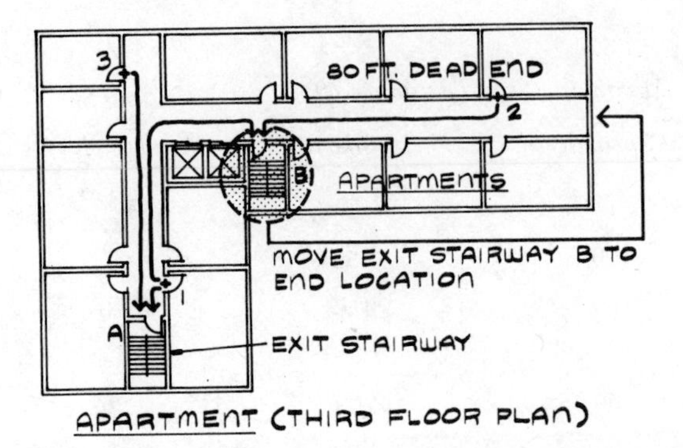

APARTMENT (THIRD FLOOR PLAN)

Assembly

In auditoriums, classrooms, and the like, many occupants may attempt to leave by the door they entered. Consequently, main exits are usually sized to handle at least 50% of the total occupant load.

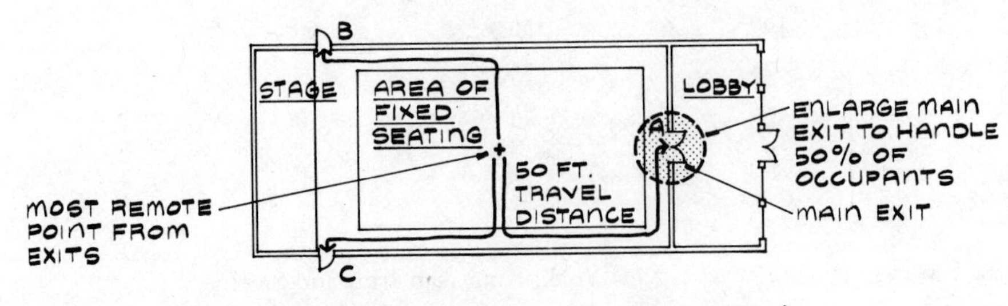

CLASSROOM (GROUND FLOOR PLAN)

Office

In offices, occupants are usually alert and familiar with the egress routes. Consequently, codes usually permit longer travel distances to exits.

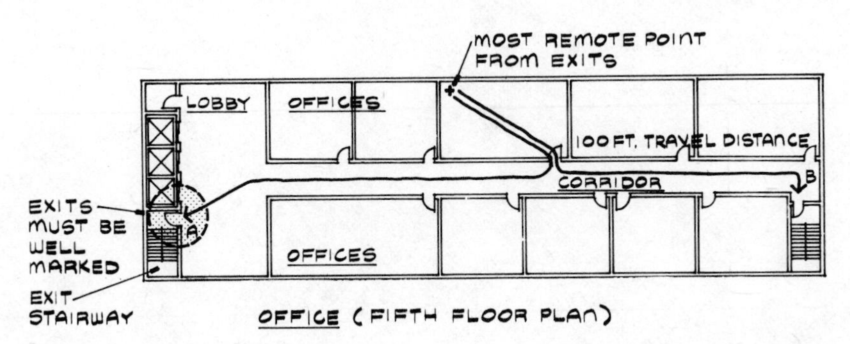

OFFICE (FIFTH FLOOR PLAN)

EXIT COMPUTATION TABLES FOR EXAMPLE LAYOUTS

Exits

Location in Building	*Occupancy*	*Area (sq ft)*	*Density (sq ft per person)*	*Total Occupants*	*Exit Capacity Factor (persons per unit of exit width)*		*Exit Unit Width*		*Number of Exits Provided*
					Horizontal	*Stairway*	*Required*	*Actual*	
Third floor	Residential (apartment)	4500	200	23	—	45	½	3	2
Ground floor	Assembly (fixed seats)	3000	15	200	100	—	2	6½	3
Fifth floor	Office	9000	100	90	—	60	1½	3	2

Travel Distances:

Location in Building	*Occupancy*	*Travel Distance (ft)*		*Dead Ends (ft)*	
		Limit	*Actual*	*Limit*	*Actual*
Third floor	Residential (apartment)	100	80	35	80*
Ground floor	Assembly (fixed seats)	150	50	—	—
Fifth floor	Office	200	100	50	None

*Occupants at position 1 have two alternate escape routes, both under a 100 ft limit for buildings without sprinklers. At positions 2 and 3, however, occupants are located in dead ends. Exit stairway B should be relocated at the end of the building to eliminate long dead ends.

ESCAPE AND REFUGE—EXIT SIGNS

Exit signs should be readily visible and should not be obscured by nearby brightly illuminated signs or displays. Studies under simulated smoke conditions at the University of Maryland indicate that yellow and orange may be the most effective colors for illuminated exit signs (cf., R. M. Brave, "A Study of the Effectiveness of Illumination for Exit Sign Lighting," Fire Protection Curriculum, University of Maryland, 1964). Avoid decorations and furnishings which can interrupt the line-of-sight to exit signs. Illumination on the exit access floor should be at least 1 footcandle (fc). Higher emergency illumination levels should be provided in occupancies involving the handicapped or the very young or very old. Where floors are illuminated by recessed wall mounted units, install units about 1 ft above the floor so smoke cannot easily occlude the light. Lighting fixtures at or near the ceiling can rapidly become hidden by smoke.

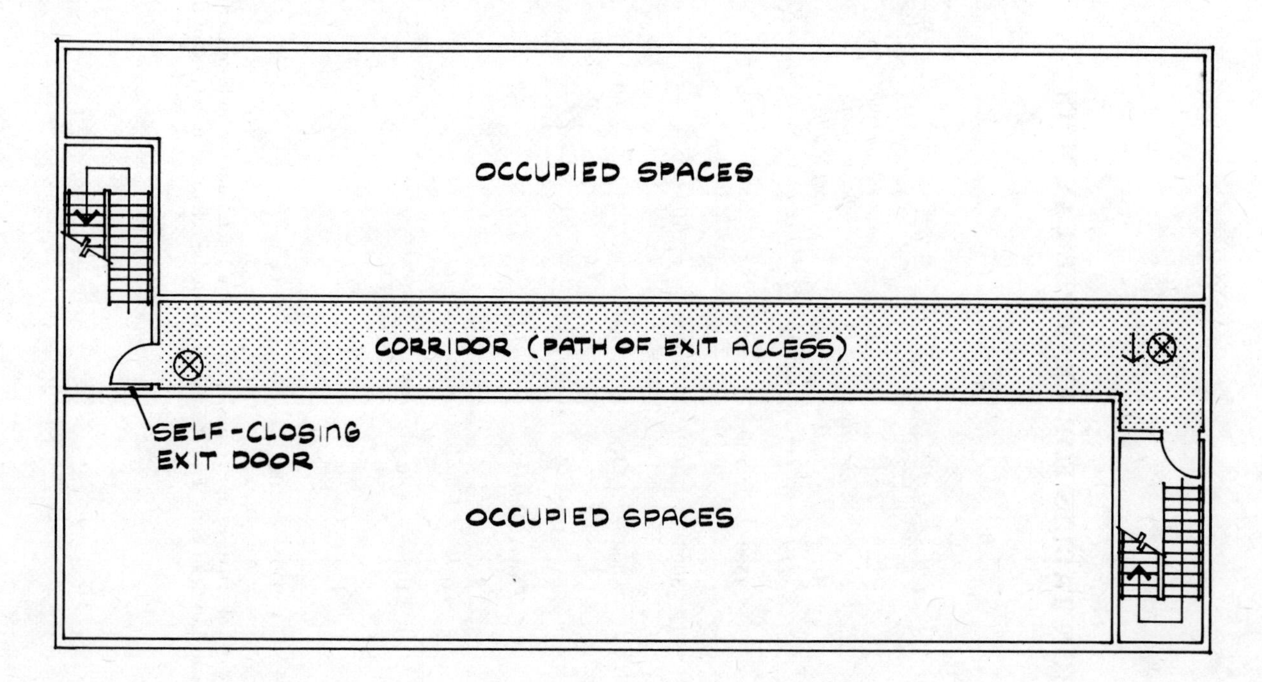

⊗ EXIT SIGN
(ILLUMINATED BY RELIABLE LIGHT SOURCE, E.G., SELF-CONTAINED BATTERY POWERED UNIT FOR USE DURING ELECTRICAL POWER FAILURE)

⊗→ DIRECTIONAL EXIT SIGN
(HAS ARROW INDICATING DIRECTION TO EXIT)

Note: Exit access also must be clearly marked (e.g., by wall graphics, patterns on floor coverings) and be free of visual obstructions.

ESCAPE AND REFUGE—VISIBILITY OF EXIT SIGNS IN SMOKE-FILLED ROOMS

Smoke can prevent occupants from locating exits and can restrict rescue and fire-fighting operations. The graph below can be used to find the visibility in feet for various illumination levels in lux (1 fc equals 10.76 lux) in rooms with burning materials. It is extremely difficult to predict the smoke density in buildings from actual fires as fire growth rate, air movement, ventilation rate, and so on can vary considerably. Nevertheless, the graph can be used to demonstrate the effect of room volume and area of burning material on visibility for various materials of specific thickness tested in the smoke density chamber (described on the following page). The smoke density factor (D_f) is found by the formula:

$$D_f = D_m \times \frac{A}{V}$$

where D_m = maximum optical density (see table on following page)
A = area of material producing smoke in sq ft
V = room volume in cu ft

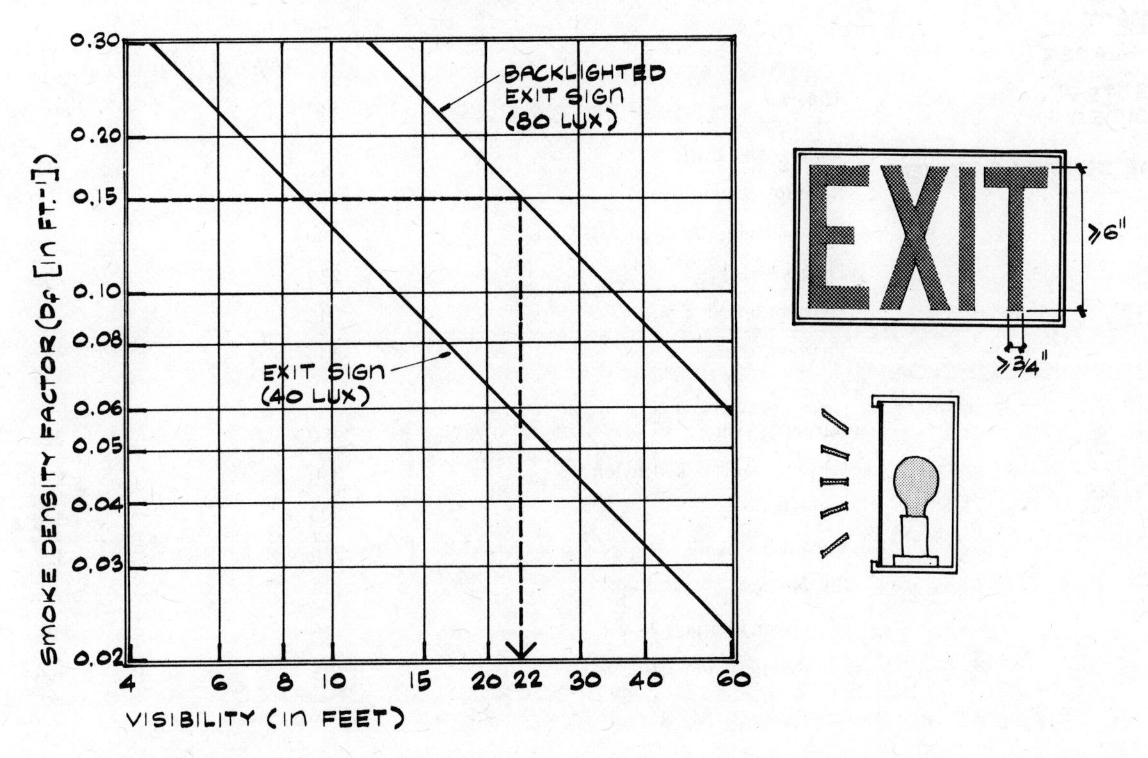

Example—Use of Graph

Given: A 60,000 cu ft room has a 40 sq ft area rug with a flaming optical density (D_m) of 220 measured by laboratory tests on sample specimens.

Find: What will be the visibility in feet for a backlighted exit sign?

First, calculate $D_f = D_m \times \frac{A}{V} = 220 \times \frac{40}{60{,}000} = \boxed{0.15}$. Next, enter graph at $D_f = 0.15$ and read opposite "backlighted exit sign" curve to a visibility distance of $\boxed{22 \text{ ft}}$. (See dashed lines on graph.)

SMOKE DENSITY

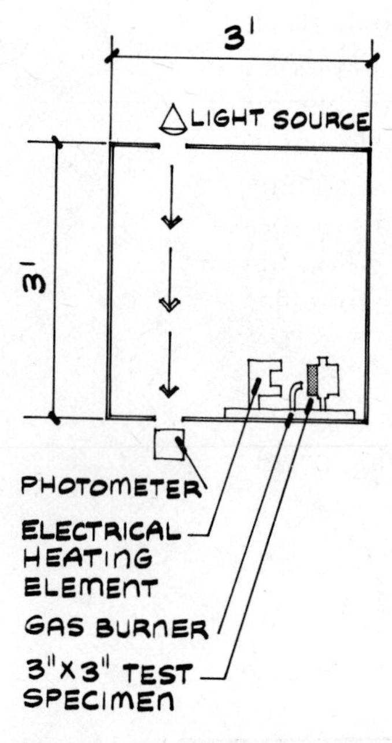

SMOKE DENSITY CHAMBER

The reduction of light transmission within a small volume test chamber can be used to measure the maximum optical density (D_m) of smoke from burning test specimens. The range of D_m measurements, covering the smoke generation potential of most building materials, is from 0 (no light obscuration) to 800. The table below shows optical density data for materials tested while "flaming" and smoldering (called "nonflaming") to represent actual burning conditions. Note that optical density data are not fundamental material properties. They are specimen properties representing smoke generation for the test specimen thickness and configuration imposed by the test conditions.

As shown by the sketch in the margin, a 3 in. × 3 in. test specimen is ignited either electrically (for smoldering test) or by a gas burner (for flaming test). Optical density in the 3 ft wide × 2 ft deep × 3 ft high test chamber is measured by a photometer.

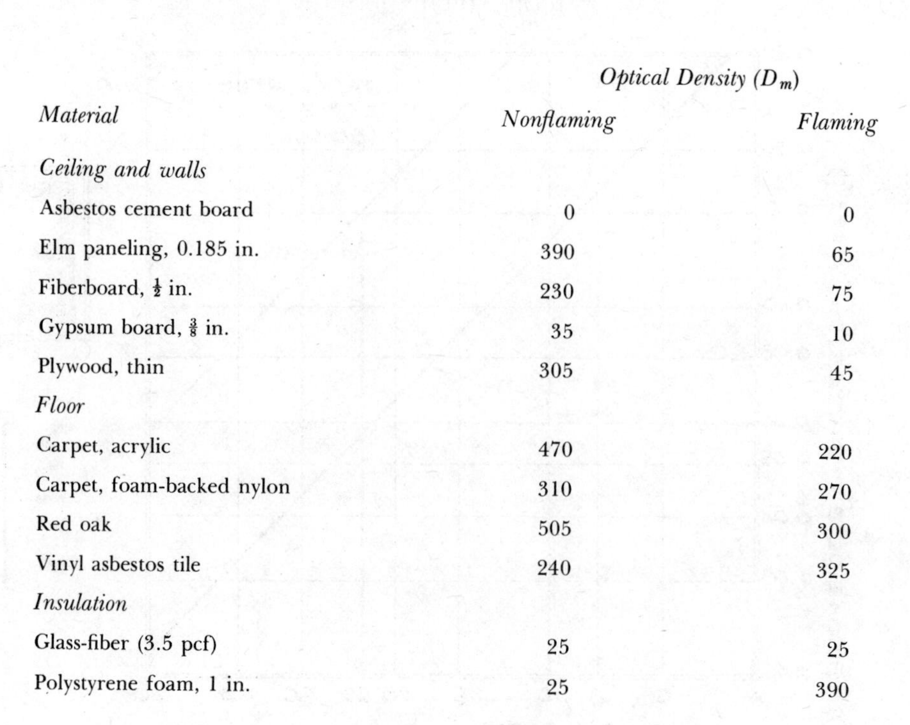

Material	*Optical Density (D_m)* Nonflaming	*Flaming*
Ceiling and walls		
Asbestos cement board	0	0
Elm paneling, 0.185 in.	390	65
Fiberboard, $\frac{1}{2}$ in.	230	75
Gypsum board, $\frac{3}{8}$ in.	35	10
Plywood, thin	305	45
Floor		
Carpet, acrylic	470	220
Carpet, foam-backed nylon	310	270
Red oak	505	300
Vinyl asbestos tile	240	325
Insulation		
Glass-fiber (3.5 pcf)	25	25
Polystyrene foam, 1 in.	25	390

SOURCE

Lee, T. G., "The Smoke Density Chamber Method for Evaluating the Potential Smoke Generation of Building Materials," U.S. National Bureau of Standards, NBS TN 757, January 1973.

ESCAPE AND REFUGE—SMOKEPROOF ENCLOSURES

"Smokeproof" enclosures are stairway exits designed to prevent untenable conditions from smoke infiltration. As a consequence, they also can provide a reliable means of access for fire fighters to upper stories of buildings. Shown below are various arrangements of smokeproof enclosures.

Perimeter Enclosure

To conserve space along the building perimeter, the vestibule can be designed to open into a large inner court.

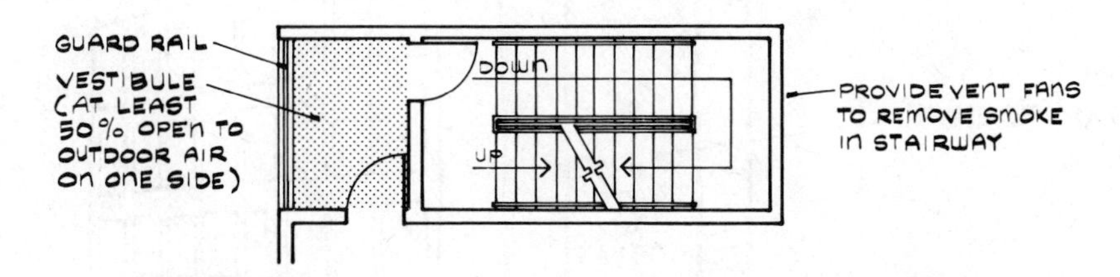

Note: Wind can retard or reverse the flow of smoke through the vestibule to the outdoors. In addition, an open vestibule (or balcony, porch) can be slippery during wet or icy weather conditions.

Isolated Enclosure

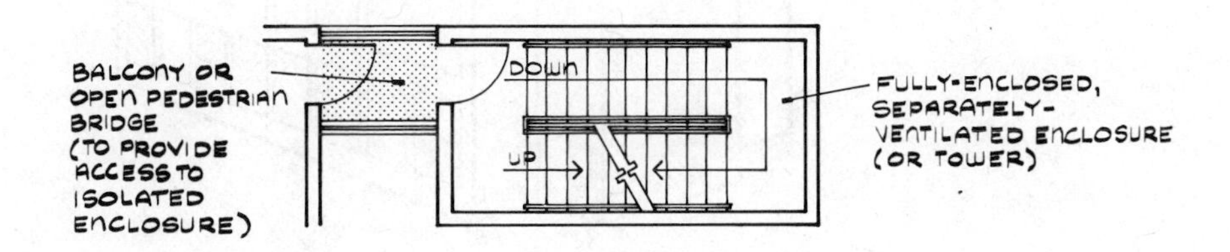

Pressurized Enclosure

Stairways can be free of smoke if a positive air pressure is established in the stairway enclosure and a negative air pressure is established in the vestibule.

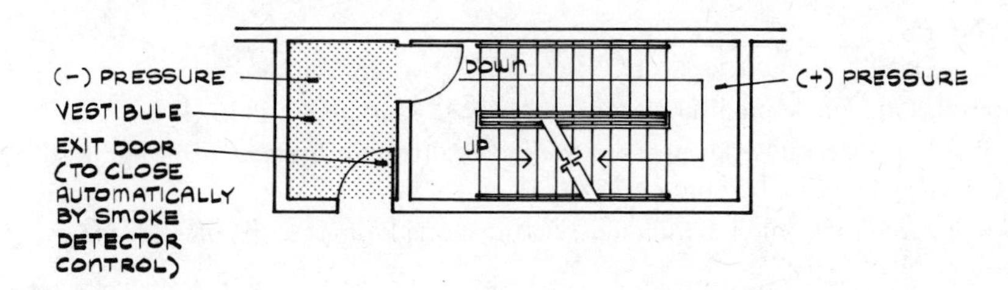

ESCAPE AND REFUGE—PRESSURIZED SMOKEPROOF TOWER

To prevent smoke infiltration, stairway towers can be pressurized by mechanically supplying more air than is exhausted. Positive pressure can be achieved by air supply into the tower at the upper or lower end, by separate air duct system (to contain the air pressures needed for vertical distribution), or by supply air fans located vertically along the tower perimeter as shown below. Towers with air supply fans at the lower end can prevent smoke movement to grade level where fire fighters may be entering and occupants leaving the building.

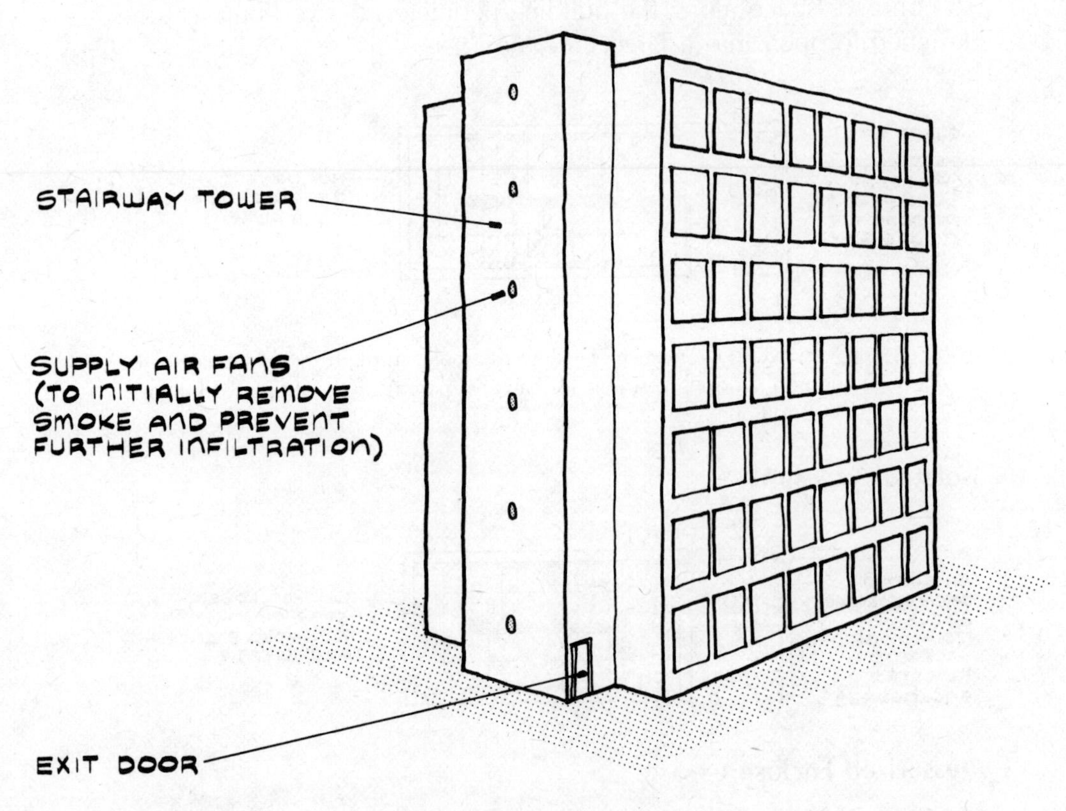

Note: Where pressurized stairway towers open directly to the outdoors at grade level, provide an additional door between grade level and the floor above to prevent loss of tower pressure from door openings.

REFERENCES

Degenkolb, J. G., "Smokeproof Enclosures," *ASHRAE Journal,* April 1971.

Erdelyi, B. J., "Test Results: Ducted Stairwell Pressurization System in a High-Rise Building," *ASHRAE Journal,* February 1976.

Webb, W. A., "Smoke Control in Buildings: A Threat or a Promise?," *ASHRAE Transactions,* Vol. 81, Part 2, 1975.

ESCAPE AND REFUGE—EMERGENCY USE OF ELEVATORS

Elevators, located in mechanically pressurized vestibules or smokeproof towers, can be used by fire fighters for rescue and suppression operations and by elderly and handicapped persons for limited emergency escape or movement to a refuge area. Elevators are normally designed to move a certain percentage of the building population during the peak 5 minutes of elevator demand, not to move everyone when they try to leave at the same time. Under these design conditions, complete evacuation of buildings by elevators typically would take 20 to 45 minutes for offices, $1\frac{1}{4}$ to $1\frac{3}{4}$ hours for apartments, and $\frac{3}{4}$ to $1\frac{1}{4}$ hours for hotels.

Smokeproof Tower

Pressurized or isolated stairway enclosures can be designed to accommodate elevators as shown below.

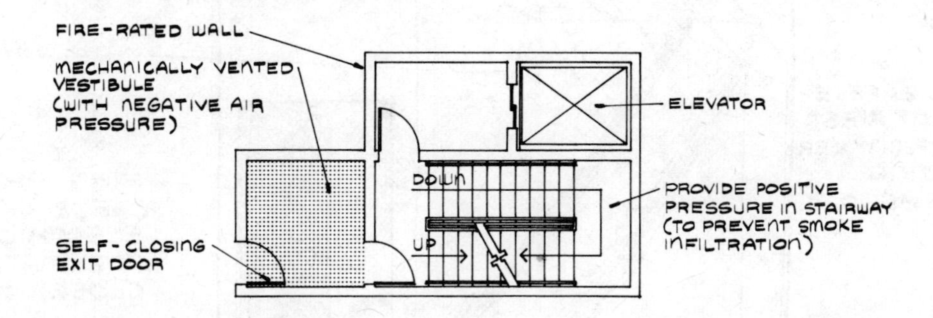

Pressurized Vestibule

Vestibules can be pressurized to create a "buffer" zone surrounding elevators. Self-closing doors are used to isolate the vestibule (or lobby) as shown below.

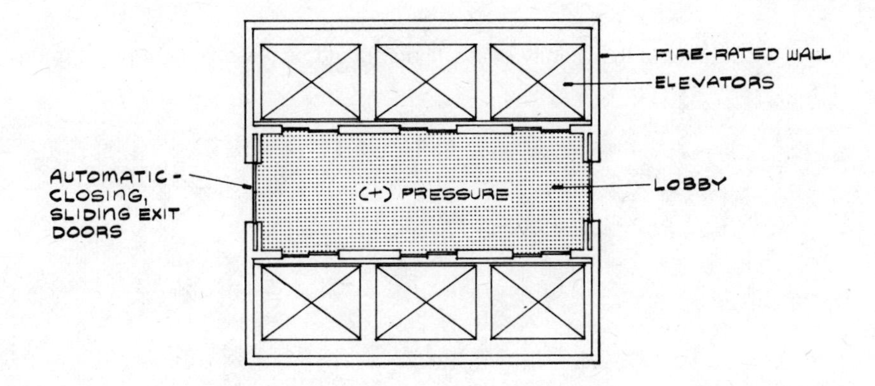

Note: Sufficient emergency electrical power should be provided to operate elevators, fans, and emergency lighting.

REFERENCE

"Tall Building Criteria and Loading," Vol. 1b in *Planning and Design of Tall Buildings,* New York: American Society of Civil Engineers, August 1972.

ESCAPE AND REFUGE—ESCALATOR FIRE PROTECTION (ENCLOSURES)

Two fire-rated enclosures, one for "down" escalators and one for "up" escalators, can be used to prevent the spread of smoke and fire between floors through escalator openings. Doors, as shown in the sketch below, can be held open by electrical or pneumatic holder-closer devices that release the doors when a fire is detected. The enclosure should have a fire-resistance rating equal to that of the floor-ceiling assemblies. It is desirable to restrict unprotected escalator penetrations to no more than two floors (i.e., no more than three floors connected by escalator openings).

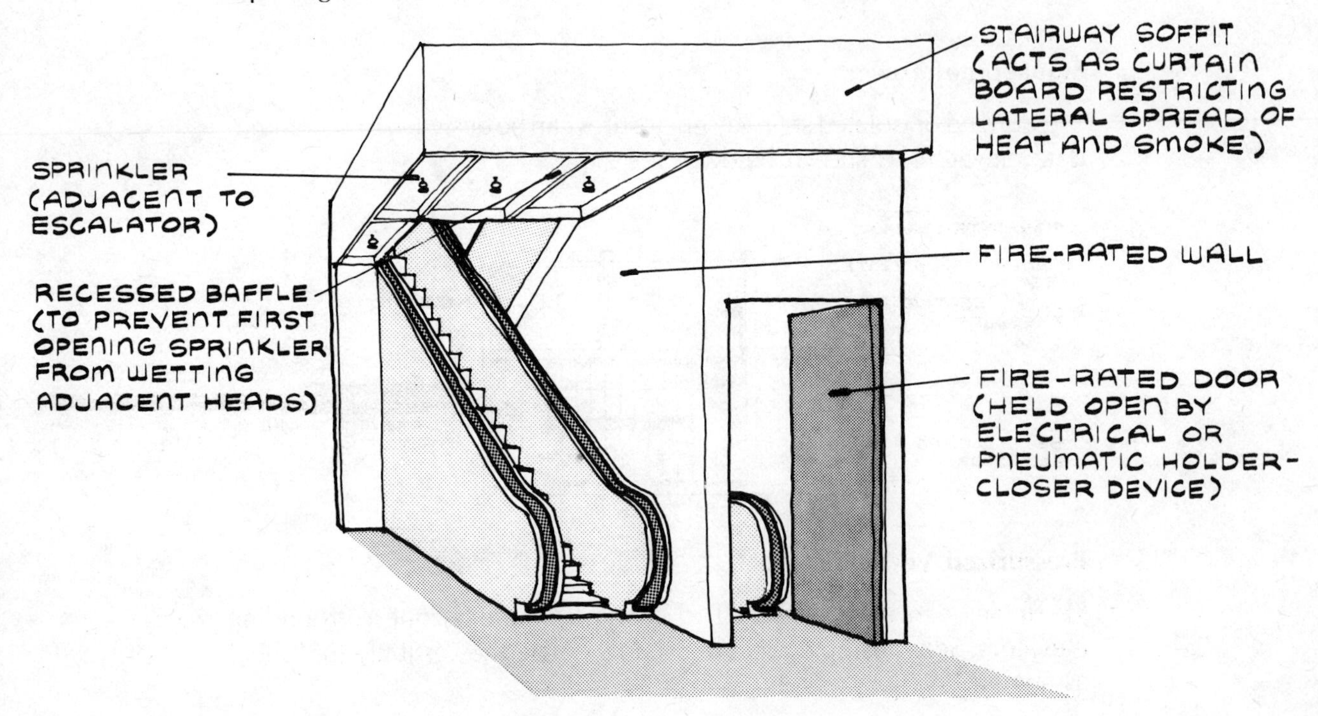

Note: Floor areas in the vicinity of escalators should be reasonably free of combustibles.

ESCAPE AND REFUGE—ESCALATOR FIRE PROTECTION (ROLLING SHUTTER)

A fire-resistant shutter device can be used to prevent the spread of smoke and fire between floors through escalator openings. When a fire is detected, the shutters should close automatically and the escalator should stop (blocking the escalator escape route). As a consequence, building occupants can only leave the fire floor by stairway exits. To avoid injury to persons in the openings, the shutters should close at a slow rate and have pressure-sensitive reversing devices along their front edges. Note that the openings where conveyors penetrate fire-rated partitions also can be protected by rolling shutter devices.

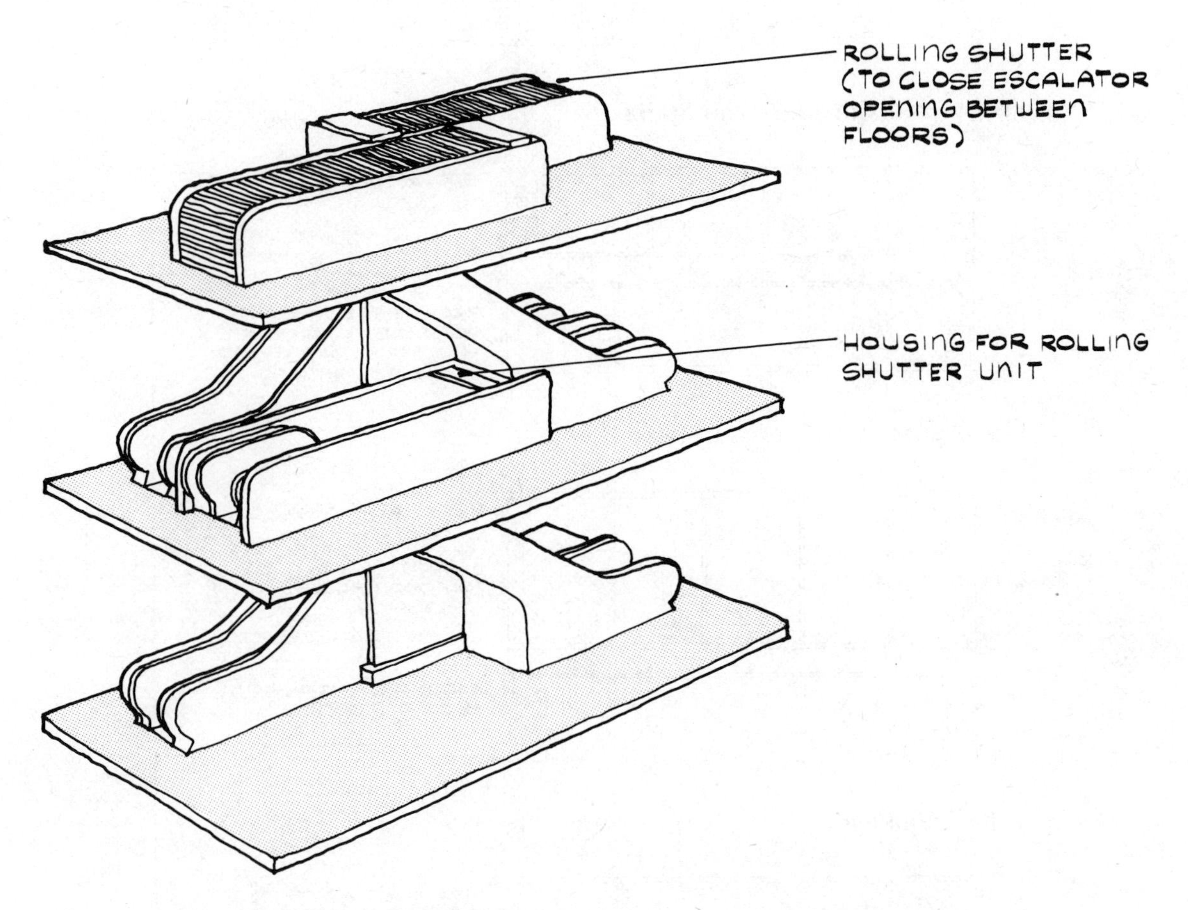

Note: See Chapter 4 for opening protection by sprinkler water spray devices.

REFERENCE

Mitchell, N. D., E. D. Bender and J. V. Ryan, "Fire Resistance of Shutters for Moving-Stairway Openings," Building Materials and Structures Report No. 129, U.S. National Bureau of Standards, March 1952.

ESCAPE AND REFUGE—EXAMPLE LIFE SAFETY MEASURES

Shown below, in order of increasing effectiveness, are example life safety measures for low-rise buildings.

Pressurized Stairs

Pressurized Corridors and Stairs

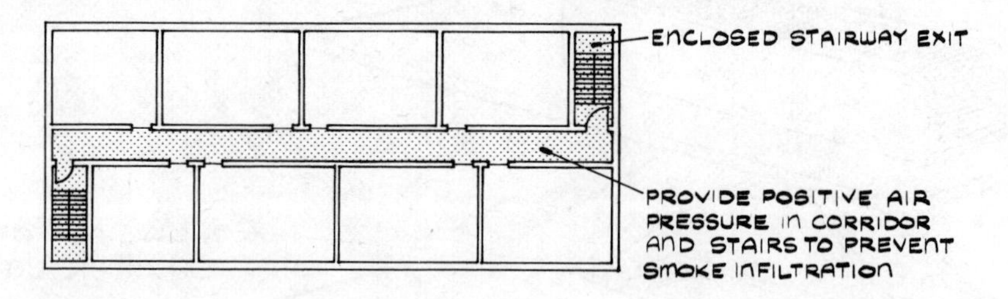

Open Corridor Access to Stairs

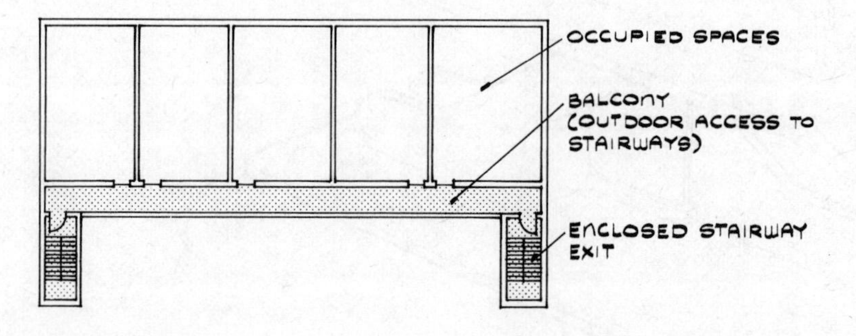

Fully Sprinklered

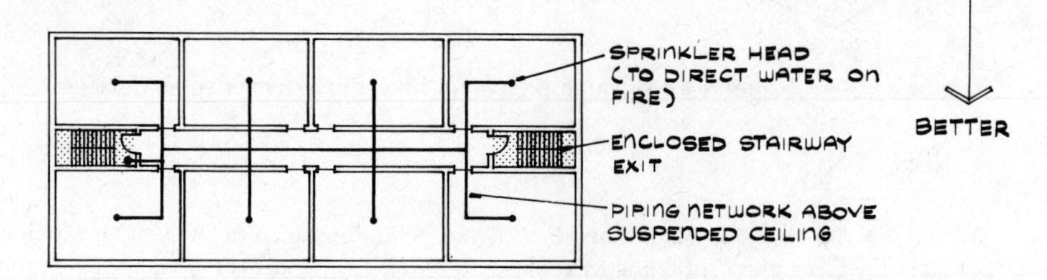

Note: For a detailed discussion of sprinkler systems, see Chapter 4.

ESCAPE AND REFUGE—EVACUATION TIMES FOR TALL BUILDINGS

The graph below, based on data from actual evacuations in Canadian buildings, shows that evacuation time depends on the number of persons being evacuated and the available stairway width. The data represent several test evacuations in 8 to 29 story buildings. For example, an actual evacuation population of 600 persons /22 in. of stairway width in an office building can be evacuated (i.e., occupants leave at ground level) in 21 minutes (see dashed lines on graph). Evacuation times in cold weather, when occupants wear or carry bulky outdoor clothing, can take considerably longer than evacuations during warm weather.

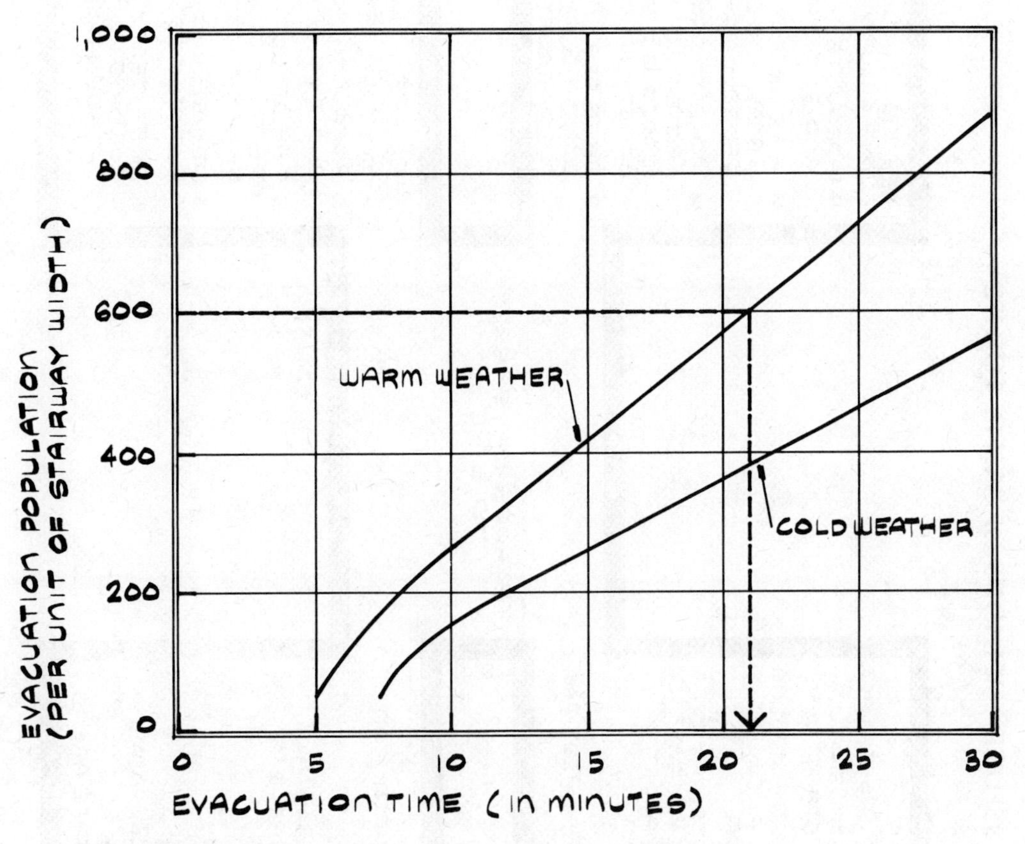

REFERENCE

Pauls, J. L., "Evacuation and Other Fire Safety Measures in High-Rise Buildings," *ASHRAE Transactions*, Vol. 81, Part 1, 1975.

ESCAPE AND REFUGE—METHODS OF SAFETY TO LIFE

METHODS OF SAFETY TO LIFE

MINIMAL DISRUPTION
(TO DEFEND OCCUPANTS IN PLACE)

- PREVENT HEAT INTRUSION WITH FIRE-RATED CONSTRUCTIONS
- PREVENT SMOKE INFILTRATION WITH BARRIERS AND PRESSURIZATION
- MANUAL OR AUTOMATIC SUPPRESSION SYSTEMS (TO DEFEND AGAINST HEAT INTRUSION)
- VENTS (TO DEFEND AGAINST SMOKE INFILTRATION)

REFUGE
(MOVE OCCUPANTS TO SAFE AREA)

- ADEQUATE NUMBER OF EXITS AND EXIT CAPACITY
- FIRE DETECTION, WARNING SIGNAL, AND CONDITIONS FOR ORDERLY MOVEMENT
- MOVE TO SAFE AREA THROUGH PROTECTED ROUTES (ENCLOSED BY FIRE-RATED CONSTRUCTIONS)
- SAFE DESTINATION IN BUILDING (OF ADEQUATE CAPACITY)

ALARM WARNING
(SEE CHAPTER ON DETECTION AND SUPPRESSION)

- FIRE DETECTION (LOCATE DETECTORS IN AREAS OF PROBABLE FIRE OCCURANCE)
- WARNING SIGNAL TO OCCUPANTS
- INSTRUCTIONS TO ENDANGERED OCCUPANTS
- INSTRUCTIONS TO OCCUPANTS IN SAFE AREAS

ESCAPE
(OCCUPANTS LEAVE BUILDING)

- ADEQUATE NUMBER OF EXITS AND EXIT CAPACITY
- FIRE DETECTION, WARNING SIGNAL, AND CONDITIONS FOR ORDERLY MOVEMENT
- LEAVE BUILDING THROUGH PROTECTED ROUTES
- SAFE DESTINATION OUTSIDE BUILDING (OF ADEQUATE CAPACITY)

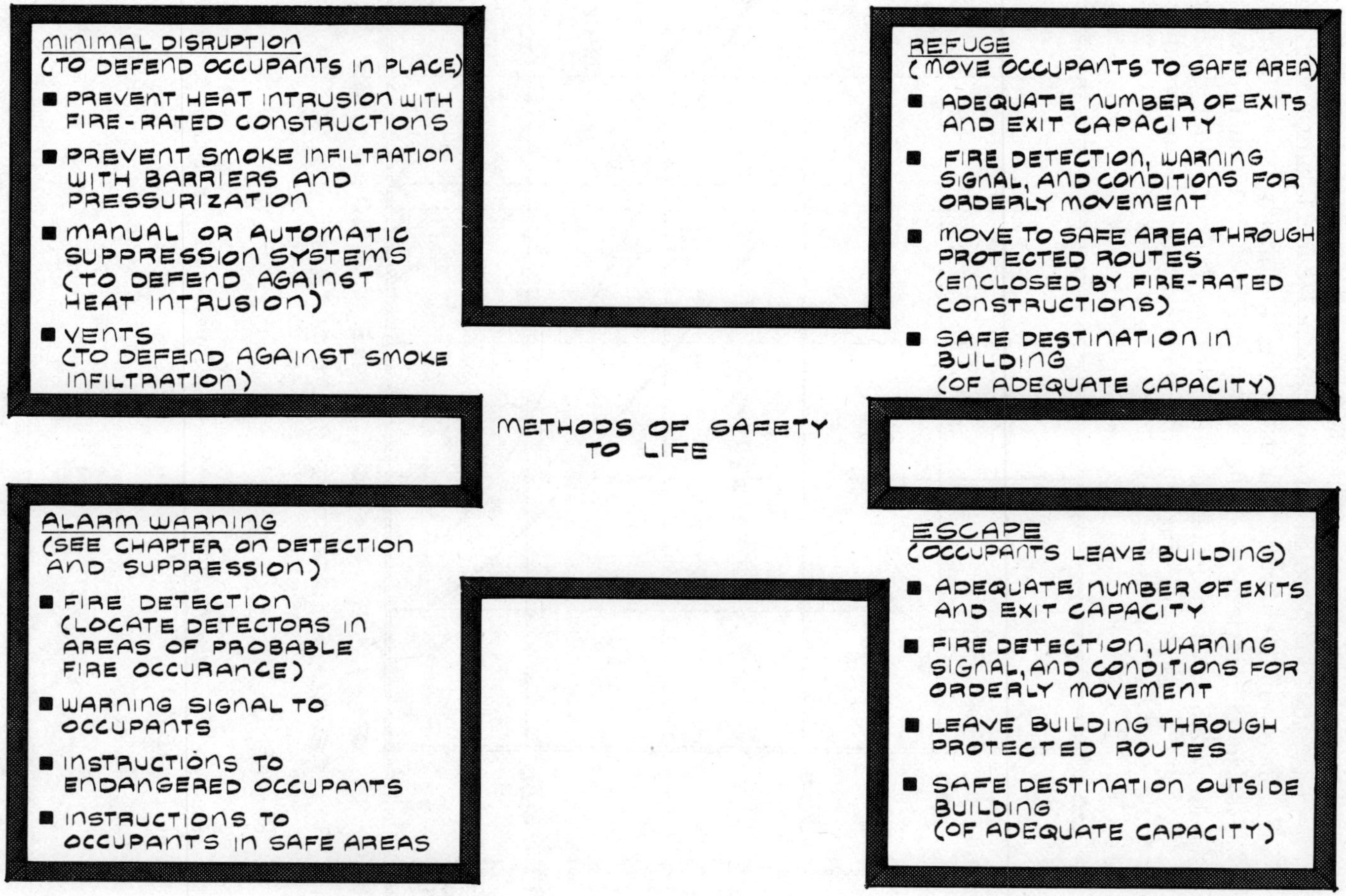

ESCAPE AND REFUGE CHECKLIST

Determine exit requirements by evaluating the building occupancy (e.g., type of office, school, store), fire hazard, and the number and mobility of the users. Access to exits should be direct, unobstructed, well-illuminated, and clearly marked. Locate manual fire alarm boxes in the natural path of egress, near exits. (Manual fire alarm systems can be designed to automatically transmit a signal to the fire department or another location.)

In general, a minimum of two exits, located remotely from each other, will be required. Exits are considered remote by some codes if the distance measured in a straight line between exits is greater than one-half of the overall diagonal dimension of the area being served.

Avoid dead-end corridors. During a fire, corridors can become smoke-filled and trap occupants in the dead-end areas beyond the exits. Occupants also can bypass exits and become trapped in the dead ends.

Occupied rooms should not be within the access routes to exits as furniture, equipment, space dividers, and the like may obstruct egress. It is especially important to avoid access to an exit through rooms subject to locking.

Do not place storage rooms, trash rooms, and so on within the exit access as fire may originate in these spaces. These hazardous areas should be isolated by fire-resistant partitions and doors.

Use automatic-closing door hardware (e.g., door held open by electromagnetic or pneumatic device actuated by smoke detector adjacent to door or mounted on the frame) on exit doors placed in the paths of everyday traffic. At these locations, self-closing doors are frequently propped open by occupants with rubber stops, blocks, and so on eliminating the intended barrier to fire spread.

Do not obstruct egress routes with furniture, water coolers, and the like and avoid exit doors that restrict corridor access to exits by swinging into the corridor.

Smokeproof stairway enclosures can provide a reliable means of egress for building occupants and access for fire fighters. However, enclosed stairways do not provide the opportunities for effective police surveillance of intruders. For improved building security, exit doors can be latched to allow entry into stairways at any floor but exit only at ground level (or at the refuge floor). Where access to floors designated as refuge areas is required in tall buildings, doors can be unlocked from an emergency control center.

Unmarked exit doors can be easily bypassed during a fire. Do not cover exits (e.g., with curtains) or disguise them (e.g., with mirrors on exit doors or on opposite walls). Use signs adjacent to stairway exit doors to indicate the floor number and the upper and lower stairway terminal points (e.g., top floor, roof, street). Doors that could be mistaken for exit doors also should be clearly marked.

Emergency lighting can be provided by (1) separate circuits to designated fixtures, (2) standby power from electric generators or central bank of storage batteries, and (3) fixture units operated by self-contained charger and battery (e.g., sealed beam incandescent or fluorescent units).

TALL BUILDINGS

TALL BUILDINGS

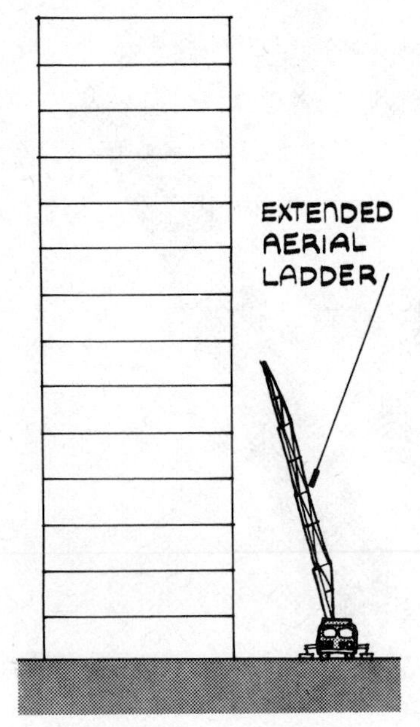

HIGH-RISE BUILDING

Low-rise buildings less than six stories are within the reach of some aerial fire ladders and fire hose streams and can be designed to be evacuated within a reasonable amount of time. However, for tall buildings from 6 to 12 stories reasonable evacuation times are only possible if considerable exit capacity is provided. In addition, tall building fires cannot be fought from the outside as upper stories are generally beyond the reach of aerial ladder equipment. The usual height limitation for aerial ladder operations is about 75 ft due to wide sidewalks, parked vehicles, and so on. Tall buildings also have the potential for significant "stack effect" due to the inside-outside temperature difference at cold climate locations. The stack effect (i.e., air infiltration at low building levels and exfiltration at high levels) can accelerate fire spread throughout the building. In addition, it is often difficult to quickly locate fires in tall buildings as smoke may have spread vertically several stories beyond the origin.

For buildings greater than about 12 stories, complete emergency evacuation is not practical and occupants must be protected internally for the duration of the fire.

From the perspective of building firesafety, the United States General Services Administration defines a high-rise (or tall) building as "one in which emergency evacuation is not practical and in which fires must be fought internally because of height. The usual characteristics of such a building are (1) it is beyond the reach of fire department aerial equipment, (2) it poses a potential for significant stack effect, and (3) it requires unreasonable evacuation time." (cf., *Public Buildings Service International Conference on Firesafety in High-Rise Buildings*, Washington, D.C.: General Services Administration, May 1971.)

TALL BUILDINGS—RESCUE BY LADDERS AND AERIAL EQUIPMENT

Shown below are typical height limitations for fully extended rescue ladder and aerial equipment. Ladder rescues from any height are normally hazardous operations! Evacuation by aerial platform (or cherry picker) is not continuous and is limited by the capacity of the platform (or basket).

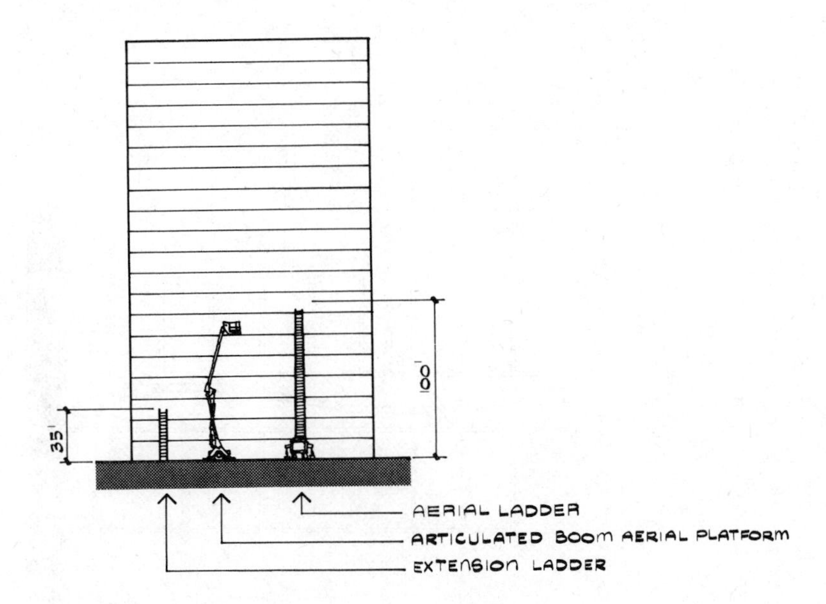

To achieve maximum vertical reach for rescue operations, the aerial platform apparatus must be positioned (called "spotted") close to the building as shown below. Proper positioning is influenced by direction of fire travel within the building, overhead wires and other obstructions, and space needed for outriggers and fire apparatus arriving later. Aerial platforms should be used on firm, level ground, and steep ground slopes should be avoided.

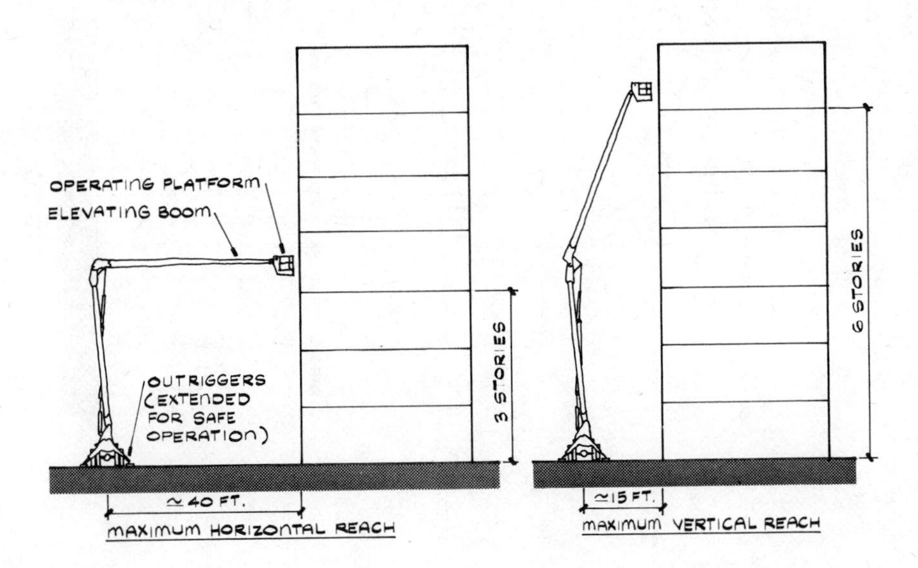

TALL BUILDINGS—STACK EFFECT

In cold weather, the stack (or "chimney") effect in tall buildings causes an upward movement of air. If a fire occurs, smoke can easily move to the upper floors through vertical shafts and stairwells. In warm weather, the stack effect is reversed with air flowing out of the building at low levels and in at upper levels. However, the airflow volumes are lower than in cold weather because the inside-outside temperature difference is generally smaller. The principal effect of wind as shown below is horizontal air movement with exfiltration on one side and infiltration on the other.

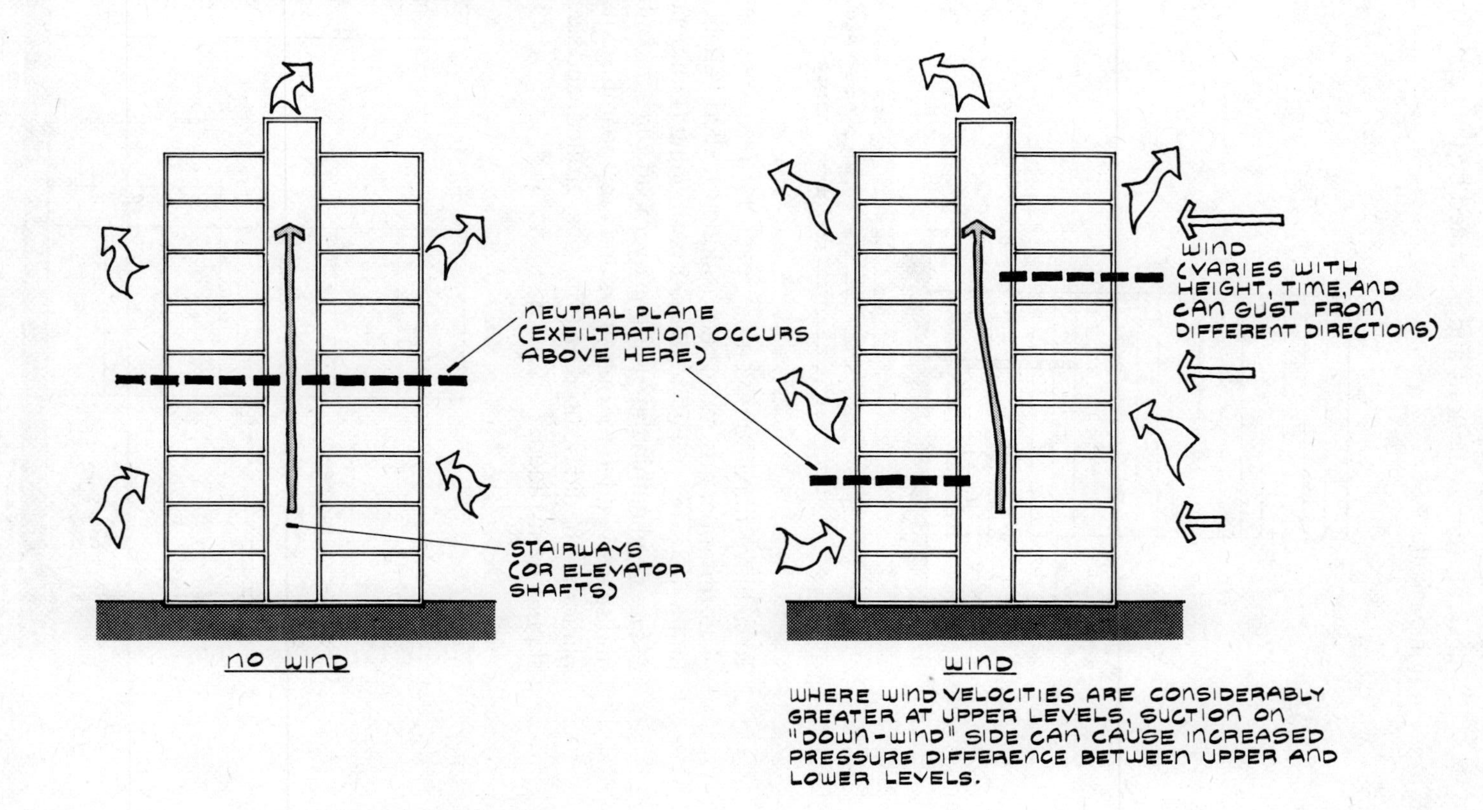

WIND SPEED AND PRESSURE

The table below describes some typical effects of wind on people. Wind speed, expressed in miles per hour (mph), is presented along with the corresponding wind pressure (for air at 70°F) in inches water gauge (in. w.g.). At elevated fire temperatures in tall buildings subject to stack effect, relative air pressures remote from the fire can be greater than 0.10 in. w.g. Air distribution systems can be designed to help control smoke movement from the fire floor by pressurizing adjacent and remote areas.

*Wind Speed (in mph)**	*Wind Pressure (in in. w.g.)*	*Typical Effect of Wind on People*
0 to 2	0 to 0.002	No noticeable wind
2 to 10	0.002 to 0.05	Wind felt on face, hair disturbed
10 to 20	0.05 to 0.20	Hair and clothing disarranged
20 to 25	0.20 to 0.30	Wind force felt on body
25 to 30	0.30 to 0.45	Umbrellas used with difficulty

* Wind velocity in fpm is 88 times a given wind speed in mph.

TALL BUILDINGS—STACK EFFECT FROM TEMPERATURE DIFFERENCE

The maximum pressure difference due to the temperature differential between inside and outside air temperatures in cold weather is shown by the graph. Indoor air temperature is 70°F for the data presented and the neutral plane is located at either the top or bottom of the building. At pressure differences of about 0.6 in. w.g. (1 in. w.g. = 0.04 psi) and greater it is difficult to open ground-level doors. Consequently, in tall buildings, vestibules are often provided. To help control smoke movement from the stack effect, walls and floor-ceiling constructions should be relatively airtight. In current construction practice, however, stairwells, mechanical service shafts, and elevator shafts are seldom completely sealed.

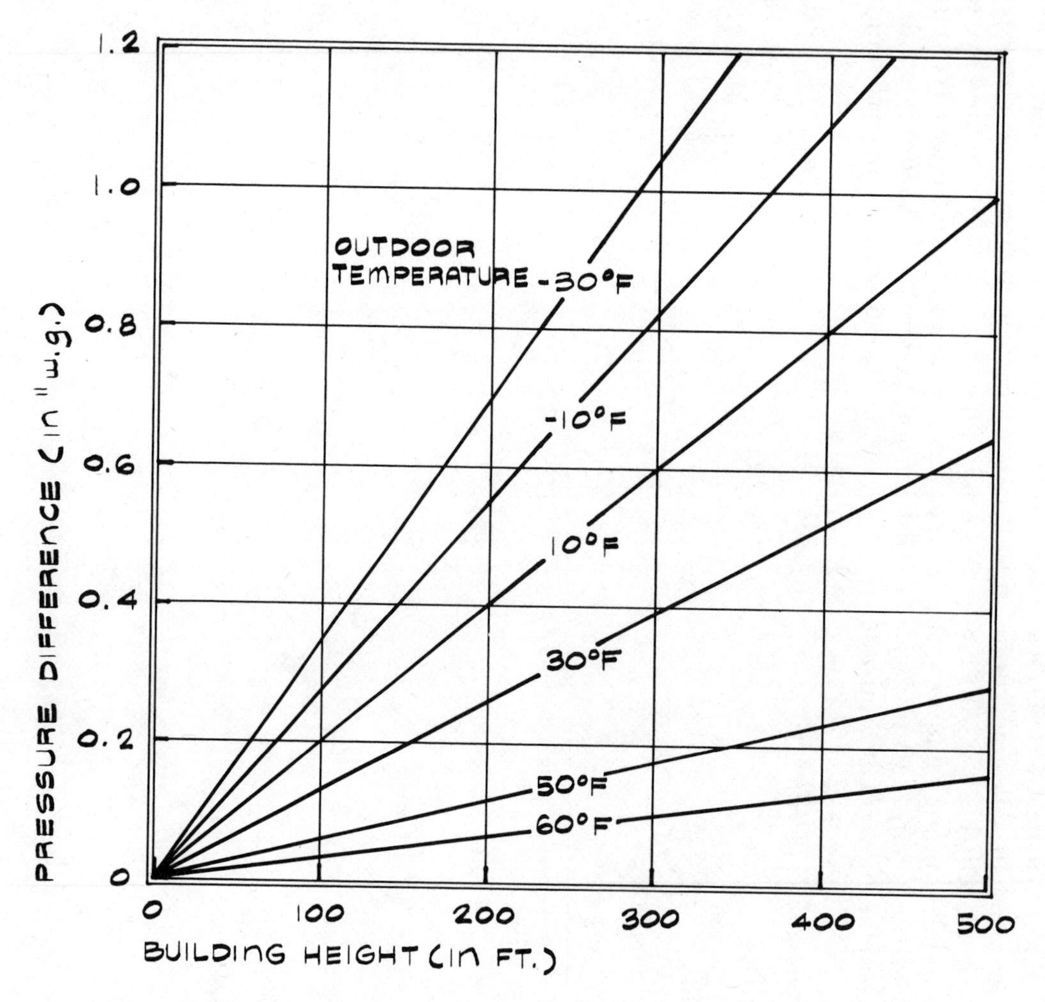

Note: For neutral planes located at building mid-height, the pressure difference would be one-half the values from the graph.

TALL BUILDINGS—EVACUATION TIMES

The graph below shows approximate evacuation times for stairway exits in tall buildings. Stair width is 44 in. and the building occupant load is 240 persons per floor. For example, a 30 story building will require 78 minutes for total evacuation through one stairway exit (see dashed lines on graph). A reduction in the occupant load would reduce the evacuation time (e.g., 120 persons per floor would require one-half the times indicated by the graph).

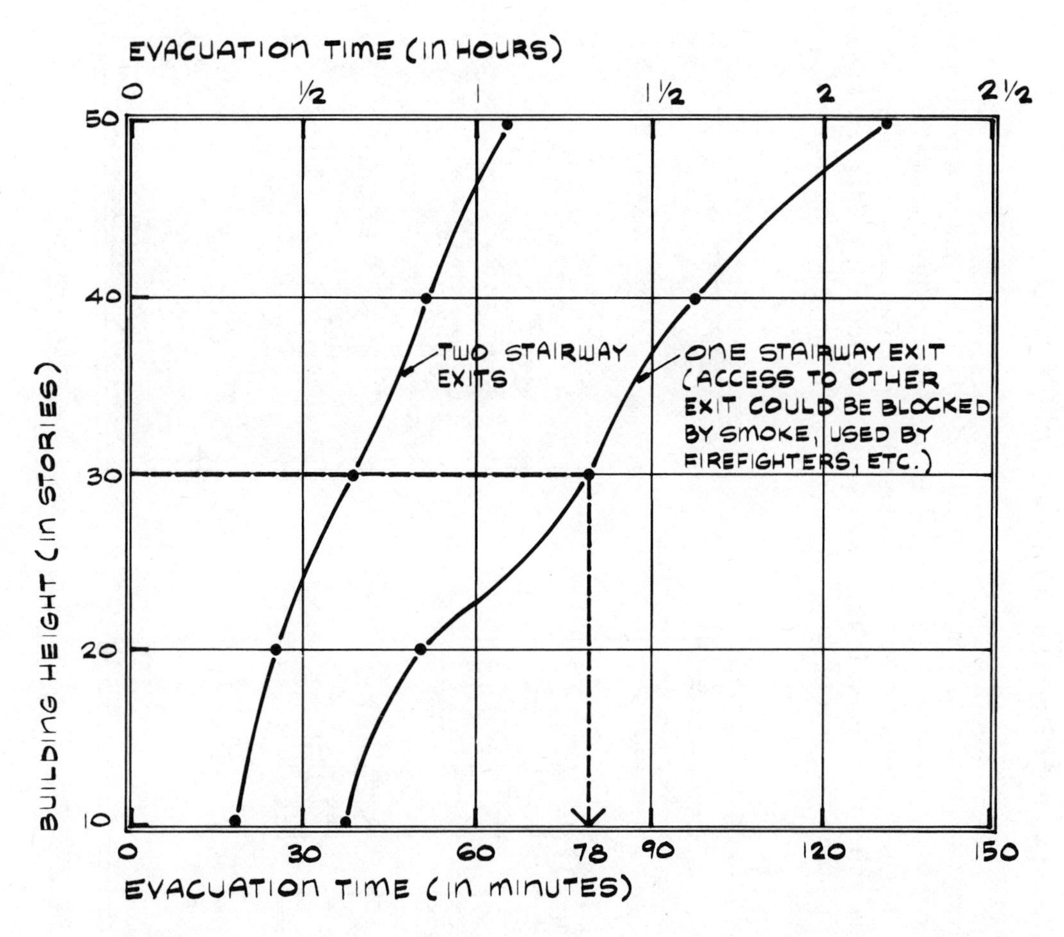

REFERENCE

Galbreath, M., "Time for Evacuation by Stairs in High Buildings," *Fire Fighting in Canada*, Vol. 13, No. 1, February 1969.

TALL BUILDINGS—SMOKE CONTROL (CONVENTIONAL ZONING)

Shown below is an example floor plan for a tall building which has separate fan rooms, each serving a zone consisting of several floor levels. In the example fire, some of the smoke would be exhausted to the outdoors and the smoke-filled return air would be diluted by the clean air from the other floor levels in the zone. As a consequence, the fans (with smoke detectors at the fan rooms) would not shut down immediately thus allowing smoke to spread to other floor levels.

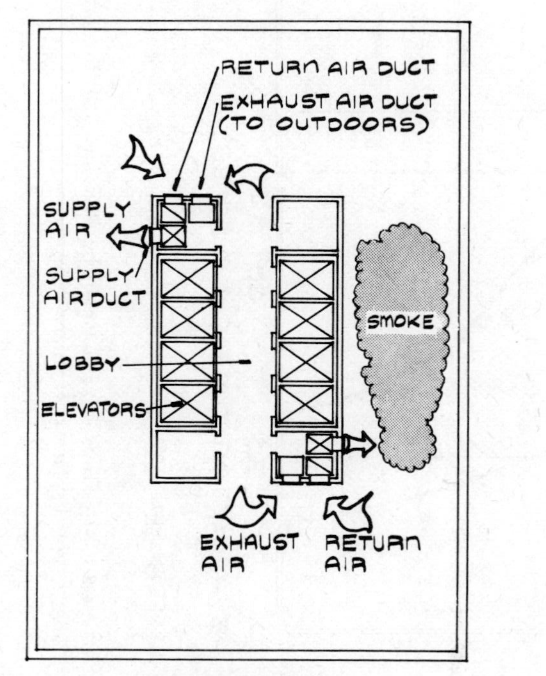

FIRE STARTS

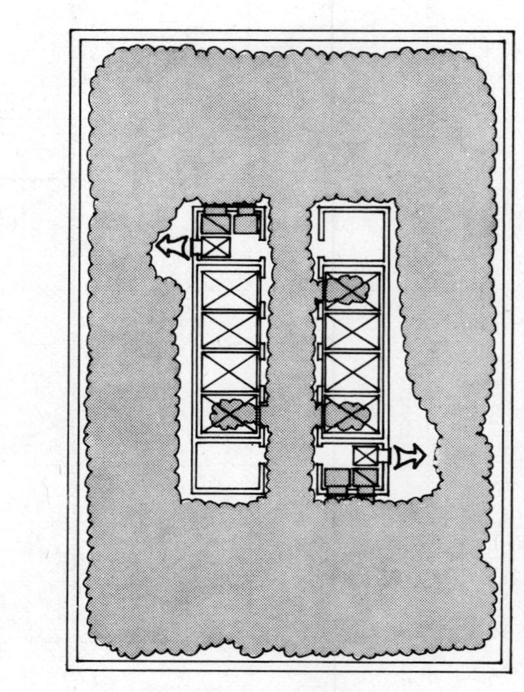

SPREAD OF SMOKE

TALL BUILDINGS—SMOKE CONTROL (SMOKE ZONING)

The example tall building from the preceding page can be divided into two or more smoke zones per floor. In the example fire, dampers would be activated to shut down the return and supply air ducts in the fire zone. Then exhaust air duct and exhaust vents into the smoke shaft remove smoke. Exterior panel vents or window vents could be provided to remove smoke. The return and exhaust air ducts would shut down in the adjacent zones to provide increased (or positive) pressure in the refuge areas. Refuge areas should have sufficient oxygen to support life and should not allow dangerous concentrations of toxic gases or intolerable temperatures to occur.

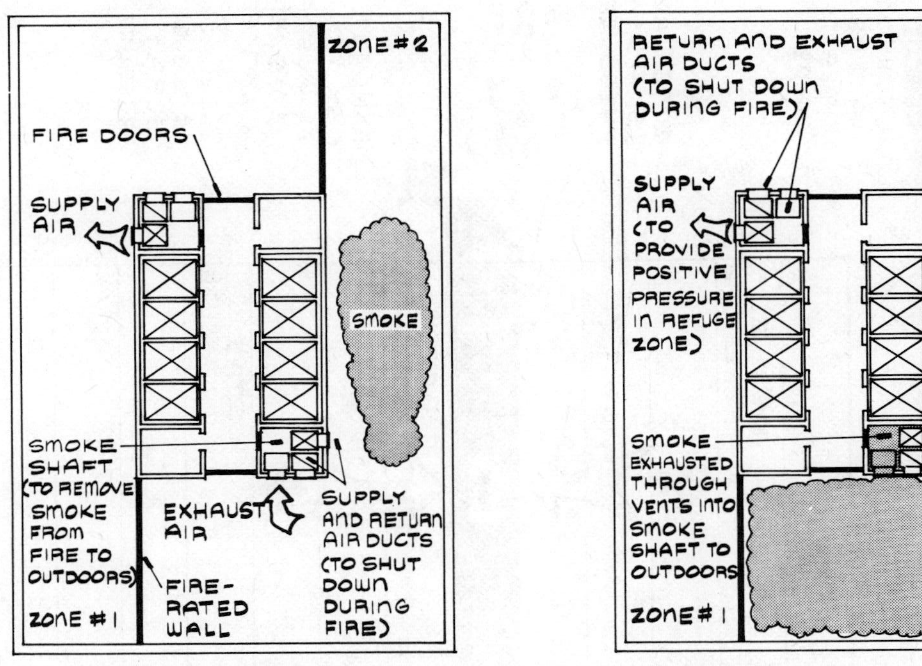

REFERENCE

Semple, J. B., "Smoke Control in High Rise Buildings," *ASHRAE Journal*, April 1971.

TALL BUILDINGS—SMOKE CONTROL AND HELICOPTER RESCUE

In the building shown below, smoke movement is controlled by the mechanical system. The fan supplying air to the zone or floor involved in the fire is turned off. The return air system from this zone is also shut down and the smoke is exhausted to the outdoors. The mechanical system serving the surrounding zones establishes an increased or positive air pressure above and below the fire zone by closing return and exhaust air dampers and keeping supply air dampers open. This confines smoke movement to the fire area.

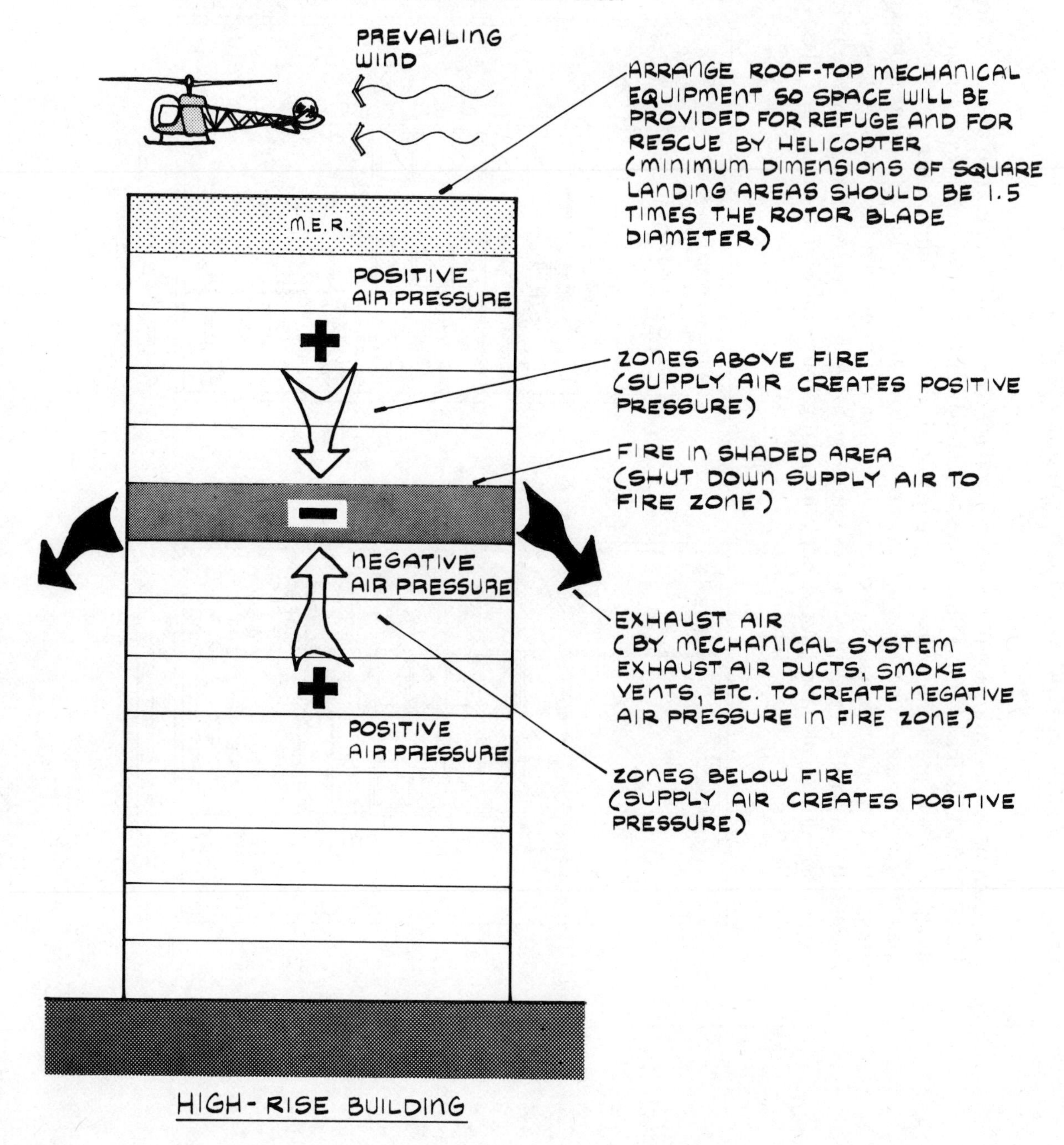

Note: Emergency helicopter rooftop landing areas should be located on the upwind side of buildings and the roof construction should be able to withstand impact loads from hard landings. In addition, the effects of adverse turbulence that could be caused by nearby tall buildings during high wind conditions must be evaluated.

TALL BUILDINGS—AIR DISTRIBUTION NETWORKS

Mechanical system air distribution networks can be divided into zones (e.g., by floors or subdivided floors) to isolate and contain fire. Shown below and on the following page is an example single-duct air distribution network which serves two zones. During a fire, the supply and return air ducts serving the fire zone are closed by fire dampers activated by smoke detectors. Exhaust vents open to vent smoke and hot gases to the outdoors. In tall buildings, air intake preferably should be below building mid-height and exhaust above mid-height so that inside-outside pressure differences will not interfere with system airflow.

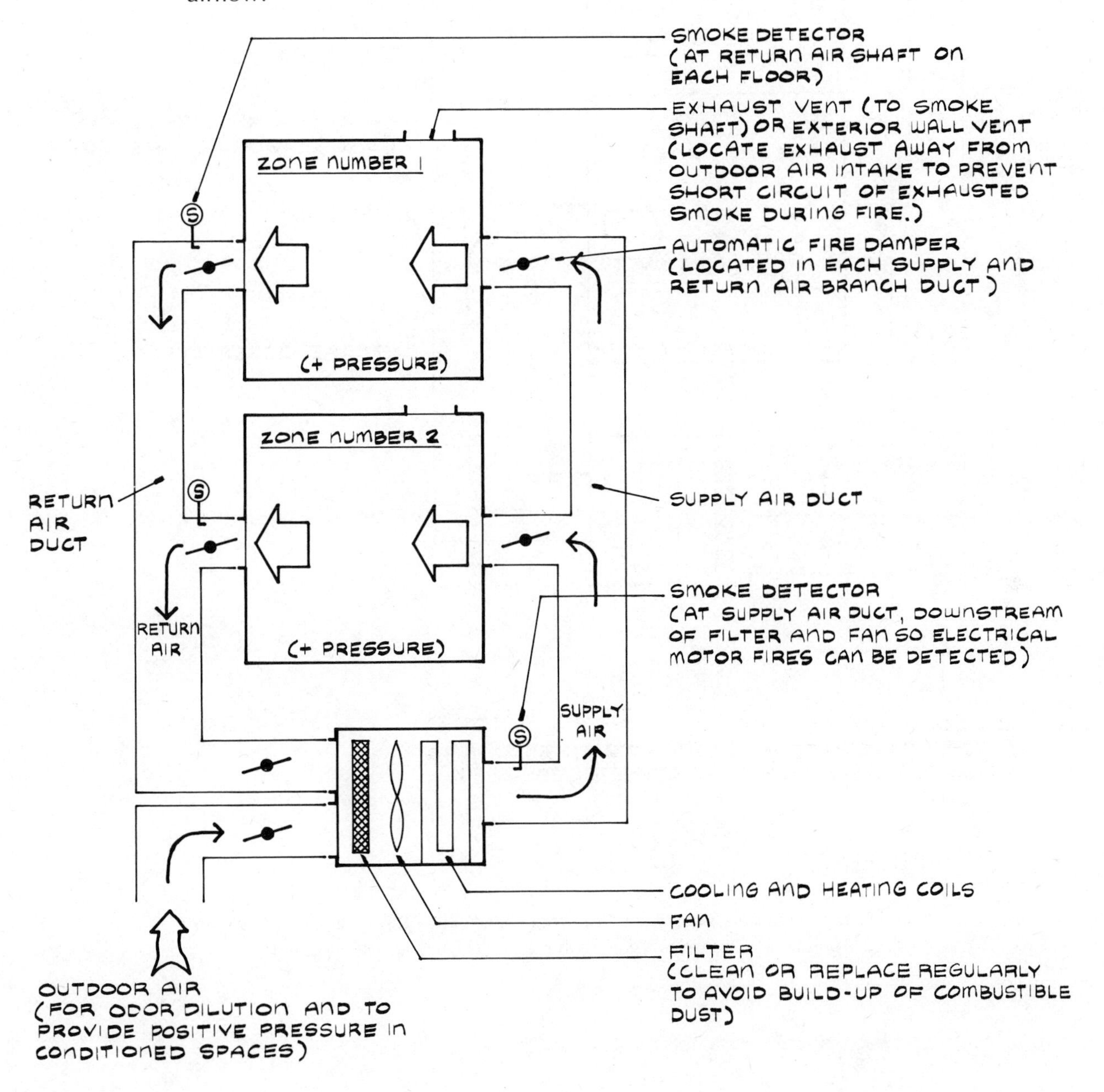

TALL BUILDINGS—AIR DISTRIBUTION NETWORKS (Continued)

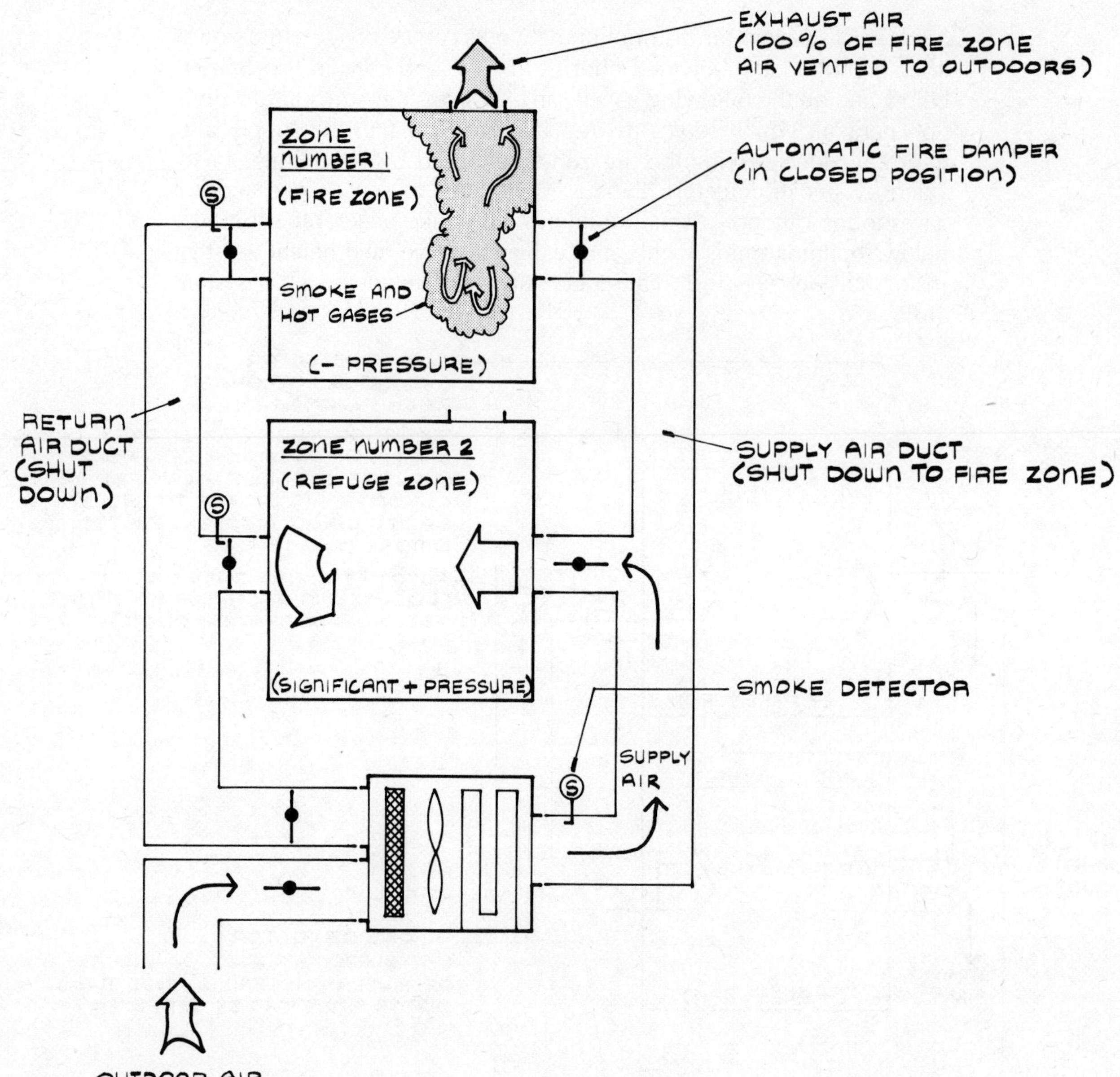

REFERENCE

Masters, R. E., "Bankers Trust Plaza-New York City: A High-Rise Office Building with a Systems Approach to Smoke and Fire Control," *ASHRAE Journal*, August 1973.

TALL BUILDINGS—FIRE AND SMOKE DAMPERS

Fire dampers, used to restrict heat flow through air ducts, also may prevent the spread of smoke if there is sufficient pressure difference on opposite sides of the damper. Single blade, multiblade, and curtain blade type fire dampers with $1\frac{1}{2}$ hour fire-resistance ratings are commercially available. However, smoke dampers are primarily for the control of smoke and are not necessarily fire dampers.

Multiblade Fire Damper

To compensate for reduction of fire resistance where air ducts penetrate fire-resistant constructions.

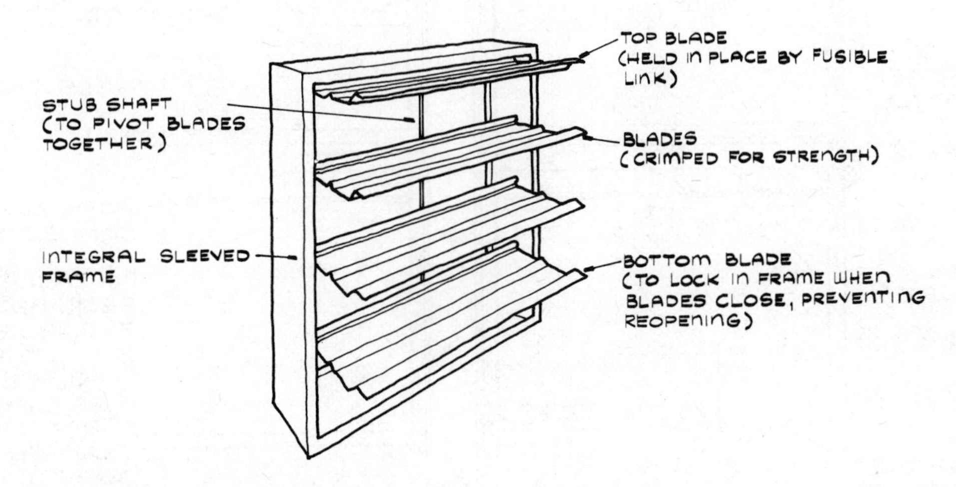

Curtain Blade Smoke Damper

To seal duct, restricting spread of smoke.

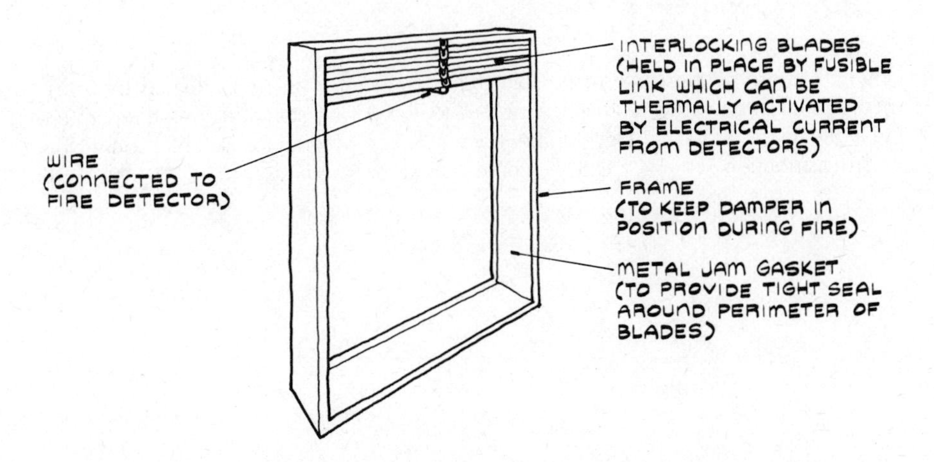

Note: An access door can be installed in a duct sleeve to allow servicing of the damper.

TALL BUILDINGS—FIRE DAMPERS

The sketch below shows examples where fire dampers are not normally required. It is adapted from "Installation of Air Conditioning and Ventilating Systems," NFPA No. 90A, 1976. This standard, which presents requirements for the location of fire dampers, attempts to compensate for the reduction in fire resistance of building constructions that is caused by the installation of air conditioning systems.

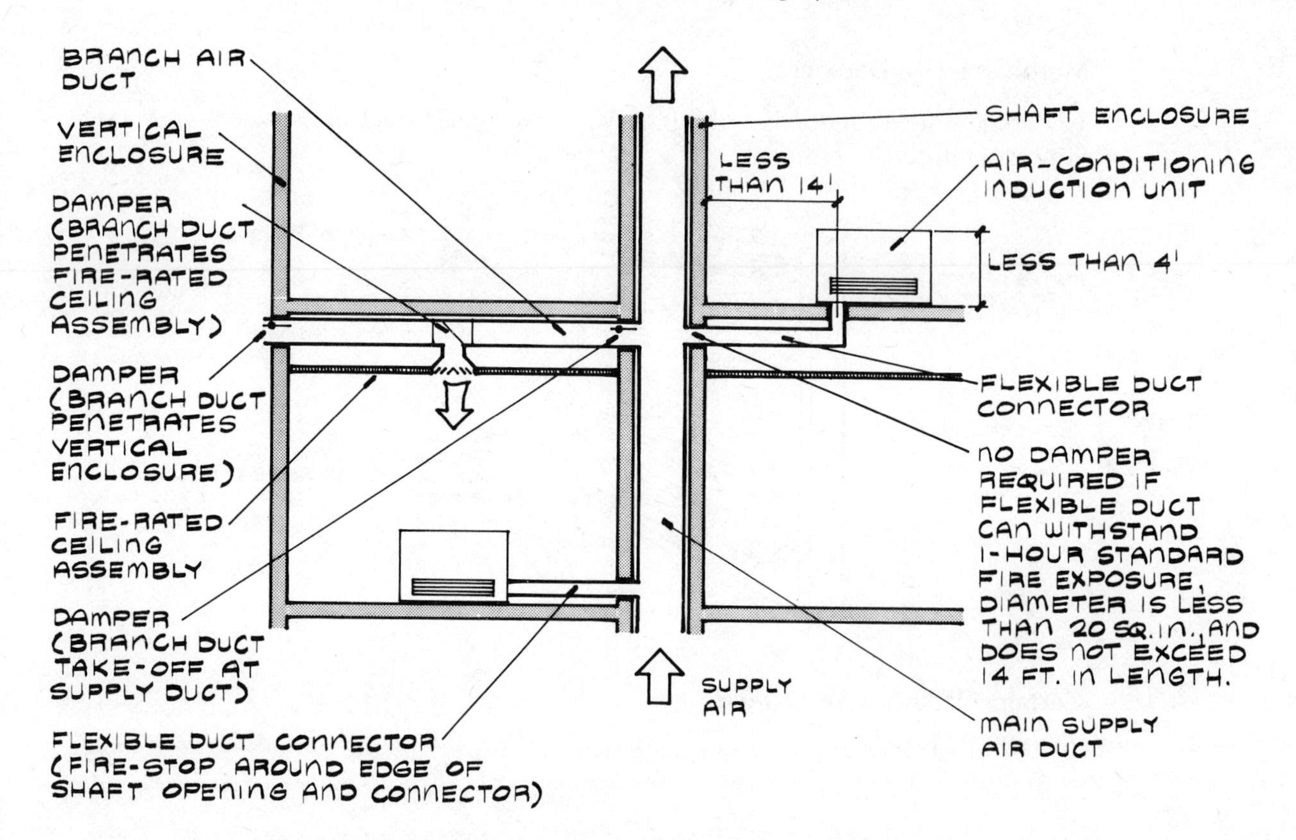

Note: The fire performance of flexible duct connectors can be evaluated by a test method that exposes the connector system to the ASTM standard time-temperature curve fire (cf., L. A. Issen, "Development of a Fire Test Method for Flexible Connectors in Air Distribution Systems," U.S. National Bureau of Standards, NBSIR 75-673, April 1975).

TALL BUILDINGS—AIRFLOW BALANCE IN BUILDINGS

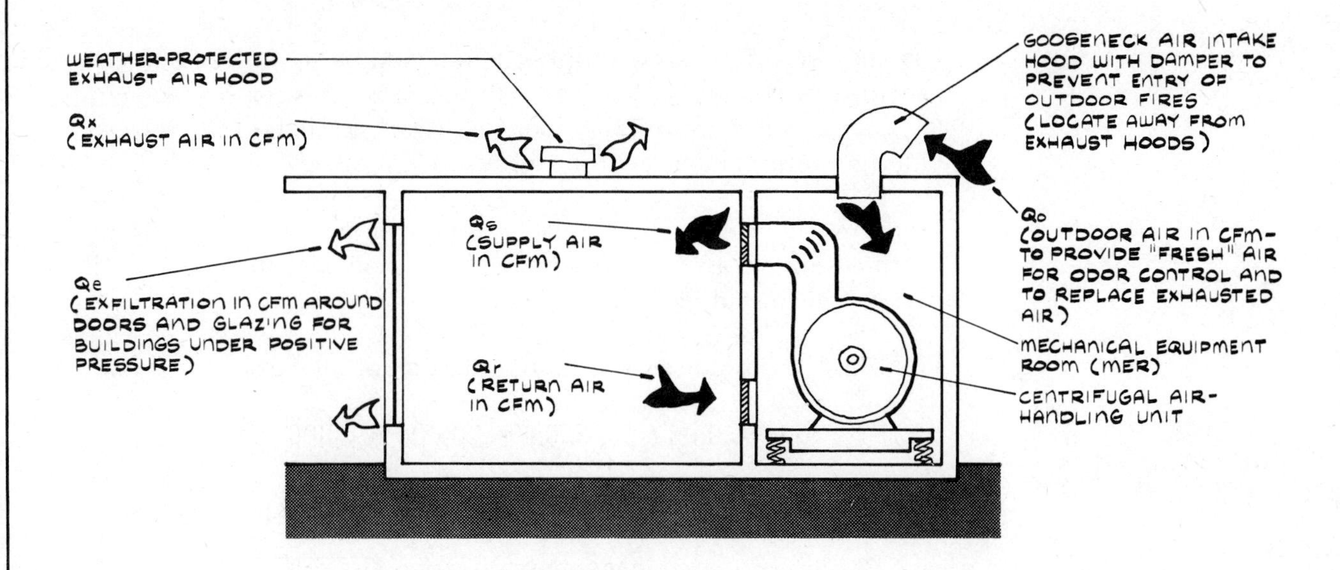

Indoor-Outdoor Air Balance

$$Q_s = Q_r + Q_o$$

where Q_s = supply airflow volume in cfm
Q_r = return airflow volume in cfm
Q_o = outdoor (or "make-up") airflow volume in cfm

To provide positive room air pressure,

$$Q_o \geqslant Q_e + Q_x$$

where Q_o = outdoor airflow volume in cfm
Q_e = exfiltration airflow volume in cfm
Q_x = exhaust airflow volume in cfm

EXAMPLE PROBLEM—NORMAL PRESSURIZATION

Given: An office has a supply airflow volume (Q_s) of 20,000 cfm for summer cooling. Exhaust fans in the restrooms remove 300 cfm (Q_x). Exfiltration around the doors and glazing will be 150 cfm (Q_e). Outdoor replacement air (or "make-up" air) is 8%.

Find: What will be the return airflow volume (Q_r) in cfm? How much additional air could be exhausted without causing negative air pressure in the office?

First, find the required outdoor replacement air at 8% (or 0.08).

$$Q_o = 0.08\ Q_s = 0.08 \times 20{,}000 = \boxed{1600 \text{ cfm}}$$

Next, find the return air from the formula, $Q_s = Q_r + Q_o$.

$$Q_r = Q_s - Q_o = 20{,}000 - 1600 = \boxed{18{,}400 \text{ cfm}}$$

Finally, to provide positive pressure, additional exhausted air (Q) should not exceed:

$$Q = Q_o - (Q_x + Q_e) = 1600 - (300 + 150) = \boxed{1150 \text{ cfm}}$$

TALL BUILDINGS—AIRFLOW REQUIRED TO COUNTERACT STACK EFFECT

The graph below shows the airflow volume required to counteract building air pressures caused by the stack effect. The size of openings at stairwell doors, elevator shafts, and air duct shafts controls the airflow volume needed for a specific air pressure condition. For example, if air pressure from the stack effect is 0.05 in. w.g., and the moving air must pass 5 sq ft of "friction-free" openings, the required airflow volume from the mechanical system to counteract it will be 4500 cfm (see dashed lines on graph).

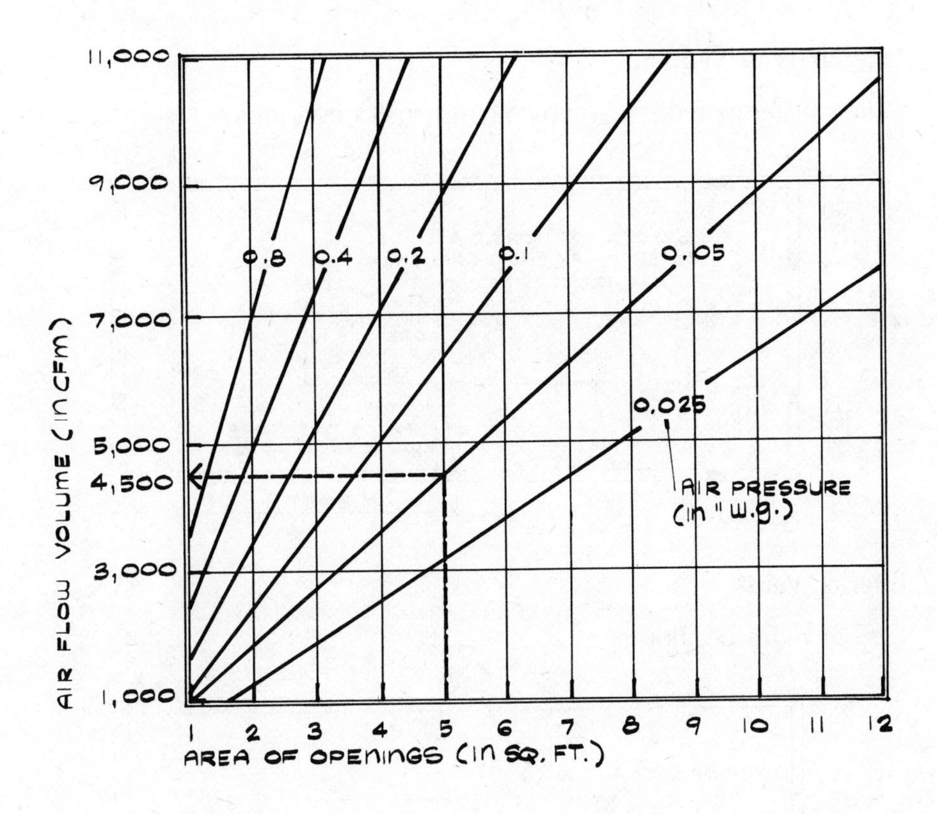

TALL BUILDINGS—EXAMPLE SMOKE VENT LAYOUTS

Fixed glazing can be shattered by fire fighters to vent smoke, but falling glass could injure pedestrians at ground level. Example exterior and interior vent layouts designed to remove smoke and heat are shown below. Panel wall vents, under fire-fighter key control, should be operable as the initial venting actions could become detrimental to fire-fighting operations (e.g., change in wind direction, change in fire spread). In addition, noncombustible wall vents should be located one above the other to protect against vertical flame spread through windows.

Exterior Wall Vents

Panels or tempered glass, typically 20 sq ft open area.

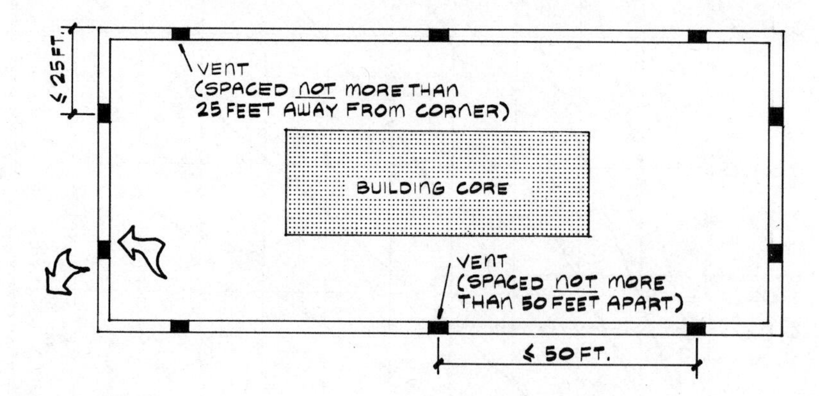

Interior Vents

One or more per floor.

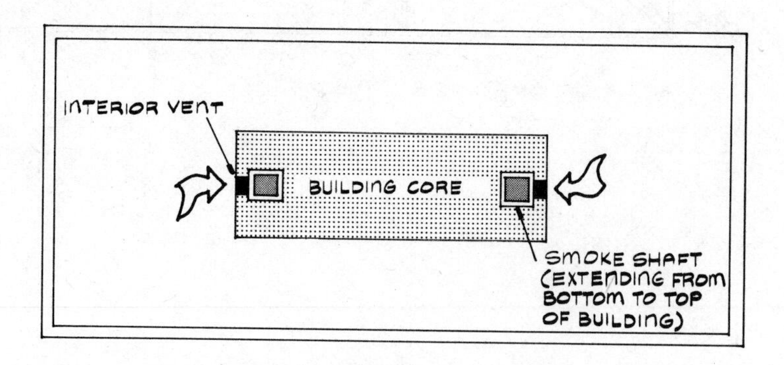

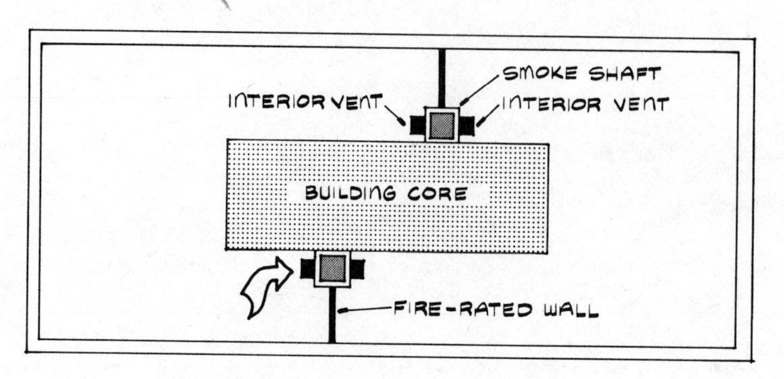

SMOKE SHAFT AREAS

Example smoke shaft cross-section areas (equal to $W \times D$ as shown by the sketch in the margin) needed to exhaust smoke utilizing the stack effect caused by cold weather conditions are given in the table. These shafts will not exhaust smoke when the inside and outside air temperatures are equal. Smoke shaft sizes are greatly influenced by shaft air leakage (e.g., through the shaft wall, vent and frame, frame and wall). The values in the table are minimum sizes based on a 2% leakage ratio of shaft leakage area per floor to open vent area. In general, smaller smoke shaft cross-section areas can be used where the shaft construction is relatively airtight and closed vents are effectively sealed. To reduce shaft wall leakage, monolithic concrete shafts or smooth metal liners can be used inside smoke shafts with sealed joints. For comprehensive smoke shaft tables, based on various percentages of leakage, see G. T. Tamura, and C. Y. Shaw, "Basis for the Design of Smoke Shafts," *Fire Technology*, Vol. 9, No. 3, August 1973.

Shaft Height	*Floor Area (in sq ft)*					
(in stories)	*2,000*	*5,000*	*10,000*	*20,000*	*30,000*	*40,000*
5	1	2	5	9	13	17
10	1	3	6	11	15	20
20	2	4	7	14	20	26
30	3	5	10	17	25	32
40	4	8	13	24	33	43
50	6	12	19	33	46	59
60	13	22	34	55	74	93

Note: An exhaust fan installed at the top of the smoke shaft can provide an effective means of exhausting smoke for year-round weather conditions.

TALL BUILDINGS—PEOPLE EVACUATION

In a tall building fire, where total evacuation of the building is impractical and dangerous, only the fire affected and adjacent areas should be evacuated. The evacuation procedure for an example tall building fire is shown below. Occupants from the fire floor and the floor below move by stairs to a "safe" area two or more floors below the fire floor and occupants of the floor above the fire floor move by stairs to a safe area one or more floors above. This creates "buffer" zones immediately above and below the fire floor for use by the fire fighters.

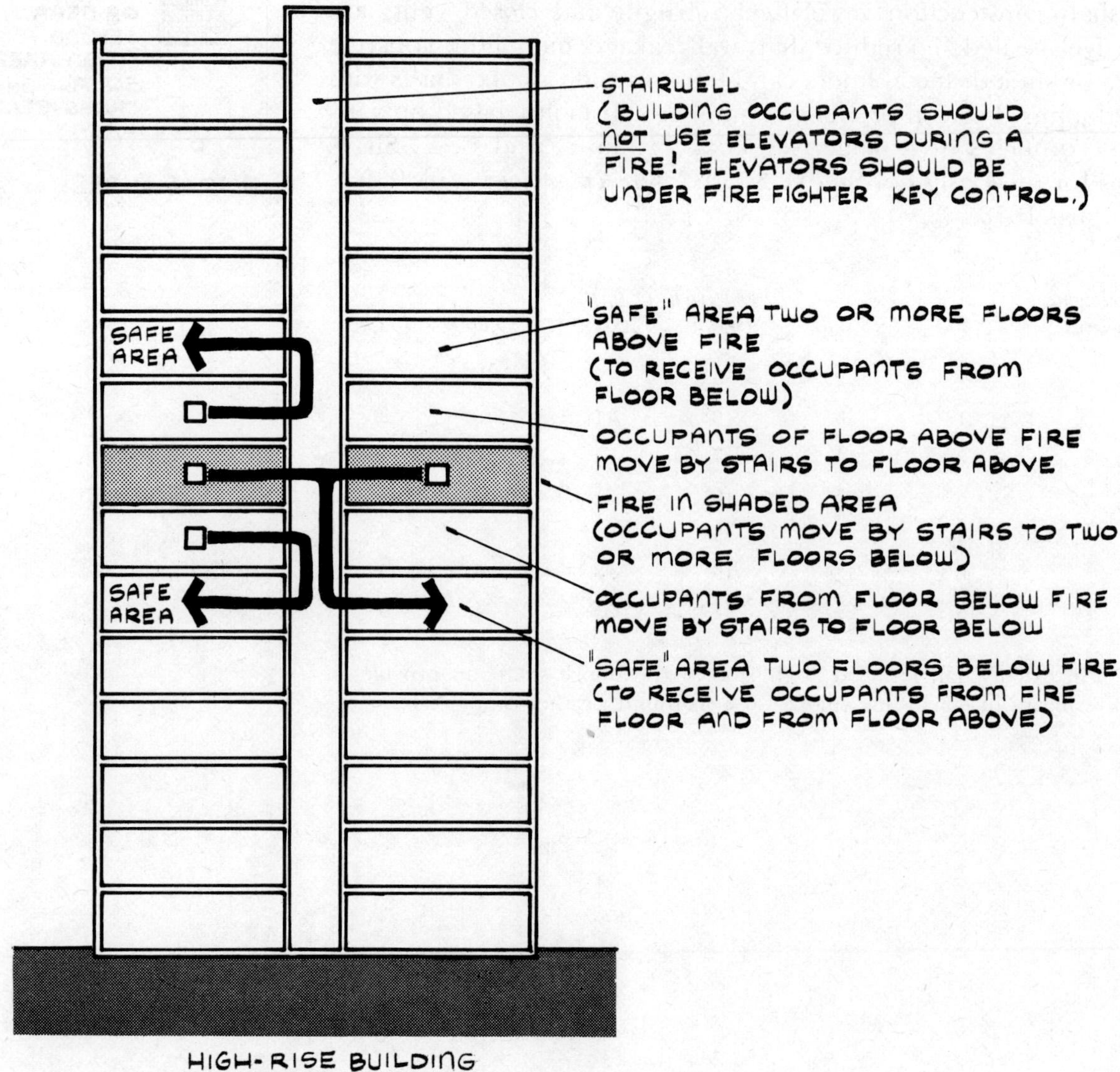

HIGH-RISE BUILDING

Note: When a fire occurs, elevators can be summoned to the fire floor by building occupants trying to escape and by elevator passengers who, unaware of the fire, push car buttons for the fire floor. The responding elevators could entrap passengers at the fire floor. Other dangers from elevator use during a fire include trapping of passengers in smoke-filled hoistways by electrical power failure, distortion of door mechanisms, and similar mechanical failures. Consequently, fire detection systems should override routine elevator operations and return all elevators to ground-level lobbies where they will be under fire-fighter key control (cf., "Safety Code for Elevators, Dumbwaiters, Escalators, and Moving Walks", ANSI A17.1).

EVACUATION MESSAGES

Prerecorded fire emergency messages over building public address systems should inform occupants in simple words what has happened, what they are expected to do, and that they should not use the elevators. In the sample evacuation messages given below for a fire on the 12th floor of an office building, note that the 13th floor occupants are asked to go up (rather than the more familiar descent to safety outdoors) specifically to "a safe area."

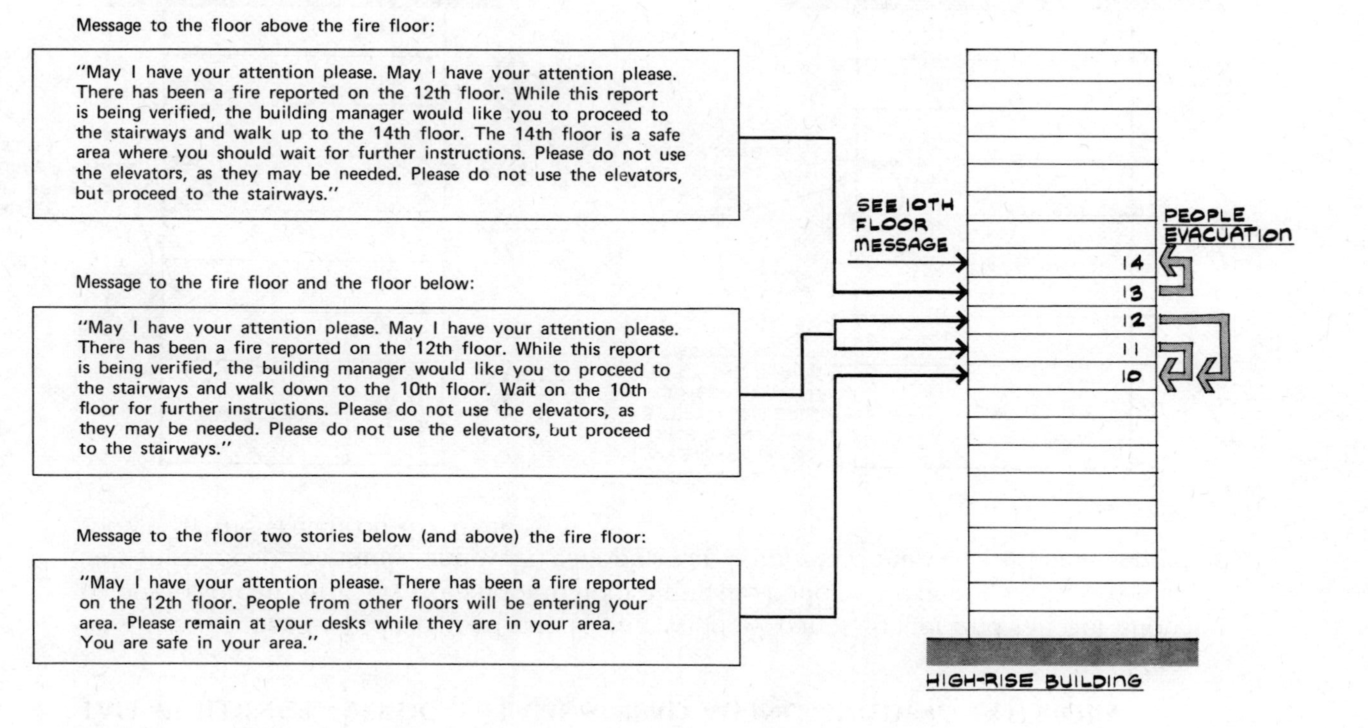

REFERENCE

Loftus, E. F. and J. P. Keating, "The Psychology of Emergency Communications" in *International Conference on Firesafety in High-Rise Buildings*, Washington, D.C.: General Services Administration, November 1974.

TALL BUILDINGS—VERTICAL FLAME SPREAD ALONG BUILDING EXTERIORS

As a fire reaches fully developed conditions, window panes shatter and fall out allowing flames to spread vertically from floor to floor along the building's exterior walls. Overhangs, balconies, deep spandrels, and wide canopies can help retard the vertical flame spread as shown by the sketch on the right.

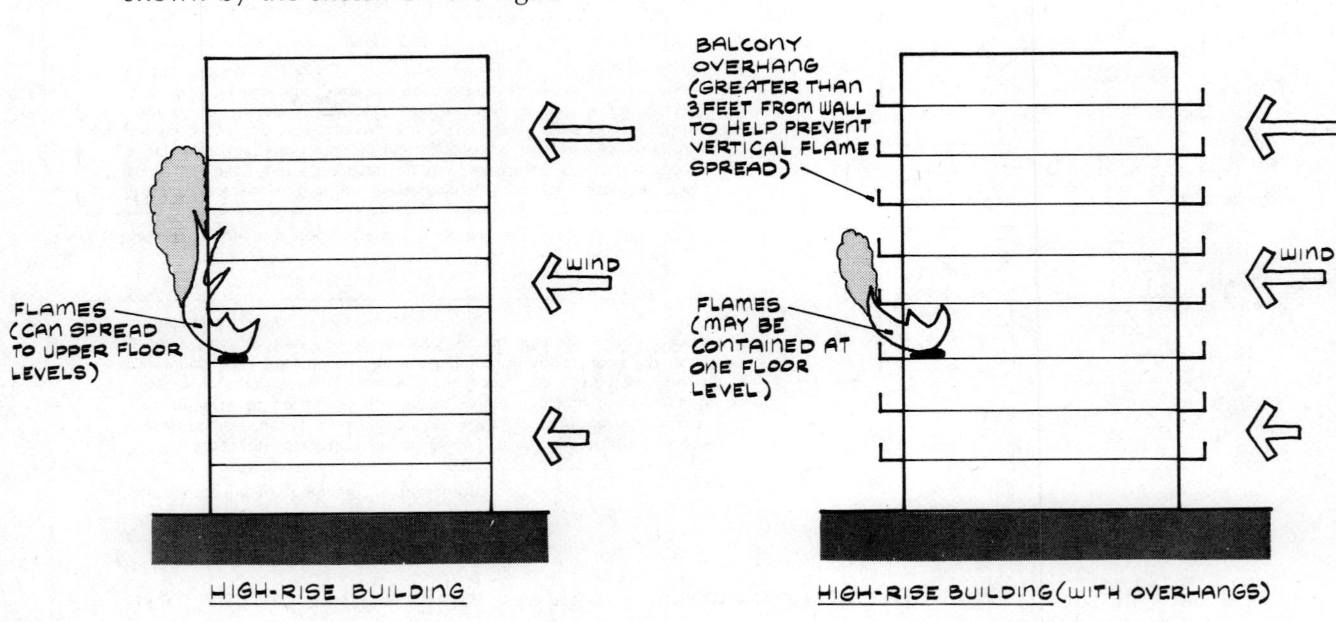

TALL BUILDINGS—FLAME BARRIERS

Flame barriers can be used to retard the vertical spread of fire from story to story. Shown below are fixed flame barriers achieved by overhang constructions (e.g., cantilever, balcony), recessed openings (with spandrel height greater than one-half floor to ceiling height), and operable panels mounted along exterior walls (called flame deflectors).

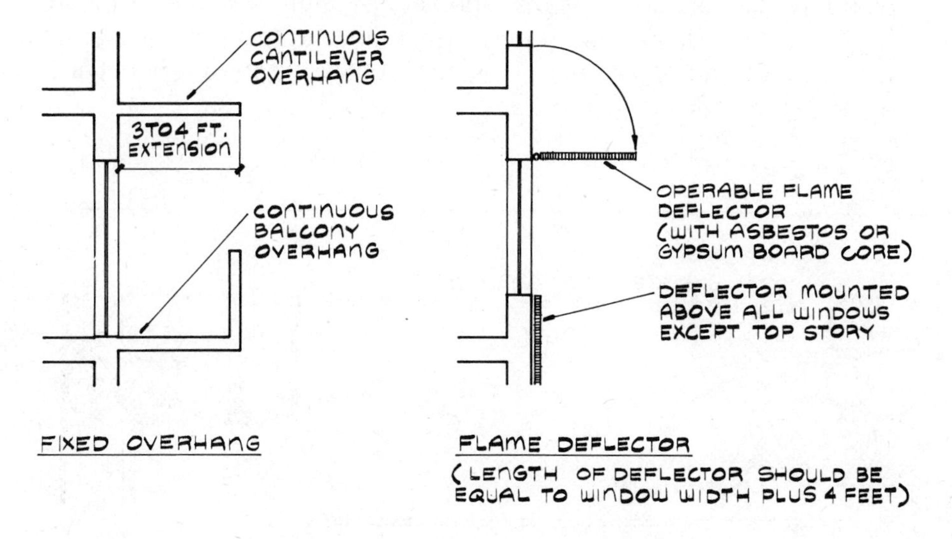

FIXED OVERHANG

FLAME DEFLECTOR (LENGTH OF DEFLECTOR SHOULD BE EQUAL TO WINDOW WIDTH PLUS 4 FEET)

REFERENCE

Harmathy, T. Z., "Flame Deflectors," National Research Council of Canada, Building Research Note No. 96, October 1974.

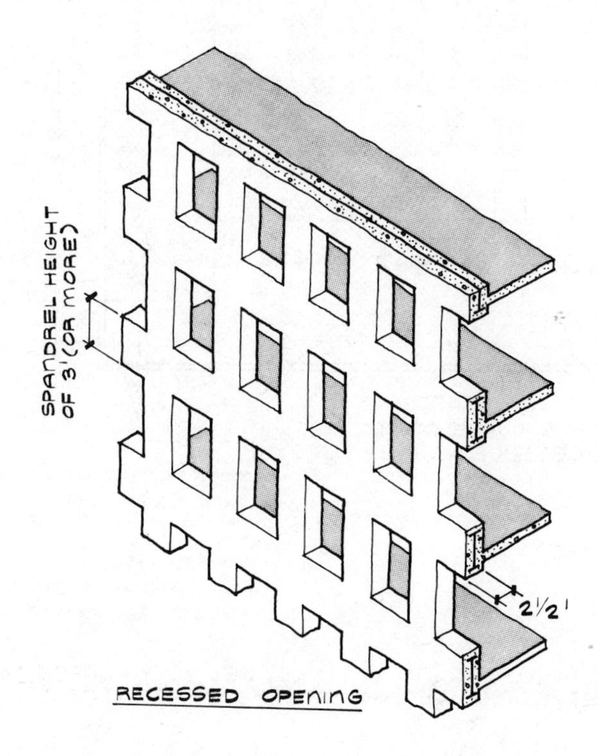

RECESSED OPENING

TALL BUILDINGS—SITE PLANNING

Effective site planning for tall buildings should incorporate many of the firesafety features shown below. Also indicated on the ground-level floor plan is the emergency control center which contains the annunciator panel to identify location of the fire; voice communication system to loudspeakers on each floor; equipment to monitor and control elevators, air-handling systems, sprinkler systems, and the like; and devices to unlock stairway doors. In addition, it is important that fire fighters have up-to-date floor plans that identify the various occupancy areas.

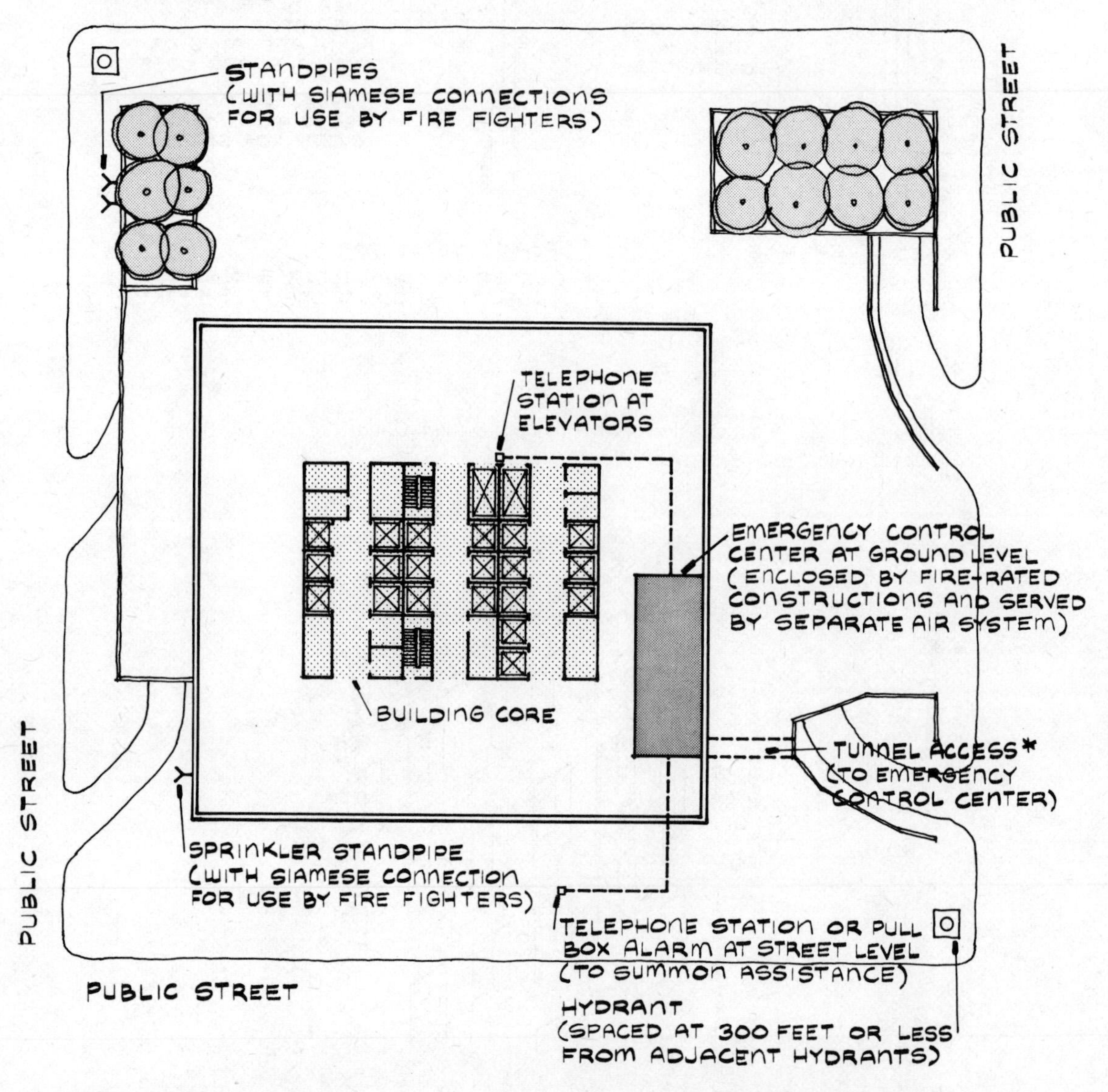

Note: Uninterruptible electrical power should be provided to operate emergency lighting, voice communication system, fire pumps, designated elevators, and other equipment.

TALL BUILDINGS—STANDPIPE SYSTEMS

In the example standpipe system shown below, the below-grade fire pump is connected to a water supply that serves the lower level zone. A second fire pump, that serves the upper level zone, is connected to a tank filled from the lower level zone. For comprehensive information on standpipe systems, refer to "Standpipe & Hose Systems," NFPA No. 14.

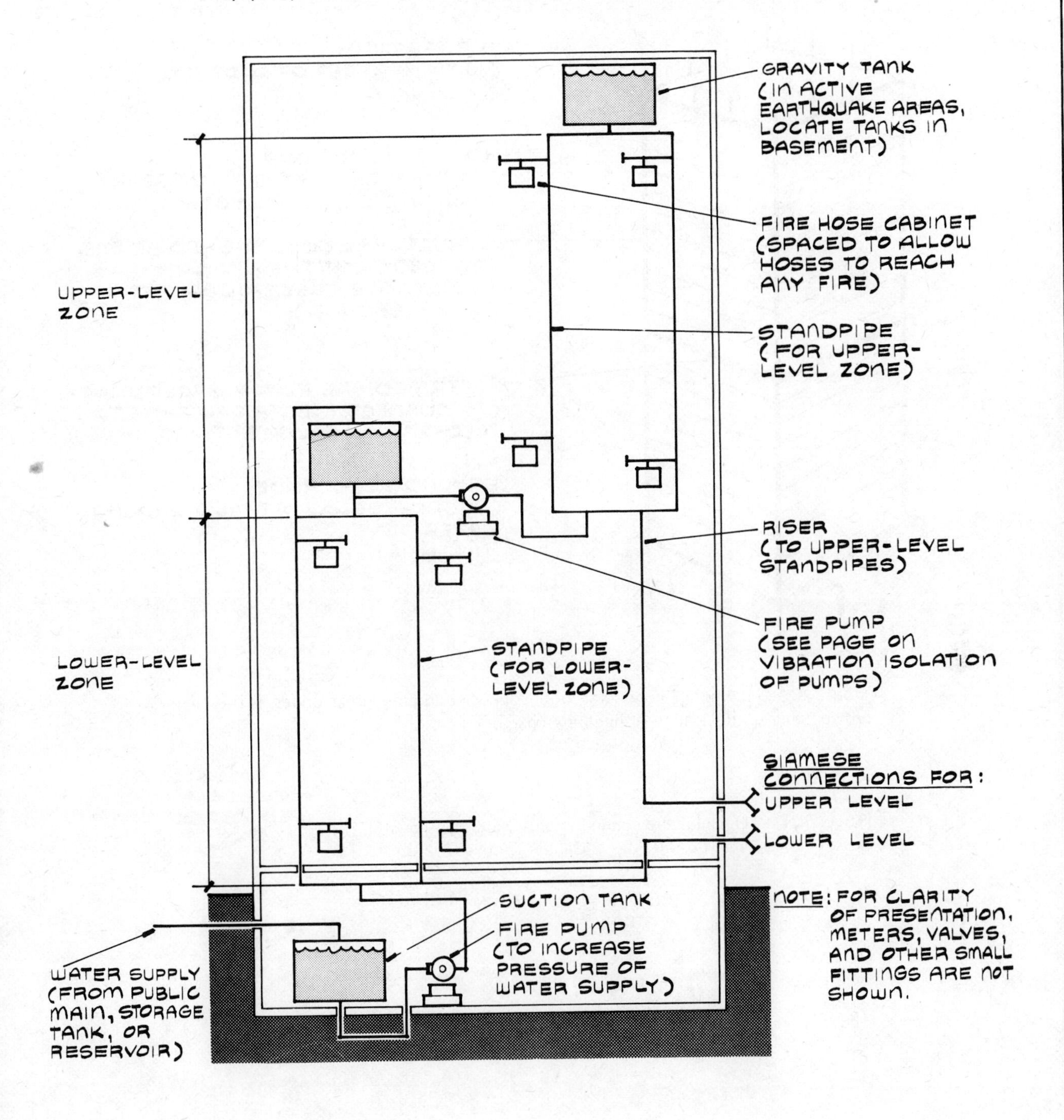

TALL BUILDINGS—LIGHTNING PROTECTION

In tall reinforced-concrete buildings, welded steel reinforcing connections can be used to provide continuous electrical paths through the concrete structural members. In tall structural-steel buildings, the steel columns can be used to provide electrical paths of low resistance to the ground electrodes as shown by the sketch.

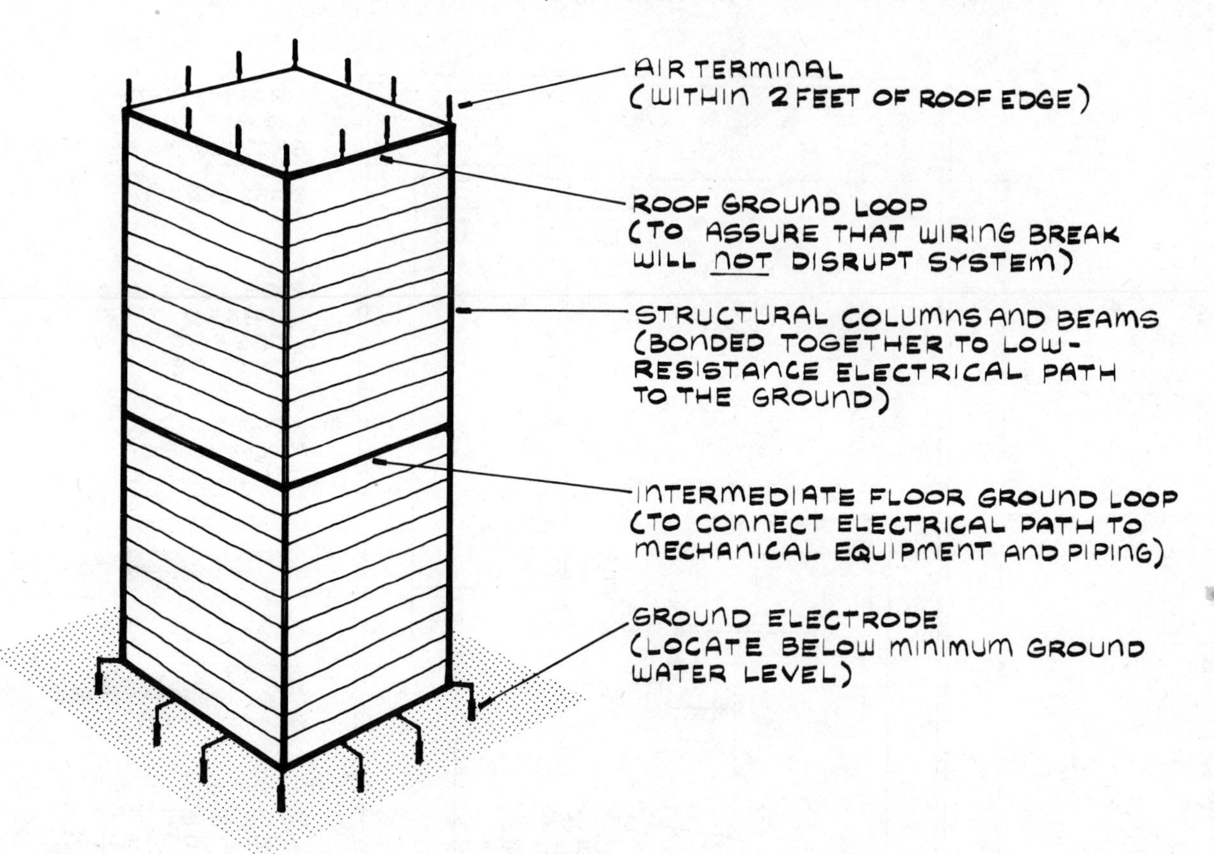

Note: A "counterpoise" conductor loop, connected to ground electrodes, is used to provide uniform dissipation of lightning charge.

REFERENCE

Harger, R. D., "Lightning Protection Systems," in R. Jensen (ed.), *Fire Protection for the Design Professional*, Boston, Mass.: Cahners Books (1975).

TALL BUILDINGS CHECKLIST

Wind and stack effects are sources of energy for moving smoke during tall building fires. Mechanical air distribution systems should be designed to control smoke movement by having the capability to establish a positive air pressure barrier surrounding the fire area. Smoke can be vented from the fire floor to the outdoors by operable windows, exterior panel vents, or interior smoke shafts.

Use fire-rated constructions to divide floors into compartments for use as refuge areas during a fire. Refuge areas must be large enough to accommodate the anticipated number of evacuees.

Enclose stairwells, elevator shafts, mechanical service shafts, and the like with fire-rated constructions. Where pipes or ducts penetrate these constructions, seal the openings airtight with fire-stop materials.

Avoid using corridors as return-air plenums that could easily become smoke-filled. Provide positive air pressure in corridors, stairwells, and shafts to remove smoke and prevent smoke infiltration.

Effective mechanical system vertical zoning in tall buildings can help neutralize the stack effect. For example, where fan rooms are provided at each floor level, the vertical ducts that can spread smoke and fire will be eliminated.

Emergency fire messages broadcasted over a communication system should advise building occupants of the necessary action to be taken. To avoid panic, the messages should convey a feeling that the situation is under control. For fire-fighter communications, locate fire department headset jacks at key points throughout the building.

Exits should be accessible from every part of the building and be remote from each other. If possible, exits should be related to elevator lobbies as occupants may tend to follow their normal egress routes during evacuations. Some codes require that fire-rated doors be used to close off lobbies during fires. In any event, building occupants must not use elevators for evacuation during fires.

Use continuous overhangs, continuous balconies, deep recesses, and other flame-deflection devices to prevent the vertical spread of fire from story to story along the exterior walls of tall buildings.

TALL BUILDINGS CHECKLIST (Continued)

During the construction of tall buildings, the amount of combustible materials (e.g., form lumber, flammable adhesives, paint, rubbish) and ignition sources (e.g., welding torches, cutting tools, electrical cables on wet floors) present are high. Consequently, firesafety measures must be integrated within the construction process. For example, as the building rises, contractors should install standpipe water supplies (temporary or permanent), complete stairway enclosures, and provide temporary construction elevators that can be used by fire fighters. In addition, portable fire extinguishers should be placed at strategic locations on each floor level.

SELECTED REFERENCES

SELECTED REFERENCES

1. *America Burning*, Report of the National Commission on Fire Prevention and Control, 1973. (Available from Superintendent of Documents, U.S. Government Printing Office, Washington, D.C. 20402.)
2. Asrow, S.P. (ed.), *Fire Protection of High Rise Buildings,* 2nd edition, Chicago Committee on High Rise Buildings, December 1971.
3. Brannigan, F. L., *Building Construction for the Fire Service*, Boston, Mass.: National Fire Protection Association (1971).
4. Bryan, J. L., *Automatic Sprinkler & Standpipe Systems,* Boston, Mass.: National Fire Protection Association (1976).
5. Bryan, J. L., *Fire Suppression and Detection Systems*, Beverly Hills, Calif.: Glencoe Press (1974).
6. Bush, L. S. and J. McLaughlin, *Introduction to Fire Science*, Beverly Hills, Calif.: Glencoe Press (1970).
7. Colburn, R. E., *Fire Protection & Suppression,* New York: McGraw-Hill (1975).
8. Erven, L. W., *Fire Company Apparatus and Procedures*, Beverly Hills, Calif.: Glencoe Press (1974).
9. *Fire Protection Through Modern Building Codes,* New York: American Iron and Steel Institute (1971).
10. *Handbook of Industrial Loss Prevention*, New York: McGraw-Hill (1967).
11. Hopf, P. S., *Designer's Guide to OSHA*, New York: McGraw-Hill (1975).
12. Jensen, R. (ed.), *Fire Protection for the Design Professional*, Boston, Mass.: Cahners Books (1975).
13. Langdon-Thomas, G. L., *Fire Safety in Buildings*, London, England: Adam and Charles Black (1972).
14. Lie, T. T., *Fire and Buildings*, London, England: Applied Science Publishers Ltd. (1972).
15. Marchant, E. W. (ed.), *A Complete Guide to Fire and Buildings*, New York: Barnes & Noble (1973).
16. McKinnon, G. P. (ed.), *Fire Protection Handbook*, Boston, Mass.: National Fire Protection Association (1976).

APPENDICES

APPENDIX A: SUMMARY OF USEFUL FORMULAS

Driveway Widths for Aerial Apparatus

1. $W = (H - 6)\cot\Theta + 4$

where W = required minimum driveway width in ft
H = building height in ft (less than 75 ft)
Θ = angle of ladder elevation in degrees

Fire-Resistance Ratings for Steel Columns

2. $R = \left(\dfrac{20w}{PG} + C\right)d$

where R = fire-resistance rating in hours
w = weight of steel in lb/linear ft
P = heated perimeter of steel in in.
G = density of insulation in lb/cu ft (pcf)
d = thickness of insulation in in.
C = protection factor: 0.5 for vermiculite, perlite, sprayed fibers, and dense mineral wool; 1.2 for asbestos cement board, plasters, and cementitious mixtures

Automatic Sprinkler Systems

3. $P = \left(\dfrac{Q}{K}\right)^2$

where P = water pressure required by sprinkler head in lb/sq in. (psi)
Q = discharge water flow rate in gal/minute (gpm)
K = sprinkler head coefficient (no units)

4. $P_a \geqslant P_d + 0.434h + P$

where P_a = available water pressure in psi
P_d = pressure drop through piping (includes fittings and valves) in psi
h = height from water source to sprinkler head in ft
P = pressure required by sprinklers in psi

Note: $\geqslant$ means greater than or equal to.

Water Flow in Pipes

5. $v = 0.4\,\dfrac{Q}{d^2}$

where v = water flow velocity in ft/second (fps)
Q = water flow rate in gpm
d = inside pipe diameter in in.

Vibration Isolation

6. $f_n = 3.13\sqrt{1/y}$

where f_n = natural frequency in hertz (Hz)
y = static deflection in in.

Roof Venting

7. $A_v \simeq 0.14Ph\sqrt{h/d}$

where A_v = required roof vent area in sq ft
P = perimeter of fire in ft
h = distance between bottom of curtain board and floor in ft
d = distance between vent and bottom of curtain board in ft

Note: $\simeq$ means approximately equal to.

Air Changes

8. $Q = \dfrac{NV}{60}$

where Q = airflow volume in cu ft/minute (cfm)
N = number of air changes per hour (ach)
V = room volume in cu ft

Natural Air Ventilation through Buildings

9. $Q \simeq 50\,Av$

where Q = airflow volume in cfm
A = open area of inlets or outlets in sq ft
v = wind velocity in mph

Stage House Venting

10. $A_v \simeq 1.6\,A_o\sqrt{h/d}$

where A_v = required stage house roof vent area in sq ft
A_o = open area of proscenium (i.e., opening between stage and auditorium) in sq ft
h = distance between bottom of proscenium curtain and stage floor in ft
d = distance between roof vent and bottom of proscenium curtain in ft

Exit Requirements

11. $W = \dfrac{A}{dc}$

where W = exit unit width
A = floor area in sq ft
d = occupant density in sq ft per person
c = capacity per unit of exit width

Smoke Density from Burning Materials

12. $D_f = \frac{D_m A}{V}$

where D_f = smoke density factor in ft^{-1}
D_m = maximum optical density (no units)
A = area of material producing smoke in sq ft
V = room volume in cu ft

Stack Effect

13. $Q \simeq 8\,A\,\sqrt{h\,(t - t_0)}$

where Q = airflow volume in cfm
A = open area of inlets or outlets in sq ft
h = height from inlets to outlets in ft
t = inside air temperature in °F
t_o = outside air temperature in °F

14. $\Delta p = 7.6h\left(\frac{1}{t_o} - \frac{1}{t}\right)$

where Δp = air pressure difference in inchés of water (in. w.g.)
h = building height in ft
t_o = outside air temperature in °R (Rankine)
t = inside air temperature in °R

Note: To convert °F to °R, add 460 to °F.

Force of Moving Air

15. $\Delta p = \frac{v^2}{3.9t}$

where Δp = air pressure difference in in. w.g.
v = air velocity in mph
t = air temperature in °R

Force on Door Due to Air Pressure

16. $F = 50\Delta p$

where F = force on door in lb
Δp = air pressure difference acting on door in in. w.g.

Building Indoor-Outdoor Air Balance

17. $Q_s = Q_r + Q_o$

where Q_s = supply airflow volume in cfm
Q_r = return airflow volume in cfm
Q_o = outdoor (or make-up) airflow volume in cfm

Water Flow Required to Extinguish Fire

18. $Q \simeq \frac{V}{25}$

where Q = fire hose water flow rate in gpm
V = room or building volume in cu ft

Flame Spread Rating from Tunnel Test

19. $FSR = \frac{550}{T}$

where FSR = flame spread rating (no units)
T = time of flame travel in min (less than $5\frac{1}{2}$ min)

Note: When flame travels $19\frac{1}{2}$ ft in more than $5\frac{1}{2}$ min, but less than 10 min, use $\frac{275}{T} + 50$. For other conditions of flame travel in the tunnel, refer to ASTM E 84.

Fire Loads

20. $FL = \frac{\Sigma qw}{8000A}$

where FL = fire (or "fuel") load in lb/sq ft (psf)
q = fuel contribution of material (e.g., furniture, equipment, finish) in Btu/lb
w = weight of material in lb
Σqw = sum of fuel contribution of combustibles times respective weight ($\Sigma qw = q_1 w_1 + q_2 w_2 + \cdots q_n w_n$)
A = floor area in sq ft

Note: The fuel contribution can be reduced where combustible materials are stored in steel containers.

APPENDIX B: BUILDING FIRESAFETY REFERENCE STANDARDS

American Society for Testing and Materials (ASTM)

The following standards on building firesafety are published by the American Society for Testing and Materials, 1916 Race Street, Philadelphia, Pa. 19103. A complete index of ASTM Standards is available from this address.

No.	*Title*
E 69	Test for Combustible Properties of Treated Wood by the Fire-Tube Apparatus
E 84	Test for Surface Burning Characteristics of Building Materials
E 108	Fire Tests of Roof Coverings
E 119	Fire Tests of Building Construction and Materials
E 136	Test for Noncombustibility of Elementary Materials
E 152	Fire Tests of Door Assemblies
E 160	Test for Combustible Properties of Treated Wood by the Crib Test
E 162	Test for Surface Flammability of Materials Using a Radiant Heat Energy Source
E 163	Fire Tests of Window Assemblies
E 176	Definition of Terms Relating to Fire Tests of Building Construction and Materials
E 286	Test for Surface Flammability of Building Materials Using an 8-ft Tunnel Furnace
D 350	Testing Flexible Treated Sleeving Used for Electrical Insulation
D 568	Test for Flammability of Flexible Plastics
D 635	Test for Flammability of Self-Supporting Plastics
D 777	Test for Flammability of Treated Paper and Paperboard
D 1360	Test for Fire Retardancy of Paints (Cabinet Method)
D 1361	Test for Fire Retardancy of Paints (Stick and Wick Method)
D 1433	Test for Flammability of Flexible Thin Plastic Sheeting
D 1692	Test for Rate of Burning or Extent of Burning of Cellular Plastics Using a Supported Specimen by a Horizontal Screen
D 1929	Test for Ignition Properties of Plastics
D 2843	Measuring the Density of Smoke from the Burning or Decomposition of Plastics
D 2859	Test for Flammability of Finished Textile Floor Covering Materials
D 2863	Test for Flammability of Plastics Using the Oxygen Index Method
D 2898	Test for Durability of Fire-Retardant Treatment of Wood
D 3014	Test for Flammability of Rigid Cellular Plastics
D 3411	Test for Flammability of Textile Materials

Note: Always refer to the latest edition of ASTM test methods, definitions, recommended practices, classifications, and specifications.

National Fire Protection Association (NFPA)

The following standards and codes on building firesafety are published by the National Fire Protection Association, 470 Atlantic Avenue, Boston, Mass. 02210. An order form (with prices and ordering information) listing over 200 NFPA codes, standards, recommended practices, and manuals is available from this address.

No.	*Title*
10	Portable Fire Extinguishers
12	Carbon Dioxide Extinguishing Systems
12A	Halon 1301 High Expansion Foam Systems
13	Installation of Sprinkler Systems
14	Standpipe and Hose Systems
17	Dry Chemical Extinguishing Systems
20	Centrifugal Fire Pumps
30	Flammable and Combustible Liquids Code
50B	Liquified Hydrogen Systems at Consumer Sites
56A	Inhalation Anesthetics
70	National Electrical Code
72E	Automatic Fire Detectors
74	Household Fire Warning Equipment
78	Lightning Protection Code
80	Fire Doors and Windows
80A	Protection of Buildings from Exterior Fire Exposures
90A	Air Conditioning and Ventilating Systems
90B	Warm Air Heating and Air Conditioning Systems
92M	Waterproofing and Draining of Floors
101	Life Safety Code
102	Tents, Grandstands and Air-Supported Structures Used for Places of Assembly
203M	Roof Coverings
204	Guide for Smoke and Heat Venting
206M	Guide on Building Areas and Heights
211	Chimneys, Fireplaces, and Vents
220	Standard Types of Building Construction
241	Building Construction and Demolition Operations
251	Fire Tests for Building Construction and Materials
252	Methods of Fire Tests for Door Assemblies
255	Tests of Surface Burning Characteristics of Building Materials
256	Methods of Fire Tests of Roof Coverings
257	Fire Tests of Window Assemblies
418	Roof-Top Heliport Construction and Protection
701	Flame-Resistant Textiles and Films
703	Fire-Retardant Treatments for Building Materials
704	Identification Systems for Fire Hazards of Materials
205M-T	Tentative Guide for Plastics in Building Construction

Note: Always refer to the latest edition and to current revisions or supplements that may be in effect. Nevertheless, the prevailing building code in a community may be based on earlier editions of NFPA standards and codes.

Underwriters Laboratories (UL)

The following standards on building firesafety are published by the Underwriters Laboratories, 333 Pfingsten Road, Northbrook, Ill. 60062. A complete catalog listing of UL Standards with prices and ordering information is available from this address.

No.	*Title*
9	Fire Tests of Window Assemblies
10A	Tin-Clad Fire Doors
10B	Fire Tests of Door Assemblies
14A	Fire Door Operators and Fire Door Operators with Automatic Closers, and Sliding Fire Door Closers
14B	Sliding Hardware for Standard Horizontally Mounted Tin-Clad Fire Doors
14C	Swinging Hardware for Standard Tin-Clad Fire Doors
33	Fusible Links for Fire Protection Service
47	Fire Hose Rack Assemblies
55A	Materials for Built-Up Roof Coverings
55B	Class C Asphalt Organic-Felt Sheet Roofing and Shingles
63	Fire Door Frames
96A	Master Labeled Lightning Protection Systems
154	Carbon Dioxide Fire Extinguishers
167	Combustion-Products Type Smoke Detectors for Fire Protective Signaling Systems
168	Photoelectric Type Smoke Detectors for Fire Alarm Service
181	Factory Made Air Duct Materials and Air Duct Connectors
193	Alarm Valves for Fire Protection Service
199	Automatic Sprinklers for Fire Protection Service
199A	Quick Response Extended Coverage Sprinklers
214	Tests for Comparative Flammability of Flame Resistant Fabrics and Films
217	Single and Multiple Station Smoke Detectors
228	Door Closers, Holders, and Integral Smoke Detectors
248	Hydrants for Fire Protection Service
260	Dry Pipe, Deluge, and Pre-Action Valves for Fire Protection Service
262	Gate Valves for Fire Protection Service
263	Tests for Fire Resistance of Building Construction and Materials
299	Dry-Chemical Fire Extinguishers
305	Panic Hardware
312	Check Valves for Fire Protection Service
346	Water Flow Indicators for Fire Protective Signaling Systems
405	Fire Department Connections
448	Pumps for Fire Protection Service
521	Fire Detection Thermostats
555	Fire Dampers
618	Concrete Masonry Units
626	$2\frac{1}{2}$ Gallon Stored Pressure Water Type Fire Extinguishers
711	Rating and Fire Testing of Fire Extinguishers
723	Tests for Surface Burning Characteristics of Building Materials
753	Alarm Accessories for Automatic Water-Supply Control Valves for Fire Protection Service

Underwriters Laboratories (UL)—(Continued)

No.	*Title*
790	Tests for Fire Resistance of Roof Covering Materials
924	Emergency Lighting Equipment
985	Household Fire Warning System Units

Note: Always refer to the current edition and most recently adopted revisions that may be issued during the life of the current edition.

APPENDIX C: CONVERSION FACTORS

The table below presents conversion factors for common firesafety units to corresponding metric system units, often referred to as Le Système International d'Unités (SI units). The basic SI units are expressed as follows: length by meter (abbreviated: m), mass by kilogram (kg), time by second (s), and temperature by degree kelvin (K). For a comprehensive presentation of metric system units, refer to "Metric Practice Guide," ASTM E 380. This publication is available from the American Society for Testing and Materials (ASTM), 1916 Race Street, Philadelphia, Pa. 19103.

To Convert	*Into*	*Multiply by**	*Conversely, Multiply by**
Btu	joule	1055	9.479×10^{-4}
Btuh	watt	0.293	3.413
Btuh/ft^2	watt/m^2	3.154	0.317
Btuh/ft^2/°F	watt/m^2/K	5.678	0.1761
Btuh/ft^2/in./°F	watt/m/K	1.442×10^{-1}	6.935
Btu/lb	calorie/g	0.5556	1.8
	joule/kg	2.326×10^{3}	4.299×10^{-4}
Btu/min	watt	17.57	0.0569
°C	°F	$(°C \times \frac{9}{5})+32$	$(°F - 32)\times\frac{5}{9}$
	K	°C + 273	K − 273
cm	in.	0.3937	2.54
	ft	3.281×10^{-2}	30.48
	mm	10	10^{-1}
	m	10^{-2}	10^{2}
cfh	m^3/s	7.865×10^{-6}	1.271×10^{5}
cfm	m^3/s	4.719×10^{-4}	2119
deg (angle)	radian	1.745×10^{-2}	57.3
°F	°C	$(°F - 32)\times\frac{5}{9}$	$(°C \times \frac{9}{5})+32$
	K	$(°F + 460)\times\frac{5}{9}$	1.8K − 460
ft	in.	12	0.0833
	mm	304.8	3.281×10^{-3}
	cm	30.48	3.281×10^{-2}
	m	0.3048	3.281
ft^2	in.^{2}	144	6.944×10^{-3}
	cm^2	9.29×10^{2}	0.01076
	m^2	9.29×10^{-2}	10.76
ft^3	in.^{3}	1728	5.787×10^{-4}
	cm^3	2.832×10^{4}	3.531×10^{-5}
	m^3	2.832×10^{-2}	35.31
fpm	m/s	5.08×10^{-3}	197
	mph	1.136×10^{-2}	88
fps	m/s	0.3048	3.281
footcandle	lux	10.76	0.0929
gal	m^3	3.785×10^{-3}	264
gpm	m^3/s	6.309×10^{-5}	1.585×10^{4}
	litres/min	3.785	0.264

To Convert	*Into*	*Multiply by**	*Conversely, Multiply by**
in.	ft	0.0833	12
	mm	25.4	0.03937
	cm	2.54	0.3937
	m	0.0254	39.37
$in.^2$	ft^2	0.006944	144
	cm^2	6.452	0.155
	m^2	6.452×10^{-4}	1550
$in.^3$	ft^3	5.787×10^{-4}	1.728×10^3
	cm^3	16.388	6.102×10^{-2}
	m^3	1.639×10^{-5}	6.102×10^4
in. w.g.	pascal (Pa)	2.4884×10^2	4.019×10^{-3}
K	°C	K −273	°C + 273
	°F	1.8K − 460	$(°F + 460) \times \frac{5}{9}$
	°R (Rankine)	1.8K	$°R \times \frac{5}{9}$
ksi	psi	10^3	10^{-3}
m	in.	39.37	0.0254
	ft	3.281	0.3048
	yd	1.0936	0.9144
	mm	10^3	10^{-3}
	cm	10^2	10^{-2}
m/s	fpm	197	5.08×10^{-3}
	fps	3.281	0.3048
mil	in.	10^{-3}	10^3
miles	ft	5280	1.894×10^{-4}
	km	1.6093	0.6214
mph	fpm	88	1.136×10^{-2}
	fps	1.47	0.6816
	km/min	2.682×10^{-2}	37.28
	km/hr	1.6093	0.6214
lb (force)	newton (N)	4.448	0.2248
lb (weight)	gm	453.6	2.205×10^{-3}
	kg	0.4536	2.205
psf (force)	psi	6.944×10^{-3}	144
	N/m^2	47.85	2.09×10^{-2}
	kg/m^2	4.878	0.205
	ksi	10^{-3}	10^3
psi (force)	psf	144	6.944×10^{-3}
	N/m^2 (or Pa)	6895	1.45×10^{-4}
	kg/cm^2	7.03×10^{-2}	14.22
	kg/m^2	703	1.422×10^{-3}
	kPa	6.895	0.145

*Round converted quantity to proper number of significant digits commensurate with the intended precision.

APPENDIX D: NFPA DECISION TREE

The firesafety "decision tree," developed by the NFPA Committee on Systems Concepts for Fire Protection in Structures, is shown on the following pages. For a detailed discussion of the decision tree, see R. J. Thompson, "The Decision Tree for Fire Safety Systems Analysis: What It Is and How To Use It," *Fire Journal*, July, September, and November 1975 issues.

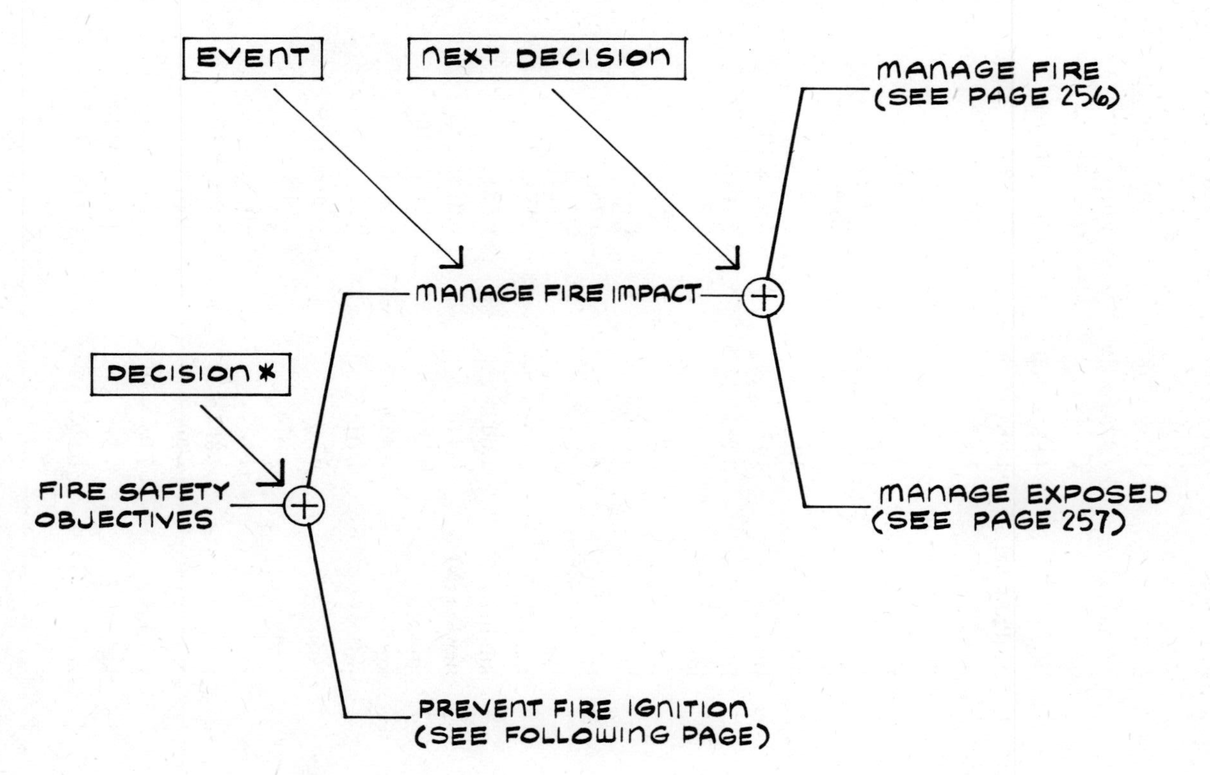

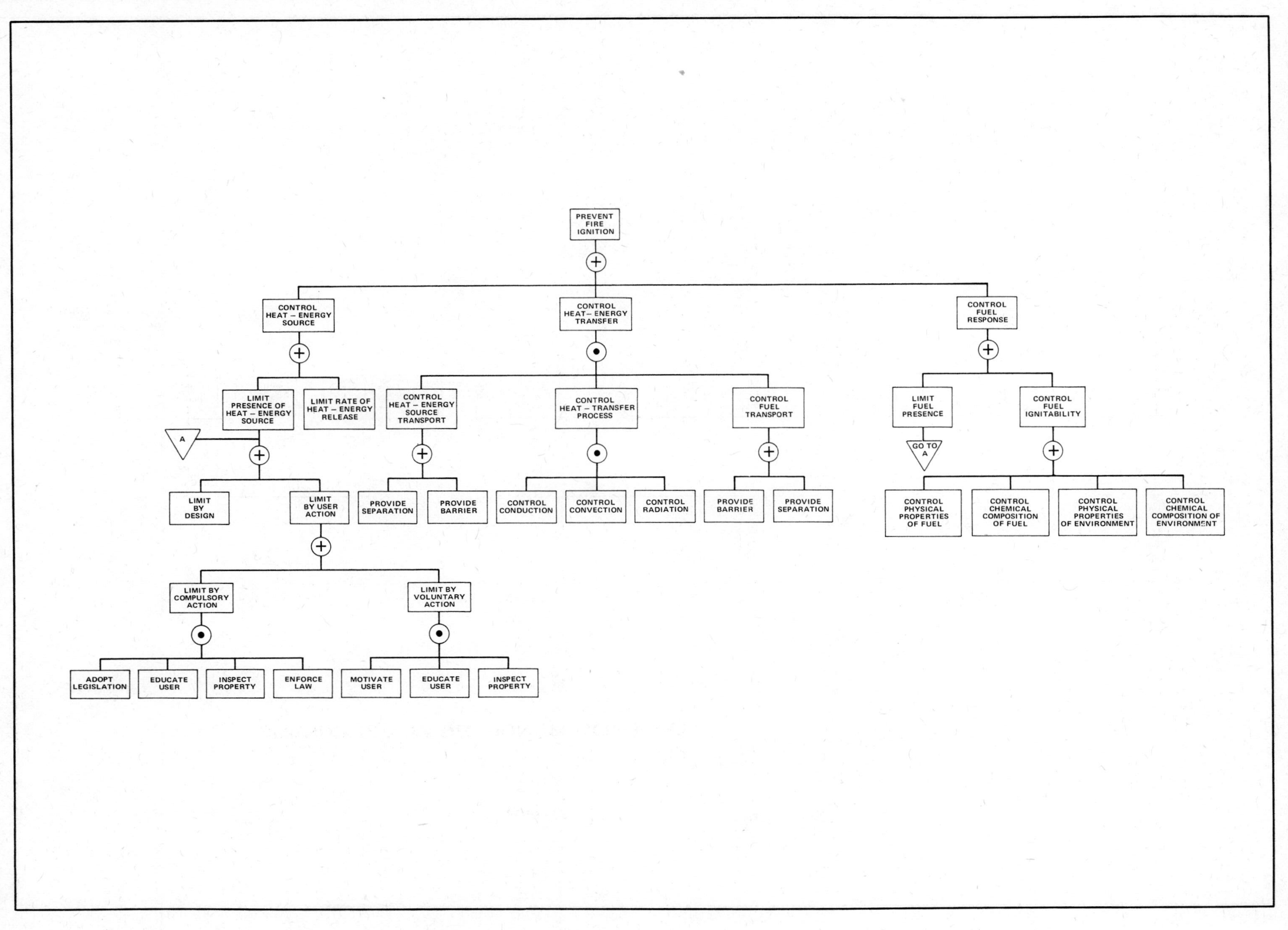

PREVENT FIRE IGNITION
CONTROL HEAT – ENERGY SOURCE
CONTROL HEAT– ENERGY TRANSFER
CONTROL FUEL RESPONSE
LIMIT PRESENCE OF HEAT – ENERGY SOURCE
LIMIT RATE OF HEAT – ENERGY RELEASE
CONTROL HEAT – ENERGY SOURCE TRANSPORT
CONTROL HEAT – TRANSFER PROCESS
CONTROL FUEL TRANSPORT
LIMIT FUEL PRESENCE
CONTROL FUEL IGNITABILITY
A
LIMIT BY DESIGN
LIMIT BY USER ACTION
PROVIDE SEPARATION
PROVIDE BARRIER
CONTROL CONDUCTION
CONTROL CONVECTION
CONTROL RADIATION
PROVIDE BARRIER
PROVIDE SEPARATION
GO TO A
CONTROL PHYSICAL PROPERTIES OF FUEL
CONTROL CHEMICAL COMPOSITION OF FUEL
CONTROL PHYSICAL PROPERTIES OF ENVIRONMENT
CONTROL CHEMICAL COMPOSITION OF ENVIRONMENT
LIMIT BY COMPULSORY ACTION
LIMIT BY VOLUNTARY ACTION
ADOPT LEGISLATION
EDUCATE USER
INSPECT PROPERTY
ENFORCE LAW
MOTIVATE USER
EDUCATE USER
INSPECT PROPERTY

APPENDIX D: NFPA DECISION TREE (Continued)

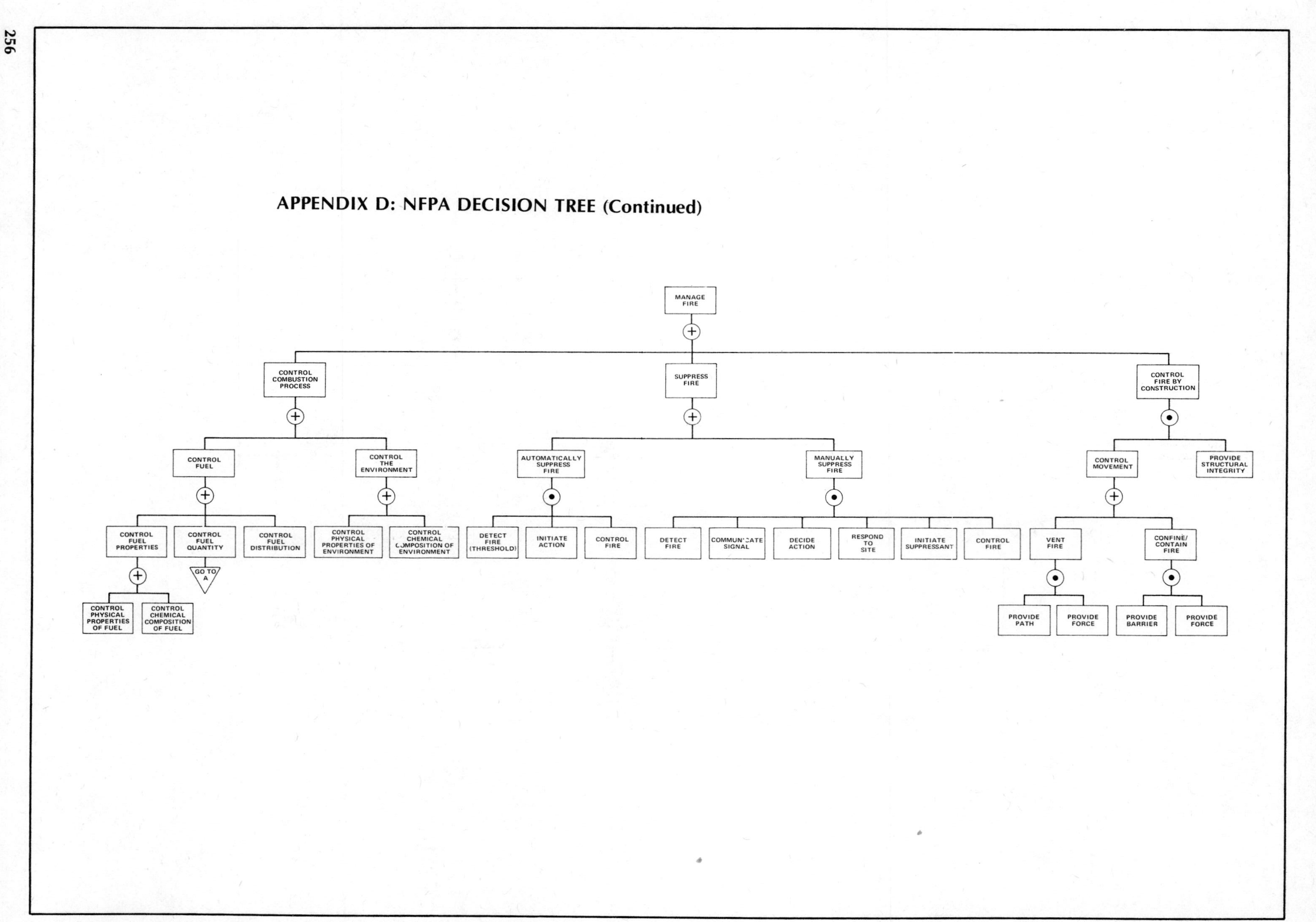

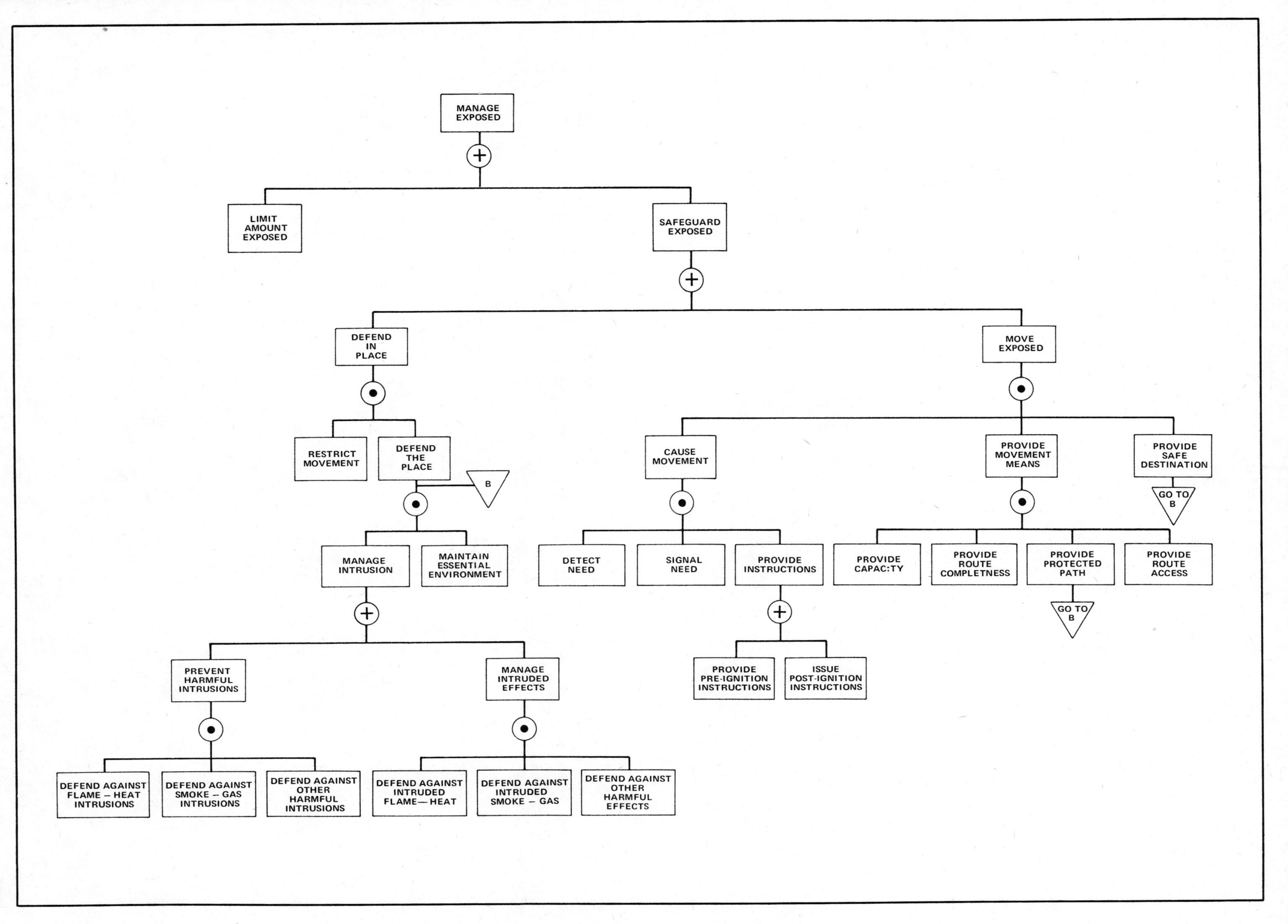

MANAGE EXPOSED
LIMIT AMOUNT EXPOSED
SAFEGUARD EXPOSED
DEFEND IN PLACE
MOVE EXPOSED
RESTRICT MOVEMENT
DEFEND THE PLACE
B
MANAGE INTRUSION
MAINTAIN ESSENTIAL ENVIRONMENT
PREVENT HARMFUL INTRUSIONS
MANAGE INTRUDED EFFECTS
DEFEND AGAINST FLAME – HEAT INTRUSIONS
DEFEND AGAINST SMOKE – GAS INTRUSIONS
DEFEND AGAINST OTHER HARMFUL INTRUSIONS
DEFEND AGAINST INTRUDED FLAME—HEAT
DEFEND AGAINST INTRUDED SMOKE – GAS
DEFEND AGAINST OTHER HARMFUL EFFECTS
CAUSE MOVEMENT
PROVIDE MOVEMENT MEANS
PROVIDE SAFE DESTINATION
GO TO B
DETECT NEED
SIGNAL NEED
PROVIDE INSTRUCTIONS
PROVIDE PRE-IGNITION INSTRUCTIONS
ISSUE POST-IGNITION INSTRUCTIONS
PROVIDE CAPAC:TY
PROVIDE ROUTE COMPLETNESS
PROVIDE PROTECTED PATH
GO TO B
PROVIDE ROUTE ACCESS

INDEXES

NAME INDEX

SUBJECT INDEX*

Building constructions, classifications of, 29
Building heights, 31
  computations, 32

*For easy reference, checklists, example problems, and tables are listed on page 269.

CHECKLISTS, EXAMPLE PROBLEMS, TABLES

Checklists

Example Problems

Tables